Java Web
框架开发技术

(Spring+Spring MVC+MyBatis)

史胜辉 王春明 ◎ 编著

清华大学出版社
北京

内 容 简 介

本书讲解 Spring、Spring MVC 和 MyBatis 3 个框架的基本知识和 3 个框架的整合技术。本书在编写过程中力求内容精简,只有 10 章,第 10 章是一个完整的项目案例。本书的宗旨是让读者用尽量少的时间掌握上述 3 个框架的技术。本书既可作为大中专教材,也可作为读者的自学用书。

本书如果作为教材,教学时数可以控制在 48～54 学时,为方便教学,每章后面都有习题和实验,习题和实验的内容都与本章教学内容紧密相关,这样便于教师组织教学。第 10 章的项目案例是在教学中使用的一个学生作业管理系统,教师完全可以将此系统用于实际教学过程中的作业管理。教材的配套资源包括 PPT、源代码、视频。

如果读者是自学,本书除提供书中示例的源码,还为一些章节录制了视频,视频的内容以教材为基础,针对教学内容的知识点和难点做进一步的讲解,特别是程序调试的技术,书中不便用文字叙述,通过视频可以让读者一目了然,目的是让读者能尽快掌握 3 个框架的核心内容,并能将 3 个框架整合起来应用于实践。

本书封面贴有清华大学出版社防伪标签,无标签者不得销售。
版权所有,侵权必究。举报: 010-62782989, beiqinquan@tup.tsinghua.edu.cn

图书在版编目(CIP)数据

Java Web 框架开发技术: Spring＋Spring MVC＋MyBatis/史胜辉,王春明编著. —北京: 清华大学出版社,2020.8(2023.1重印)
ISBN 978-7-302-55095-2

Ⅰ. ①J… Ⅱ. ①史… ②王… Ⅲ. ①JAVA 语言-程序设计-高等学校-教材 Ⅳ. ①TP312.8

中国版本图书馆 CIP 数据核字(2020)第 046192 号

责任编辑: 袁勤勇　常建丽
封面设计: 杨玉兰
责任校对: 李建庄
责任印制: 宋　林

出版发行: 清华大学出版社
　　　网　　址: http://www.tup.com.cn, http://www.wqbook.com
　　　地　　址: 北京清华大学学研大厦 A 座　　邮　编: 100084
　　　社 总 机: 010-83470000　　邮　购: 010-62786544
　　　投稿与读者服务: 010-62776969, c-service@tup.tsinghua.edu.cn
　　　质量反馈: 010-62772015, zhiliang@tup.tsinghua.edu.cn
　　　课件下载: http://www.tup.com.cn, 010-83470236
印 装 者: 三河市铭诚印务有限公司
经　　销: 全国新华书店
开　　本: 185mm×260mm　　印　张: 23.25　　字　数: 537 千字
版　　次: 2020 年 9 月第 1 版　　印　次: 2023 年 1 月第 4 次印刷
定　　价: 69.80 元

产品编号: 082539-01

前言

PREFACE

在当今的软件开发中，Java 是热门的编程语言之一。Java Web 开发的高效性和便利性是 Java 开发流行的原因之一。在 Java Web 开发中，SSH 和 SSM 框架技术的流行有功不可没的作用。SSM 3 个框架开发的灵活性和高效性可能更适合现在的互联网应用，已经得到越来越多程序员的认可。在本书中，对 SSM 3 个框架开发技术做了全面系统的概述，同时更侧重于实践操作，教材中对理论的叙述并不太多，主要是通过一些实例讲述相关的概念和一些框架技术的使用方法。3 个框架在书中占的比重是不同的，其中 Spring MVC 和 MyBatis 占的比重要大一些，Spring 占的比重少一些，这主要是从教学学时受限考虑的。每章后都有习题和一个实验，这样既便于读者掌握教材的内容，也便于教师组织教学。

本书的编写宗旨是对 SSM 3 个框架进行精简，以够用为原则，主要讲解各个框架最基本的知识和技术，同时又给读者扩展 3 个框架所需技术提供网址和参考资料。这样，读者可以在最短的时间内掌握 SSM 3 个框架技术，为后续的进一步开发打下良好的基础。

各章的具体内容如下：

第 1 章主要讲解 Spring 框架入门的一些基础知识，内容包括 Spring 框架简介、Spring IoC 容器、依赖注入的 3 种方式、Bean 的作用域、Spring 中 Bean 的装配方式。

第 2 章主要讲解 Spring MVC 框架的基础知识，内容包括 MVC 设计模式、Spring MVC 的架构、Spring MVC 的工作机制、Spring MVC 基于注解的开发、请求处理方法的返回类型。

第 3 章主要讲解 Spring MVC 的组件开发，内容包括拦截器、文件的上传与下载、Spring MVC 的表单标签库。

第 4 章主要讲解 Spring MVC 的数据转换与表单验证，内容包括数据绑定过程、数据类型转换、基于注解格式化数据、JSON 数据格式的转换、表单验证。

第 5 章主要讲解 MyBatis 基础知识，内容包括 ORM 与 MyBatis、MyBatis 的开发环境、MyBatis 中的 API、MyBatis 的配置文件、MyBatis 映射器。

第 6 章主要讲解结果映射与动态 SQL，内容包括结果映射、动态 SQL。

第 7 章主要讲解关联映射，内容包括一(多)对一的关联操作、一对多的关联操作、多对多的关联操作、MyBatis 的缓存机制。

第 8 章主要讲解 MyBatis 的注解开发，内容包括常用注解、用注解完成数据库中单表的 CRUD 操作、一对多的双向关联操作、用注解完成多对多的关联操作、注解中的动态 SQL。

第 9 章主要讲解基于 SSM 3 个框架的整合技术，内容包括基于 MyBatis 映射文件的整合开发、基于 MyBatis 注解的整合开发。

第 10 章主要讲解一个作业管理系统案例，内容包括系统的实体类图、功能框图、每个功能模块的类图、主要功能模块的流程图、每个功能模块的多层体系结构。

本书可作为计算机专业本科生或大专生的教材，也可作为有一定 JSP 和 Web 开发基础的计算机编程爱好者的学习用书。

本书由史胜辉负责编写，王春明、陆培军、王进、张晓峰、沈学华、王则林、马海英、严燕、王丹丹、何鹏、朱浩、魏晓宁等参与了本书的编写及代码测试。

由于作者水平有限，书中难免会有不足之处，敬请读者批评指正。

作　者

2020 年 7 月

目 录

第1章 Spring 框架 1

1.1 Spring 框架简介 1
1.1.1 Spring 的基本概念 1
1.1.2 Spring 的下载和安装 2
1.1.3 在 Eclipse 中配置 Spring 应用程序 3

1.2 Spring IoC 容器 5
1.2.1 IoC 容器 5
1.2.2 Spring IoC 容器的设计 6
1.2.3 Spring 中的依赖注入 8

1.3 依赖注入的3种方式 9
1.3.1 构造器注入 9
1.3.2 Setter 注入 10
1.3.3 接口注入 11

1.4 Bean 的作用域 12
1.4.1 作用域的分类 12
1.4.2 singleton 的作用域 12
1.4.3 prototype 的作用域 13

1.5 Spring 中 Bean 的装配方式 13
1.5.1 基于 XML 装配 Bean 14
1.5.2 基于注解装配 Bean 16
1.5.3 基于组件扫描注解装配 Bean 17
1.5.4 基于注解@Autowired 自动装配 19

习题 20
实验1 Spring IoC 中 Bean 的装配 21

第2章 Spring MVC 框架的基础知识 23

2.1 MVC 设计模式 23

2.2 Spring MVC 的架构 ······ 24
2.3 开发一个 Spring MVC 简单应用示例 ······ 25
2.4 Spring MVC 的工作机制 ······ 30
2.5 Spring MVC 基于注解的开发 ······ 31
　2.5.1 @Controller 注解 ······ 31
　2.5.2 @RequestMapping 注解 ······ 32
　2.5.3 @SessionAttribute 和 @SessionAttributes 注解 ······ 34
　2.5.4 控制器处理请求方法的参数类型 ······ 36
2.6 请求处理方法的返回类型 ······ 37
　2.6.1 Model 类型的使用 ······ 38
　2.6.2 ModelAndView 类型的使用 ······ 38
　2.6.3 返回类型为 String ······ 40
2.7 一个基于注解开发的示例 ······ 41
习题 ······ 46
实验 2　Spring MVC 基于注解开发 ······ 47

第 3 章　Spring MVC 的组件开发 ······ 49

3.1 拦截器 ······ 49
　3.1.1 Spring MVC 拦截器的设计 ······ 49
　3.1.2 单个拦截器的使用 ······ 50
　3.1.3 多个拦截器的使用 ······ 53
　3.1.4 拦截器应用——用户权限验证 ······ 55
3.2 文件的上传与下载 ······ 59
　3.2.1 文件的上传 ······ 59
　3.2.2 文件的下载 ······ 64
3.3 Spring 的表单标签库 ······ 66
　3.3.1 form 标签 ······ 67
　3.3.2 input 标签 ······ 67
　3.3.3 checkboxes 标签 ······ 68
　3.3.4 radiobuttons 标签 ······ 71
　3.3.5 select 标签 ······ 73
　3.3.6 标签应用示例 ······ 74
习题 ······ 79
实验 3　组件开发 ······ 79

第 4 章　Spring MVC 的数据转换与表单验证 ······ 81

4.1 数据绑定过程 ······ 81
4.2 数据类型转换 ······ 82

	4.2.1	ConversionService ···	82
	4.2.2	Spring 支持的转换器 ··	84
	4.2.3	自定义数据转换器 ··	84
4.3	基于注解格式化数据 ···		87
	4.3.1	@DateTimeFormat 注解 ····································	87
	4.3.2	@NumberFormat 注解 ······································	88
	4.3.3	基于注解格式化数据示例 ····································	88
4.4	JSON 数据格式的转换 ···		91
	4.4.1	JSON 格式简介 ··	92
	4.4.2	JSON 数据格式转换 ··	92
4.5	表单验证 ··		97
	4.5.1	JSR 303 校验规则 ··	97
	4.5.2	校验规则示例 ···	99

习题 ··· 103
实验 4 数据转换与表单验证 ··· 104

第 5 章 MyBatis 基础知识 ··· 107

5.1	ORM 与 MyBatis ··		107
5.2	MyBatis 的开发环境 ···		108
	5.2.1	MyBatis 框架的 JAR 包下载 ······························	108
	5.2.2	日志信息配置 ···	109
5.3	MyBatis 中的 API ···		112
	5.3.1	SqlSessionFactoryBuilder ································	112
	5.3.2	SqlSessionFactory ···	113
	5.3.3	SqlSession ···	114
5.4	MyBatis 的配置文件 ···		116
	5.4.1	<properties>元素 ···	117
	5.4.2	<settings>元素 ··	118
	5.4.3	<typeAliases>元素 ···	119
	5.4.4	<typeHandlers>元素 ·······································	120
	5.4.5	<environments>元素 ······································	122
	5.4.6	<mappers>元素 ··	124
5.5	MyBatis 映射器 ···		125
	5.5.1	XML 映射文件的主要元素 ································	125
	5.5.2	<select>元素 ··	125
	5.5.3	<insert>元素 ··	130
	5.5.4	<update>和<delete>元素 ··································	133
	5.5.5	<sql>元素 ··	134

习题 ··· 136
实验 5　用 MyBatis 完成单表的增、删、改、查操作 ······················· 136

第 6 章　结果映射与动态 SQL ··· 137

6.1　结果映射(<resultMap>元素) ··· 137
6.2　动态 SQL ··· 140
　　6.2.1　<if>元素 ··· 140
　　6.2.2　<choose>元素 ··· 142
　　6.2.3　<where>元素 ·· 144
　　6.2.4　<set>元素 ·· 145
　　6.2.5　<foreach>元素 ·· 146
习题 ··· 148
实验 6　用动态 SQL 完成单表的修改和查询操作 ··························· 148

第 7 章　关联映射 ·· 149

7.1　一(多)对一的关联操作 ·· 149
7.2　一对多的关联操作 ·· 154
　　7.2.1　一对多关联操作示例 ·· 155
　　7.2.2　影响关联操作性能的相关配置 ································· 158
7.3　多对多的关联操作 ·· 161
7.4　MyBatis 的缓存机制 ··· 166
　　7.4.1　一级缓存(SqlSession 级别) ··································· 166
　　7.4.2　二级缓存(mapper 级别) ······································· 168
习题 ··· 171
实验 7　表的关联操作 ·· 171

第 8 章　MyBatis 的注解开发 ··· 175

8.1　常用注解 ··· 175
8.2　单表的操作 ·· 176
8.3　一对多的双向关联操作 ·· 180
8.4　多对多的关联操作 ·· 184
8.5　注解中的动态 SQL ··· 186
习题 ··· 189
实验 8　基于注解的开发 ··· 189

第 9 章　SSM 框架整合 ··· 191

9.1　基于 MyBatis 映射文件的整合开发 ······································ 191

9.1.1 创建 Web 项目 …… 191
9.1.2 编写配置文件 …… 193
9.1.3 创建映射文件与接口 …… 197
9.1.4 创建 Service 及其实现类 …… 198
9.1.5 创建 Controller …… 202
9.1.6 创建 JSP 页面 …… 203
9.1.7 运行程序 …… 204
9.2 基于 MyBatis 注解的整合开发 …… 204
9.2.1 创建 Web 项目 …… 205
9.2.2 编写配置文件 …… 205
9.2.3 创建接口与注解 …… 205
9.2.4 创建 Service 及其实现类 …… 209
9.2.5 创建 Controller …… 212
9.2.6 创建 JSP 页面 …… 213
9.2.7 运行程序 …… 215
习题 …… 216
实验 9 SSM 整合开发 …… 216

第 10 章 项目案例：作业管理系统 …… 217

10.1 系统简介 …… 217
  10.1.1 系统用例图 …… 217
  10.1.2 系统功能框图 …… 217
10.2 系统设计 …… 219
  10.2.1 数据库设计 …… 219
  10.2.2 实体类的设计 …… 222
  10.2.3 系统结构设计 …… 226
10.3 系统环境的搭建 …… 227
  10.3.1 所需 JAR 包 …… 227
  10.3.2 创建数据库 …… 228
  10.3.3 创建 Web 项目 …… 232
10.4 功能模块实现 …… 237
  10.4.1 教师管理模块 …… 237
  10.4.2 班级管理模块 …… 264
  10.4.3 学生管理模块 …… 272
  10.4.4 课程管理模块 …… 289
  10.4.5 习题管理模块 …… 296

10.4.6　作业管理模块 …………………………………………………………… 310
10.4.7　批改作业模块 …………………………………………………………… 330
10.4.8　学生端作业管理模块 …………………………………………………… 340
10.5　单元测试 ………………………………………………………………………… 357
10.6　发布运行系统 …………………………………………………………………… 358

参考文献 …………………………………………………………………………………… 360

第 1 章 Spring 框架

本章目标
1. 了解 Spring IoC 和 DI 的概念。
2. 了解并掌握 Spring Bean 的装配方式。
3. 理解 Spring Bean 的生命周期。

1.1 Spring 框架简介

Spring 是一个开源框架，Spring 框架由 Rod Johnson 开发，2004 年发布了 Spring 框架的第一个版本。经过十多年的发展，Spring 已经成为 Java Web 开发中最重要的框架之一。对于一个 Java 开发者来说，Spring 已经成为其必须掌握的技能之一。以 Spring 为核心还衍生出了一系列框架，如 Spring Boot、Spring Cloud、Spring Cloud Data Flow、Spring Security 等（具体查看可登录 Spring 官网 www.spring.io）。本章介绍的是 Spring 框架的基本内容。

1.1.1 Spring 的基本概念

Spring 当初是为了解决企业应用开发的复杂性而创建的。Spring 使用基本的 JavaBean 完成以前只可能由 EJB 完成的事情。然而，Spring 的用途不仅限于服务器端的开发。从简单性、可测试性和松耦合的角度来看，任何 Java 应用都可以从 Spring 中受益。Spring 是一个轻量级的控制反转(IoC)和面向切面(AOP)的容器框架。

轻量是指 Spring 从大小与开销两方面而言都是轻量的。完整的 Spring 框架可以在一个大小只有 1MB 多的 JAR 文件里发布，并且 Spring 所需的处理开销也是微不足道的。此外，Spring 是非侵入式的，Spring 应用中的对象不依赖于 Spring 的特定类。

IoC 是指 Spring 通过一种称作 IoC 的技术促进了松耦合。当应用了 IoC，一个对象依赖的其他对象会通过被动的方式传递进来，而不是这个对象自己创建或者查找依赖对象。可以认为IoC与JNDI相反——不是对象从容器中查找依赖,而是容器在对象初始化时不等对象请求就主动将依赖传递给它。

AOP是指Spring提供了面向切面编程的丰富支持,允许通过分离应用的业务逻辑与系统级服务(例如审计和事务管理)进行内聚性的开发。应用对象只实现它们应该做的业务逻辑。它们并不负责其他的系统级关注点,如日志或事务支持。

Bean的生命周期,在Spring框架中,Spring包含并管理应用对象的配置和生命周期,在这个意义上它是一种容器,你可以配置基于一个可配置原型(prototype)的bean,也可以创建一个单独(singleton)的Bean。Spring可以将简单的组件配置、组合成为复杂的应用。在Spring中,应用对象可以声明方式进行组合,这些组合提供了很多基础功能(如事务管理、持久化框架集成等)并以XML文件方式进行配置。

1.1.2 Spring 的下载和安装

Spring是一个独立的框架,它不需要依赖任何Web容器,它既可在独立的Java SE项目中使用,也可以在Java Web项目中使用。下面先介绍如何为Java项目和Java Web项目添加Spring的支持。

下载和安装Spring框架可按如下步骤进行。

(1) 登录https://repo.spring.io/libs-release-local/org/springframework/spring/这个网址,该页面显示一个目录列表,从列表中可以看到现在最新的版本是5.1.8,单击这个链接即可下载。

(2) 下载完成后得到一个名字为spring-framework-5.1.8.RELEASE-dist的压缩包,解压此压缩包得到一个spring-framework-5.1.8.RELEASE的文件夹。此目录下有3个子目录。

docs:该目录中存放的相关文档,包含开发指南和API参考文档。

libs:该目录中的JAR包分为3类:

- Spring框架class文件的JAR包;
- Spring框架源文件的压缩包,文件名以-sources结尾;
- Spring框架API文档的压缩包,文件名以-javadoc结尾。

整个Spring框架由21个模块组成,在该目录下可看到Spring为每个模块都提供了3个压缩包。

Schema:该目录下包含了Spring各种配置文件的XML Schema文档。

(3) 将libs目录下所需要模块的JAR包复制添加到项目的类加载路径中——既可通过添加环境变量的方式添加,也可使用IDE工具管理应用程序的类加载路径。如果需要发布该应用,将这些JAR包一同发布即可。

经过上面3个步骤,即可在Java应用程序中使用Spring框架。

解压后的docs目录中有一个spring-framework-reference子目录,打开此目录下的index.html网页,可在浏览器中查看Spring框架的详细信息,特别是后面要用到的XML配置文件的相关的Schema可以从这里得到。

1.1.3 在 Eclipse 中配置 Spring 应用程序

在工程中使用 Spring 框架非常简单,只要将 Spring 所需的 JAR 包配置到 classpath 中即可。在 Eclipse 中可以通过构建路径配置,在 Web 工程中将所需的 JAR 包复制到工程的 lib 目录中即可。下面以一个简单的程序演示应用 Spring 框架的配置步骤。

(1) 在 Eclipse 中创建一个名为 chap1 的 Web 项目,将 Spring 的 4 个基础 JAR 包以及 commons-logging 的 JAR 包复制到 lib 目录中,并发布到类路径下。

(2) 在 src 目录下创建一个 com.ssm.ioc 包,并在包中创建一个 Customer 类,如代码清单 1-1 所示。

代码清单 1-1:Customer 类

```
1.  package com.ssm.ioc;
2.  public class Customer {
3.      private Long id;
4.      private String username;
5.      private String password;
6.      /* setter and getter */
7.      public Customer() {
8.          id=1L;username="admin"; password="123456";
9.      }
10.     @Override
11.     public String toString() {
12.         return "Customer [id=" +id +", username=" +username +", password=" +password +"]";
13.     }
14. }
```

(3) 在 src 目录下创建 Spring 的配置文件 applicationContext.xml,并在配置文件中创建一个 id 为 Customer 的 Bean。

applicationContext.xml 配置文件:

```
1.  <?xml version="1.0" encoding="UTF-8"?>
2.  <beans xmlns="http://www.springframework.org/schema/beans"
3.      xmlns:xsi="http://www.w3.org/2001/XMLSchema-instance"
4.      xsi:schemaLocation="http://www.springframework.org/schema/beans
5.          https://www.springframework.org/schema/beans/spring-beans.xsd">
6.  <!--此处的 bean 相当于创建一个对象,id="customer"表示创建一个 Customer 类的对象变量,-->
7.      <bean id="customer" class="com.ssm.Customer">
8.      </bean>
9.  </beans>
```

代码分析：第 2～5 行代码是 Spring 的约束配置。该配置信息无须手工输入，可以在 Spring 的帮助文档中找到。

（4）在 com.ssm.ioc 包下创建测试类 TestIoc，并在类中编写 main() 方法。在 main() 方法中，首先需要初始分 Spring 容器，并加载 Spring 配置文件，然后通过 Spring 容器获取 customer 实例，最后调用对象的 toString() 方法输出对象的属性值。

TestIoc 类的代码如下：

```java
1.  package com.ssm.ioc;
2.
3.  import org.springframework.context.ApplicationContext;
4.  import org.springframework.context.support.ClassPathXmlApplicationContext;
5.
6.  public class TestIoc {
7.
8.      public static void main(String[] args) {
9.          //读取配置文件的信息
10.         ApplicationContext context =new ClassPathXmlApplicationContext
             ("applicationContext.xml");
11.
12.         //获取 customer 实例
13.         Customer customer =context.getBean("customer",Customer.class);
14.
15.         //调用 Customer 中的 toString() 方法将对象的属性打印到控制台
16.         System.out.println(customer);
17.
18.      }
19.
20.  }
```

执行程序后，控制台输出结果如图 1-1 所示。

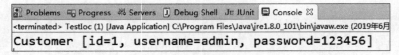

图 1-1　控制台输出结果

从图 1-1 可以看出，控制台已成功输出了 Customer 类中的 toString() 的内容。在 TestIoc 类中并没有通过 new 关键字创建 Customer 对象，而是通过 Spring 容器获取实现类的对象，这就是 Spring IoC 容器的工作机制。

（5）配置好的包结构和 JAR 包如图 1-2 所示。

图 1-2　配置好的包结构和 JAR 包

1.2　Spring IoC 容器

1.2.1　IoC 容器

　　日常的 Java 项目开发都是由两个或多个类的彼此合作实现业务逻辑的，这使得每个对象都需要与其合作的对象的引用（称为所依赖的对象），如果合作的对象的引用或依赖关系由具体的对象实现，这对复杂的面向对象系统的设计与开发是非常不利的，由此，如果能把这些依赖关系和对象的注入交给框架实现，让具体对象交出手中对于依赖对象的控制，就能很大程度上解耦代码，这显然是极有价值的，而这就是"依赖反转"，即反转对依赖的控制，把控制权从具体的对象中转交到平台或者框架。

　　依赖反转的实现有很多种，在 Spring 中，IoC 容器就是实现这个模式的载体，它可以在对象生成或初始化的过程中，直接将数据或者依赖对象的引用注入对象的数据域中，从而实现方法调用的依赖。而这种依赖注入是递归的，依赖对象会被逐层注入，从而建立起一套有序的对象依赖关系，简化了对象依赖关系的管理，把面向对象过程中需要执行的如对象的创建和对象引用赋值等操作交由容器统一管理，极大程度上降低了面向对象编程的复杂性。

　　在 Spring 中，Spring IoC 提供了一个基本 JavaBean 容器，通过 IoC 模式管理依赖关系，并通过依赖注入和 AOP 切面对类似 POJO 这样的对象提供了事务管理、生命周期管理等功能。在应用开发中，设计组件往往需要引入和调用其他组件的服务时，这种依赖关系如果固化在组件设计中，就会导致组件之间的耦合和维护难度增大，这时如果使用 IoC

容器,把资源获取的方式反转,让 IoC 容器主动管理这些依赖关系,将依赖关系注入到组件中,那么这些依赖关系的适配和管理就会更加灵活。

应用管理依赖关系时,如果在 IoC 实现依赖反转的过程中能通过可视化的文本完成配置,并且通过工具对这些配置信息进行可视化的管理和浏览,那么肯定能提高依赖关系的管理水平,而且如果耦合关系变动,并不需要重新修改和编译 Java 代码,这符合在面向对象过程中的开闭原则。

Spring 倡导的开发方式就是如此,所有的类都会在 Spring 容器中登记,告诉 Spring 你是一个什么组件,需要哪些组件,然后 Spring 会在系统运行到适当的时候,把你要的组件或对象主动给你,同时也把你交给其他需要你的组件或对象。所有对象的创建、销毁都由 Spring 控制,也就是说,控制对象生存周期的不再是引用它的对象,而是 Spring。对于某个具体的对象而言,以前是它控制其他对象,现在是所有对象都被 Spring 控制,所以称控制反转。

IoC 的一个重点是在系统运行中,动态地向某个对象提供它所需要的其他对象。这一点是通过 DI(Dependency Injection,依赖注入)实现的。例如,对象 A 需要操作数据库,以前我们总是要在 A 中自己编写代码获得一个 Connection 对象,有了 Spring,就只需要告诉 Spring,A 中需要一个 Connection,至于这个 Connection 怎么构造,何时构造,A 不需要知道。系统运行时,Spring 会在适当的时候制造一个 Connection,然后像打针一样注射到 A 中,这样就完成了对各个对象之间关系的控制。A 需要依赖 Connection 才能正常运行,而这个 Connection 是由 Spring 注入 A 中的,依赖注入的名字就是这么来的。

1.2.2 Spring IoC 容器的设计

Spring IoC 容器的设计主要基于 BeanFactory 和 ApplicationContext 两个接口,其中 ApplictionContext 是 BeanFactory 的子接口。也就是说,BeanFactory 是 Spring IoC 容器定义的最底层接口,而 ApplictionContext 是其高级接口之一,并且对 BeanFactory 功能做了许多有用的扩展,所以,在绝大部分的工作场景下,都会使用 ApplictionContext 作为 Spring IoC 容器,其接口设计如图 1-3 所示。

可以看到,ApplicationContext 是 BeanFactory 的子类,所以 ApplicationContext 可以看作是更强大的 BeanFactory,它们两个之间的区别如下:

- BeanFactory,基础类型 IoC 容器,提供完整的 IoC 服务支持。如果没有特殊指定,默认采用延迟初始化策略(lazy-load)。只有当客户端对象需要访问容器中的某个受管对象的时候,才对该受管对象进行初始化以及依赖注入操作。所以,相对来说,容器启动初期速度较快,所需要的资源有限。对于资源有限,并且功能要求不是很严格的场景,BeanFactory 是比较合适的 IoC 容器选择。
- ApplicationContext,在 BeanFactory 的基础上构建,是相对高级的容器实现,除了拥有 BeanFactory 的所有支持,ApplicationContext 还提供了其他高级特性,如事件发布、国际化信息支持等,ApplicationContext 管理的对象在该类型容器启动之后,默认全部初始化并绑定完成。所以,相对于 BeanFactory 来说,

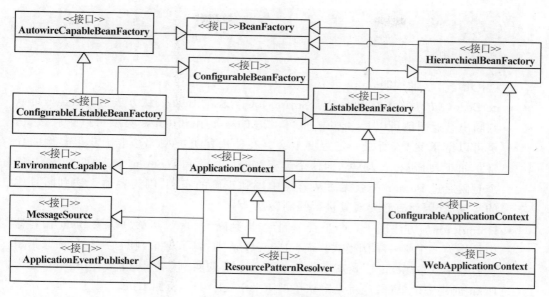

图 1-3　Spring IoC 容器的接口设计

ApplicationContext 要求更多的系统资源，同时，因为在启动时就完成所有初始化，容器的启动时间较 BeanFactory 会长一些。在系统资源充足，并且要求更多功能的场景中，ApplicationContext 类型的容器是比较合适的选择。

为了更好地理解 BeanFactory 接口，可参考 BeanFactory 类的源代码。

BeanFactory 类的源代码：

```
1.   public interface BeanFactory
2.   {
3.       String FACTORY_BEAN_PREFIX = "&";
4.       Object getBean(String name) throws BeansException;
5.       <T> T getBean (String name, Class<T> requiredType) throws BeansException;
6.       Object getBean(String name, Object... args) throws BeansException;
7.       <T>T getBean(Class<T> requiredType) throws BeansException;
8.       <T> T getBean (Class<T> requiredType, Object... args) throws BeansException;
9.       <T>ObjectProvider<T>getBeanProvider(Class<T>requiredType);
10.      <T>ObjectProvider<T>getBeanProvider(ResolvableType requiredType);
11.      boolean containsBean(String name);
12.      boolean isSingleton(String name) throws NoSuchBeanDefinitionException;
13.      boolean isPrototype(String name) throws NoSuchBeanDefinitionException;
14.      boolean isTypeMatch (String name, ResolvableType typeToMatch) throws NoSuchBeanDefinitionException;
15.      boolean isTypeMatch (String name, Class<?> typeToMatch) throws NoSuchBeanDefinitionException;
```

```
16.     @Nullable
17.     Class<?> getType(String name) throws NoSuchBeanDefinitionException;
18.     String[] getAliases(String name);    }
```

对类中几个方法的说明：

- getBean()方法用于获取配置给Spring IoC容器的Bean。从方法的返回类型可以判断出其返回值可以是String，也可以是Class。由于Class类型可以扩展接口，也可以继承父类，所以一定程度上会存在使用父类类型无法准确获得实例的异常。如获取Person(人)类，但这个类有Man(男人)和Woman(女人)两个子类，这个时候通过Person类就无法从容器中得到正确的实例，因为容器无法判断具体的实现类，因此有必要对其进行强制转换。
- isSingleton()方法用于判断是否单例，如果判断为"真"，其意思是该Bean在容器中是作为一个唯一单例模式。而isPrototype则相反，如果判断为"真"，意思是当从容器中获取Bean，容器就为你生成了一个新的实例。默认情况下，Spring会为Bean创建一个单例，也就是默认情况下isSingleton()返回true。
- 关于type的匹配是一个按Java类型匹配的方式。
- getAliases()方法是获取别名的方法。

这就是Spring IoC最底层的设计，所有关于Spring IoC的容器，都将会遵循它所定义的方法。

1.2.3　Spring中的依赖注入

依赖注入(Dependency Injection，DI)与IoC的含义相同，只不过这两个称呼是从两个角度描述同一个概念。对于一个Spring初学者来说，这两个概念很难理解。下面对这两个概念进行简单的介绍。

当某个Java对角(调用对角)需要调用另一个Java对象(被调用者，即被依赖对象)时，在传统模式下，调用者通常会采用new运算符创建一个对象，这种方式会导致调用者与被调用者之间的耦合性增加，不利于后期项目的升级和维护。调用者创建被调用者如图1-4所示。

图1-4　调用者创建被调用者

在使用Spring框架之后，对象的实例不再由调用者创建，而是由Spring容器创建，Spring容器负责控制对象之间的关系，取代了由调用者控制对象之间的关系。这样，对象之间的这种依赖关系由程序转移到IoC容器，控制权发生了反转，这就是Spring的控制反转。

从Spring容器的角度看，Spring容器负责将被依赖对象赋值给调用者的成员变量，这相当于调用者注入了它依赖的实例，这就是Spring的依赖注入，如图1-5所示。

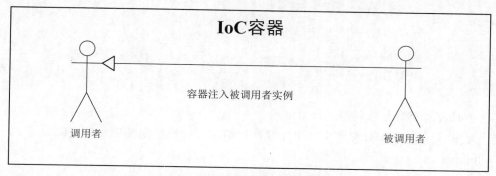

图1-5　将被调用者注入调用者对象

1.3　依赖注入的3种方式

在实际环境中实现IoC容器的方式主要分为两大类：一类是依赖查找，它是通过资源定位把对应的资源查找回来；另一类是依赖注入，Spring主要使用的是依赖注入。一般而言，依赖注入可以分为3种方式。

- 构造器注入；
- Setter注入；
- 接口注入。

构造器注入和Setter注入是主要的方式，接口注入是从别的地方注入的方式。例如，在Web工程中，配置的数据源往往是通过服务器（如Tomcat）配置的，这时可以用JNDI的形式通过接口将它注入Spring IoC容器中。下面对它们进行详细讲解。

1.3.1　构造器注入

构造器注入依赖于构造方法实现，而构造方法可以是有参数的或者是无参数的。大部分情况下，我们都是通过类的构造方法创建类的对象，Spring也可以采用反射的方式通过使用构造方法完成注入，这就是构造器注入的原理。

为了让Spring完成对应的构造注入，我们有必要描述具体的类、构造方法并设置对应的参数，这样Spring就会通过相应的信息用反射的形式创建对象，如下面的Customer类，其代码清单如下。

```
1.   package com.ssm;
2.   public class Customer {
3.       private Long id;
4.       private String username;
```

```
5.     private String password;
6.     /*setter and getter*/
7.     public Customer(Long id,String username,String password) {
8.         this.id=id;
9.         this.username=username;
10.        this.password=password;
11.    }
12. }
```

这个时候是没有办法利用无参数的构造方法创建对象的,为了使Spring能够正确创建这个对象,可以在XML配置文档中完成属性的注入。构造器配置文件代码如下。

applicationContext.xml代码:

```
1. <bean id="customer" class="com.ssm.chap1.Customer">
2.     <constructor-arg index="0" value="1" />
3.     <constructor-arg index="1" value="admin"/>
4.     <constructor-arg index="2" value="123456"/>
5. </bean>
```

constructor-arg元素用于定义类构造方法的参数,其中index用于定义参数的位置,而value则是设置值,通过这样的定义,Spring便知道如何使用Customer的构造方法创建对象了。

这样注入比较简单,但是缺点很明显。由于这里的参数比较少,所以可读性还可以,但是如果参数很多,那么这种构造方法就比较复杂了,这时候应该考虑Setter注入。

1.3.2 Setter注入

Setter注入是Spring中最主流的注入方式,它利用Java Bean规范定义的setter()方法完成注入,灵活且可读性好。它削除了使用构造器注入时出现多个参数的可能性,首先可以把构造方法声明为无参数的,然后使用Setter注入为其设置的值,其实也是通过Java反射技术得以实现的。这里假设先在前面的Customer类加入一个没有参数的构造方法,然后做如下配置。

chap1/src/applicationContext.xml配置文件代码:

```
1. <bean id="customer2" class="com.ssm.Customer">
2.     <property name="id" value="1" />
3.     <property name="username" value="admin"/>
4.     <property name="password" value="123456"/>
5. </bean>
```

这样,Spring就会通过反射调用没有参数的构造方法生成对象,同时通过Setter注入属性的值。这种方式是Spring最主要的注入方式,在实际工程中广泛使用。

1.3.3 接口注入

有些时候资源并非来自自身系统,而是来自外界。例如,数据库连接资源完全可以在 Tomcat 下配置,然后通过 JNDI 的形式获取它,这样数据库连接资源是属于开发工程外的资源,这个时候我们可以采用接口注入的形式获取它,如在 Tomcat 中可以配置数据源。又如,在 Eclipse 中配置了 Tomcat 后,可以打开服务器的 context.xml 文件,代码清单如下。

context.xml 代码:

```
1.    <Context>
2.    <!--name 为 JNDI 名称
3.        url 是数据库的 jdbc 连接
4.        username 是用户名
5.        password 是密码
6.        -->
7.     <Resource name="jdbc/ssm" auth="Container" type="javax.sql.
        DataSource"
8.        driverClassName="com.mysql.jdbc.Driver"
9.        url=" jdbc: mysql://localhost: 3306/ssm? zeroDatetimeBehavior =
        converToNull"
10.       username="root" password="123456"/>
11.
12.   </Context>
```

如果已经配置了相应的数据库连接,那么 Eclipse 会把数据库的驱动包复制到对应的 Tomcat 的 lib 文件夹中,否则就需要自己手工将对应的驱动包复制到 Tomcat 的工作目录中,它位于{Tomcat_Home}\lib。然后启动 Tomcat,这个时候数据库资源也会在 Tomcat 启动的时候被其加载进来。

如果 Tomcat 的 Web 工程使用了 Spring,那么可以通过 Spring 的机制,用 JNDI 获得 Tomcat 启动的数据库连接池,如下面的配置文件所示。

applicationContext.xml 配置文件代码:

<!--通过 JNDI 获取的数据源,通过 Spring 的接口注入实现 -->

```
1.    <bean id="dataSource"class="org.springframework.jndi.JndiObjectFactoryBean">
2.    <property name="jndiName">
3.    <value>java:comp/env/jdbc/netshop</value>
4.    </property>
5.    </bean>
```

这样就可以在 Spring 的 IoC 容器中获得 Tomcat 所管理的数据库连接池了,这是一种接口注入的形式。

1.4 Bean 的作用域

通过 Spring 容器创建一个 Bean 的实例时,不仅可以完成 Bean 的实例化,还可以为 Bean 指定特定的作用域。本节主要讲解 Spring 中的作用域的分类和含义。

1.4.1 作用域的分类

Spring 5.1 中一共定义了 6 种作用域,具体内容见表 1-1。

表 1-1 Bean 的作用域

作用域	说明
singleton(单例)	在 IoC 容器中始终只有一个实例,无论有多少个 Bean 引用它,始终指向同一个对象。这是 Spring 默认的作用域
prototype(原型)	IoC 获取 Bean 时,容器每次都为它创建一个新的 Bean 实例
request	这是对应于 HTTP 请求中的 request
session	这是对应于 HTTP 请求中的 session
application	这是对应于 HTTP 请求中的 application
websocket	为每个 websocket 对象创建一个实例

在表 1-1 的 6 种作用域中,singleton 和 prototype 是用得最多的两种,下面重点介绍这两种作用域。

1.4.2 singleton 的作用域

Spring 的 IoC 容器默认的是 singleton 模式,这一点可以通过一个示例说明。针对前面的 applicationContext.xml 配置文件,通过以下类创建两个 Bean。

chap1/src/com/ssm/TestSingleton 类代码:

```
1.   package com.ssm;
2.   import org.springframework.context.ApplicationContext;
3.   import org.springframework.context.support.ClassPathXmlApplicationContext;
     public class TestSingleton {
4.
5.     public static void main(String[] args) {
6.       //读取配置文件的信息
7.       ApplicationContext context=new ClassPathXmlApplicationContext("
         applicationContext.xml");
8.       //获取第一个 customer 实例
9.       Customer customer1=context.getBean("customer2",Customer.class);
10.      //获取第二个 customer 实例
11.      Customer customer2=context.getBean("customer2",Customer.class);
```

```
12.        //判断customer1 和 customer2 是否是同一个实例
13.        boolean result=customer1==customer2;
14.        System.out.println("customer1 与 customer2 是否相同:"+result);
15.    }
16. }
```

singleton 模式运行结果如图 1-6 所示。

图 1-6 singleton 模式运行结果

从图 1-6 中可以看出,虽然代码中获取了二次 Bean 的实例,但它们是同一个对象,默认情况下,Spring IoC 容器只会生成一个实例,而不是多个。

1.4.3 prototype 的作用域

对需要保持会话状态的 Bean 应该使用 prototype 作用域。使用 prototype 作用域时,Spring 容器会为每个对该 Bean 的请求创建一个新的对象,修改 applicationContext 配置文件。

chap1/src/applicationContext.xml 代码如下:

```
1. <bean id="customer2" class="com.ssm.Customer" scope="prototype">
2.     <property name="id" value="1"/>
3.     <property name="username" value="admin"/>
4.     <property name="password" value="123456"/>
5.     <!--collaborators and configuration for this bean go here -->
6. </bean>
```

prototype 模式运行结果如图 1-7 所示。

图 1-7 prototype 模式运行结果

从图 1-7 可以看到,两个 Customer 实例确实不是同一个对象,说明在 prototype 作用域下,Spring IoC 创建了两个不同的实例。

1.5 Spring 中 Bean 的装配方式

Spring 中,Bean 的装配可以理解为依赖关系注入,Bean 的装配方式即 Bean 依赖注入的方式。Spring IoC 容器支持 3 种 Bean 的装配方式:基于 XML 装配 Bean、基于注解装配 Bean 和基于组件扫描注解装配 Bean。

1.5.1　基于 XML 装配 Bean

Spring 的配置文件是基于 XML 格式的，Spring 1.0 的配置文件采用 DTD 格式，Spring 2.0 以后使用 Schema 的格式，后者让不同类型的配置拥有了自己的命名空间，使配置文件更具有扩展性。采取基于 Schema 的配置格式，文件头的声明会复杂一些，前面的配置文件就是一个简单值的装配的示例。在这段代码中，通过 XML 配置文件将 3 个基本类型的变量的值注入对象的 3 个属性中。

上面这种简单值的注入是一种常见的注入方式，除此之外，还有另外一种常见的属性值，即对象。也就是说，某一个类的属性是另一个类的引用，注入时要先定义引用的 Bean，然后再通过 Bean 的 ref 属性注入进来。现有一个学生类 Student，代码如下所示。

/chap1/src/com/ssm/Student 类代码：

```
1.  package com.ssm;
2.  //学生类
3.  public class Student {
4.      private Integer id;                //学生 ID
5.      private String loginname;          //学生登录名
6.      private String password;           //密码
7.      private String username;           //学生姓名
8.      private Clazz clazz;               //学生所在的班级
9.  /**setter and getter **/
10. }
```

这个类属性中有一个代表班级的引用变量 clazz，Clazz 类的代码如下所示。

/chap1/src/com/ssm/Clazz 类代码：

```
1.  package com.ssm;
2.  班级类
3.  public class Clazz implements Serializable {
4.      private Integer id; //班级 ID
5.      private String cname;    //班级名称
6.  /** setter and getter **/
7.  }
```

现在通过 Spring 的 XML 配置文件的方式将属性 clazz 的值注入 Student 的一个 Bean 中。配置文件代码如下所示。

/chap1/src/springXmlConfig.xml 配置文件：

```
1.  <?xml version="1.0" encoding="UTF-8"?>
2.  <beans xmlns="http://www.springframework.org/schema/beans"
3.      xmlns:xsi="http://www.w3.org/2001/XMLSchema-instance"
```

```
4.      xsi:schemaLocation="http://www.springframework.org/schema/beans
5.      https://www.springframework.org/schema/beans/spring-beans.xsd">
6.      <!--此处的bean相当于创建一个对象,id="clazz"表示创建一个Clazz对象变量,
        同时为变量的属性赋值 -->
7.      <bean id="clazz" class="com.ssm.ioc.Clazz">
8.         <property name="id" value="1" />
9.         <property name="cname" value="计算机191"/>
10.     </bean>
11.     <!--此处的bean相当于创建一个对象,id="student"表示创建一个student对象变
        量,在这个Bean中有一个引用变量clazz,这个变量的赋值是通过ref属性引用前面已
        定义好的Bean的 -->
12.     <bean id="student" class="com.ssm.ioc.Student" >
13.        <property name="id" value="1"/>
14.        <property name="loginname" value="admin"/>
15.        <property name="password" value="123456"/>
16.        <property name="username" value="孙悟空"/>
17.        <property name="clazz" ref="clazz"/>
18.     </bean>
19.  </beans>
```

在代码17行,id为student的bean中有一个clazz属性,通过property的ref属性将第7行定义的id为clazz的bean注入student的clazz属性中。其他的属于简单类型值的注入。通过下面的测试类可验证配置文件的正确性。

测试类代码/chap1/src/com/ssm/test/TestXmlConfig：

```
1.  package com.ssm.test;
2.  import org.springframework.context.ApplicationContext;
3.  import org.springframework.context.support.ClassPathXmlApplicationContext;
4.  public class TestXmlConfig {
5.     public static void main(String[] args) {
6.        //读取配置文件springXmlConfig.xml的信息
7.        ApplicationContext context = new ClassPathXmlApplicationContext("
          springXmlConfig.xml");
8.        //获取Student实例
9.        Student stu=context.getBean("student",Student.class);
10.       //打印学生的信息
11.       System.out.println(stu);
12.    }
13. }
```

程序运行结果如图1-8所示。

从运行结果分析,在实例化Student对象的同时也将其属性clazz进行了初始化,Spring通过解析XML配置文件完成了对象属性的注入。如果类的属性是一个集合,同

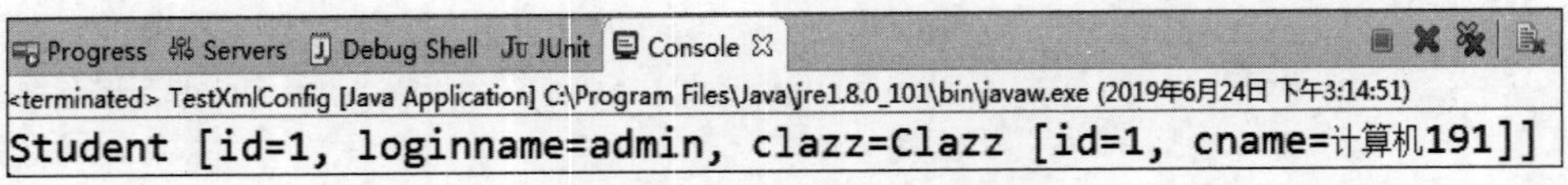
Progress
Servers
Debug Shell
JUnit
Console
<terminated> TestXmlConfig [Java Application] C:\Program Files\Java\jre1.8.0_101\bin\javaw.exe (2019年6月24日 下午3:14:51)
Student [id=1, loginname=admin, clazz=Clazz [id=1, cname=计算机191]]

图 1-8 测试类 TestXmlConfig 的运行结果

样可以通过配置文件对其进行初始化，并完成依赖注入，在此不再介绍。

1.5.2 基于注解装配 Bean

通过前面的学习，读者已经知道如何使用 XML 的方式装配 Bean，但利用 XML 配置 Bean 最大的问题是当有大量 Bean 时，配置文件会变得非常臃肿，不便于管理，因此，Spring 3.0 以后开始引入基于注解的配置方式，即 Bean 的定义信息可以通过在 Bean 的实现类上标注注解(annotation)实现。使用注解的方式可以减少 XML 的配置，注解功能更加强大，它既能实现 XML 的功能，也提供了自动装配的功能。采用自动装配后，程序员的工作量大为减少，有利于提高程序开发的效率。Spring 中提供了两种注解方式让 Spring IoC 容器装配 Bean。

- 组件扫描：通过定义资源的方式让 Spring IoC 容器扫描对应的包，从而把 Bean 装配起来。
- 自动装配：通过注解定义，使得一些依赖关系可以通过注解完成。

Spring 5.0 中的注解很多，这里只介绍几种常用的注解和使用方法。

1. @Bean 注解

@Bean 标识一个用于配置和初始化一个由 Spring IoC 容器管理的新对象的方法，类似于 XML 配置文件的<bean/>，一般与@Configration 注解配合使用。

2. @Component 注解

把普通 POJO 类实例化到 Spring 容器中，相当于配置文件中的<bean id="" class=""/>泛指各种组件，也就是说，当我们的类不属于各种归类的时候（不属于@Controller、@Services 等的时候），就可以使用@Component 标注这个类。

3. @Repository 注解

@Repository 注解用于标注数据访问组件，即 DAO 组件。

4. @Controller 注解

当组件属于控制层时，使用@Controller 注解被 Controller 标记的类就是一个控制器，这个类中的方法就是相应的动作。

5. @Autowired 注解

它可以对类成员变量、方法及构造函数进行标注，完成自动装配的工作。通过

@Autowired 的使用消除 set()、get() 方法。@Autowired 注解可用于为类的属性、构造器、方法进行注值。默认情况下,其依赖的对象必须存在(Bean 可用),如果容器中包含多个同一类型的 Bean,那么启动容器时会报找不到指定类型 Bean 的异常,解决办法是结合 @Qualifier 注解进行限定,指定注入的 Bean 名称。

6. @Resource 注解

@Resource 依赖注入时查找 Bean 的规则既不指定 name 属性,也不指定 type 属性,而是自动按 byName 方式查找。如果没有找到符合的 Bean,则回退为一个原始类型进行查找;如果找到,就注入。只有指定了 @Resource 注解的 name,才按 name 后的名字去 Bean 元素里查找有与之相等的 name 属性的 Bean。只指定 @Resource 注解的 type 属性,则从上下文中找到类型匹配的唯一 Bean 进行装配,找不到或者找到多个,都会抛出异常。既指定了 @Resource 的 name 属性,又指定了 type,则从 Spring 上下文中找到唯一匹配的 Bean 进行装配,若找不到,则抛出异常。

1.5.3 基于组件扫描注解装配 Bean

下面通过一个案例演示如何通过组件扫描这些注解来装配 Bean,具体步骤如下。

(1) 创建一个服务接口 CustomerService。

/chap1/src/com/ssm/service/CustomerService 类代码如下:

```
1.  package com.ssm.service;
2.  public interface CustomerService {
3.      public void save();
4.  }
```

(2) 创建 CustomerService 接口的实现类 CustomerServiceImpl。

/chap1/src/com/ssm/service/CustomerServiceImpl 类代码如下:

```
1.  package com.ssm.service;
2.  import org.springframework.stereotype.Service;
3.  //@Service("customerService")
4.  public class CustomerServiceImpl implements CustomerService {
5.      @Override
6.      public void save() {
7.          System.out.println("saving customer!");
8.      }
9.  }
```

(3) 创建控制器类 CustomerController。

/chap1/src/com/ssm/service/CustomerController 类代码如下:

```
1.  package com.ssm.service;
2.  import javax.annotation.Resource;
```

```
3.    import org.springframework.stereotype.Controller;
4.    //@Controller("customerController")           //控制层的 Bean
5.    public class CustomerController {
6.        @Resource(name="customerService")         //数据访问层的 Bean
7.        private CustomerService customerService;
8.        public void save() {
9.            this.customerService.save();
10.       }
11.   }
```

在 CustomerController 中引用了 CustomerService 接口作为其成员变量,在此并没有实例化,而通过第 6 行的@Resource 注解相当于在 Spring 的配置文件中定义了一个 Bean,其 id 为 customerService,对应的 class 为 CustomerServiceImpl。

(4) 创建 Spring 配置文件。

/chap1/src/springAnnotationConfig.xml 配置文件:

```
1.    <?xml version="1.0" encoding="UTF-8"?>
2.    <beans xmlns="http://www.springframework.org/schema/beans"
3.        xmlns:xsi="http://www.w3.org/2001/XMLSchema-instance"
4.        xmlns:context="http://www.springframework.org/schema/context"
5.        xsi:schemaLocation="http://www.springframework.org/schema/beans
6.         http://www.springframework.org/schema/beans/spring-beans.xsd
7.         http://www.springframework.org/schema/context
8.         http://www.springframework.org/schema/context/spring-context.xsd">
9.        <!--使用 context 命名空间,在配置文件中开启相应的注解处理器 -->
10.       <context:annotation-config/>
11.       <!--此处的 customerService 相当于创建一个服务对象-->
12.       <bean id="customerService" class="com.ssm.service.CustomerServiceImpl"/>
13.       <!--此处的 customerController 相当于创建一个控制器对象-->
14.       <bean id="customerController" class="com.ssm.service.CustomerController"/>
15.   </beans>
```

此处的配置文件与前面的配置文件不同的是多了第 4、7、8 行,这 3 行用来配置命名空间,第 10 行表示启用注解处理器处理代码中的注解。

(5) 创建测试类,测试获取 CustomerController 的实例 customerController,并调用其方法 save(),代码如下。

/chap1/src/com/ssm/service/TestAnnotationConfig 类代码:

```
1.    public class TestAnnotationConfig {
2.        public static void main(String[] args) {
3.            //读取配置文件的信息
4.            ApplicationContext context = new ClassPathXmlApplicationContext("
              springAnnotationConfig.xml");
```

```
5.      //获取一个 CustomerTroller 实例
6.      CustomerController cus = context.getBean("customerController",
        CustomerController.class);
7.      //打印 Customer 的信息
8.      cus.save();
9.   }
10. }
```

程序运行后的结果显示："saving customer!"，说明 IoC 容器正确地解析了配置文件的信息，得到 CustomerController 实例，并调用 save()方法，得到我们期待的结果。

（6）添加 Spring 的包扫描机制。

这个示例通过@Resource 注解完成了属性的注入，但还存在一个很大的问题，就是如果程序中有很多类，必然要在配置文件中定义大量的 Bean，这样非常不利于管理。Spring 为了解决这个问题，给出了一个很好的解决方案，就是通过包扫描组件自动创建 Bean 以及它们之间的依赖关系，完成属性的自动注入。为了测试 Spring 包扫描的功能，需对代码作如下修改：

将 CustomerController 类的注释去掉，启用@Service("customerService") 注解。
修改 Spring 配置文件的代码如下所示。
/chap1/src/springAutoScanConfig.xml 配置文件：

```
1.  <?xml version="1.0" encoding="UTF-8"?>
2.  <beans xmlns="http://www.springframework.org/schema/beans"
3.      xmlns:xsi="http://www.w3.org/2001/XMLSchema-instance"
4.      xmlns:context="http://www.springframework.org/schema/context"
5.      xsi:schemaLocation="http://www.springframework.org/schema/beans
6.      http://www.springframework.org/schema/beans/spring-beans.xsd
7.      http://www.springframework.org/schema/context
8.      http://www.springframework.org/schema/context/spring-context.xsd">
9.      <!--使用 context 命名空间,在配置文件中开启相应的注解处理器 -->
10.     <context:annotation-config/>
11.     <!--通知 Spring 扫描指定包下所有的 Bean -->
12.     <context:component-scan base-package="com.ssm.service"/>
13. </beans>
```

在配置文件中添加第 12 行包扫描的路径，Spring 会到指定的包下扫描所有的注解，并解析为相应的 Bean，这样省去了文件中 Bean 的定义。运行原来的测试类代码可得到同样的结果。

1.5.4 基于注解@Autowired 自动装配

通过学习 Spring IoC 容器，我们知道 Spring 是先完成 Bean 的定义和生成，然后寻找需要注入的资源。也就是当 Spring 生成所有的 Bean 后，如果发现这个注解，它就会在 Bean 中查找，然后找到对应的类型，将其注入进来，这样就完成了依赖注入。自动装配技术是一种由 Spring 自己发现对应的 Bean,自动完成装配工作的方式,它会应用到一个十分常用的注解@Autowired,当将这个注解添加到某一类型的属性上时,Spring 会根据类型寻找配置文件中已定义的 Bean,或者通过注解定义的 Bean,如果找到了相同类型的 Bean,就将其注入。如果没找到,可以忽略,也可以抛出异常,这取决于@Autowired 的属性值 required 是 true,还是 false。如果设置成@Autowired(required=true),表示这个属性不能为空,是必须注入的,这也是默认值。如果设置成@Autowired(required=false),表示这个属性值可以为空,即使不注入,这个属性值也不会抛出异常。

下面开始测试自动装配,修改 CustomerController 类代码。

/chap1/src/com/ssm/autowired/CustomerController 类修改后的代码如下:

```
1.   package com.ssm.autowired;
2.   import org.springframework.beans.factory.annotation.Autowired;
3.   import org.springframework.stereotype.Controller;
4.   @Controller("customerController")
5.   public class CustomerController {
6.       @Autowired //自动装配注解
7.       private CustomerService customerService;
8.       public void save() {
9.           this.customerService.save();
10.      }
11.  }
```

同时要在 CustomerController 类中启用注解@Service("customerService"),在 Spring 配置文件中启用注解处理器和包扫描功能,再次运行测试类 TestAutoScanConfig,可得到前面运行相同的结果。

习　题

1. 下面关于 AOP 的说法,错误的是(　　)。
   A. AOP 将散落在系统中的"方面"代码集中实现
   B. AOP 有助于提高系统的可维护性
   C. AOP 已经表现出了将要代替面向对象的趋势
   D. AOP 是一种设计模式,Spring 提供了一种实现
2. 下面关于 Spring 的说法中,错误的是(　　)。
   A. Spring 是一个轻量级的框架
   B. Spring 中包含一个"依赖注入"模式的实现
   C. 使用 Spring 可以实现声明式事务
   D. Spring 提供了 AOP 方式的日志系统
3. 关于声明事务的说法,下面说法错误的是(　　)。
   A. Spring 采取 AOP 的方式实现声明式事务

B. 声明式事务是非侵入式的,可以不修改原来代码就给系统增加事务
C. 配置声明式事务需要 tx 和 aop 两个命名空间的支持
D. 配置声明式事务主要关注"在哪儿"和"采取什么样的事务策略"

4. (　　)不是依赖注入的方式。

  A. 构造器   B. Setter   C. 接口   D. AOP

5. Spring 的 IoC 容器默认 Bean 的作用域是(　　)。

  A. singleton B. prototype C. request D. session

实验 1　Spring IoC 中 Bean 的装配

1. 实验目的

掌握在 Spring IoC 容器中基于 XML 和基于注解装配 Bean 的方法。

2. 实验内容

现有 Product 产品类,根据业务需求要对产品进行保存操作,为了遵守 3 层框架体系结构,设计了如下的类结构。

Product 实体类:

```
1.  package com.po;
2.  public class Product {
3.      private Integer id;            //商品的 id
4.      private String name;           //商品名称
5.      private String subTitle;       //商品的描述信息
6.      private float price;           //商品价格
7.      public Integer getId() {
8.          return id;
9.      }
10. /***setter and getter ###/
11.     @Override
12.     public String toString() {     //当打印对象信息时会调用此方法
13.         return "product [id=" + id + ", name=" + name + ", subTitle=" +
            subTitle +", price=" +price +"]";
14.     }
15. }
```

业务层接口 ProductService:

```
1.  package com.service;
2.  import com.po.Product;
3.  public interface ProductService {
4.      public void save(Product produce);      //保存商品操作
5.  }
```

业务层实现类 ProductServiceImpl：

```
1.  package com.service.impl;
2.  import com.po.Product;
3.  import com.service.ProductService;
4.  public class ProductServiceImpl implements ProductService{
5.      @Override
6.      public void save(Product product) {           //实现接口中定义的方法
7.          System.out.println(product);
8.      }
9.  }
```

控制器类 ProductController：

```
1.  package com.controller;
2.  import com.po.Product;
3.  import com.service.ProductService;
4.  public class ProductController {
5.      private ProductService productService;
6.      public void save(Product product) {
7.          productService.save(product);              //调用业务服务层的保存方法
8.      }
9.  }
```

具体要求：

(1) 在 Spring IoC 容器中基于 XML 完成对 Bean 的装配,创建测试类,获取 ProductController 的实例,调用其 save()方法,在控制台上打印 Product 的信息。

(2) 在 Spring IoC 容器中基于注解(@Autowired)完成对 Bean 的装配,创建测试类,获取 ProductController 的实例,调用其 save()方法,在控制台上打印 Product 的信息。

3. 实现思路及步骤

对于要求(1)可参照 1.5.1 中示例的内容,主要是完成其配置文件中 Bean 的定义和依赖注入。对于要求(2)要在类中相关类和属性上添加注解,并在配置文件中添加注解解析器和包扫描器,具体步骤为

(1) 创建项目,并添加 Spring 相关支持。

(2) 创建相关的接口和类。

(3) 配置 Spring IoC 容器,即编写 applicatinContext.xml 配置文件。

(4) 创建测试类,通过 applicationContext 获取 ProductController 类的实例,调用其 save()方法。

第 2 章 Spring MVC 框架的基础知识

本章目标

1. 了解 Spring MVC 框架的基本构成。
2. 掌握 Spring MVC 框架的开发流程。
3. 理解几个常用注解的作用和使用方法。

常用的注解包括@Controller、@RequestMapping、@SessionAttribute、@SessionAttributes。

常用的两个模型数据为 Model 和 ModelAndView。

Spring Web MVC 是基于 Servlet API 构建的 Web 框架,从一开始就包含在 Spring Framework 中。正式名称 Spring Web MVC 来自其模块(spring-web mvc)的名称,但它通常被称为 Spring MVC。

2.1 MVC 设计模式

MVC 设计模式的任务是将包含业务数据的模块与显示模块的视图解耦。为了实现这个目标,在模型和视图之间引入重定向层可以解决问题。此重定向层是控制器,控制器将接收请求,执行更新模型的操作,然后通知视图关于模型更改的消息。其设计思想用图形表示如图 2-1 所示。

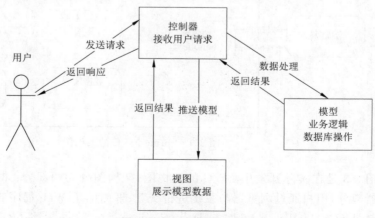

图 2-1 MVC 模式示意图

MVC 有如下特点：
- 多个视图可以对应一个模型。按 MVC 设计模式，一个模型对应多个视图，可以减少代码的复制及代码的维护量，这样，一旦模型发生改变，也易于维护。
- 模型返回的数据与显示逻辑分离。模型数据可以应用任何显示技术，例如，使用 JSP 页面、Velocity 模板或直接生成 Excel 文档等。
- Spring MVC 是对 MVC 模式的一个很好的实现。

2.2 Spring MVC 的架构

在介绍 Spring MVC 架构之前，先看看 Spring 的基本架构，如图 2-2 所示。

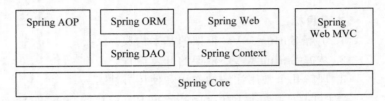

图 2-2　Spring 的基本架构

可以看到，在 Spring 的基本架构中，Spring Web MVC 是 Spring 基本架构里的一个组成部分，属于 Spring Frame Work 的后续产品，已经融合在 Spring Web Flow 里面，所以我们在后期和 Spring 进行整合的时候，几乎不需要别的配置。

Spring MVC 类似于 Struts2 的一个 MVC 框架，在实际开发中，接收浏览器的请求响应，对数据进行处理，然后返回页面进行显示，但是开发难度却比 Struts2 简单多了。而且由于 Struts2 暴露的安全问题，Spring MVC 已经成为大多数企业优先选择的框架。如图 2-3 所示为 Spring MVC 的架构图。

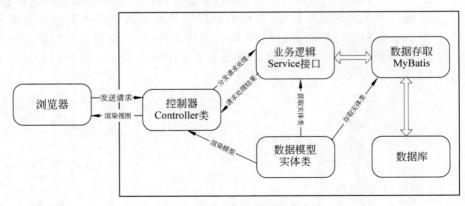

图 2-3　Spring MVC 的架构图

图 2-3 是在实际开发中 MVC 的架构图，架构图分为两部分：框外的是 Web 程序的浏览器部分，用户通过浏览器与系统进行交互；框内的是 Web 程序的后台部分，这部分包括控制器（Controller 类）、业务逻辑（Service 类）、数据模型（实体类）、数据持久层

（MyBatis 框架）和 MySQL 数据库管理系统。

在 MVC 架构中，JSP 页面就是视图，用户通过 JSP 页面发出请求后，Spring MVC 会根据请求路径将请求发给与请求路径对应的 Controller 类，Controller 类调用 Service 类对请求进行处理，Service 类会调用数据持久层 MyBatis 完成对实体类的存取和查询工作，并将处理结果返回 Controller 类，Controller 类将处理结果转换为 ModelAndView 对象，JSP 接收 ModelAndView 对象并进行渲染。

通过以上分析，总结出 Spring MVC 的如下特点：
- Spring MVC 是 Spring 框架的一部分，可以方便地与 Spring 框架集成。
- 灵活性强，易于与其他框架集成。
- 提供了一个前端控制器 DispatcherServlet，使开发人员无须开发额外的控制器。
- 可自动绑定用户输入，并能正确地转换数据类型。
- 内置了常见的校验器，可以校验用户输入。
- 支持国际化，可以根据用户区域显示多国语言。
- 支持多视图技术，如 JSP、JSON、XML、PDF、Velocity 和 FreeMarker 等。
- 使用基于 XML 的配置文件，编辑后无须重新编译应用程序。
- 支持注解开发，可以通过注解取代 XML 配置文件。

以上只是列出了 Spring MVC 的一些常见的特点，其实在应用这个框架的过程中，会发现框架更多的优点和特性，这也是 Spring 直到现在都一直深受程序员喜爱的重要原因。

2.3 开发一个 Spring MVC 简单应用示例

例 2-1 利用 Spring MVC 框架创建一个简单的 helloWorld 应用示例。

1. 创建项目，规划包的结构

在 Eclipse 中创建一个 Web 项目 chap2，在项目的 lib 目录中添加 Spring 的核心 JAR 包和 Spring MVC 所需的 JAR 包。同时创建 com.po 包和 com.controller 包，在 WEB-INF 下创建 JSP 文件夹用来存放 JSP 文件。将 JSP 文件放在 WEB-INF 下可以防止从浏览器中直接访问 JSP 页面，对 Web 资源起到一定的安全保护作用。这里，Spring 的版本是 5.1.8，项目添加的 JAR 包如图 2-4 所示。

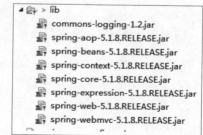

图 2-4 Spring MVC 所需 JAR 包

2. 配置前端控制器

在 web.xml 文件中配置 Spring MVC 的核心控制器 DispathcerServlet，如代码清单 2-1 所示。

代码清单 2-1：/chap2/WebContent/WEB-INF/web.xml

```xml
1.  <?xml version="1.0" encoding="UTF-8"?>
2.  <web-app xmlns:xsi="http://www.w3.org/2001/XMLSchema-instance"
3.   xmlns="http://xmlns.jcp.org/xml/ns/javaee"
4.   xsi:schemaLocation="http://xmlns.jcp.org/xml/ns/javaee
5.   http://xmlns.jcp.org/xml/ns/javaee/web-app_3_1.xsd" id="WebApp_ID" version="3.1">
6.    <display-name>chap2</display-name>
7.    <listener>
8.      <listener-class>org.springframework.web.context.ContextLoaderListener</listener-class>
9.    </listener>
10.   <context-param>
11.     <param-name>contextConfigLocation</param-name>
12.     <param-value></param-value>
13.   </context-param>
14.   <servlet>
15.     <servlet-name>app</servlet-name>
16.     <servlet-class>org.springframework.web.servlet.DispatcherServlet</servlet-class>
17.     <init-param>
18.       <param-name>contextConfigLocation</param-name>
19.       <param-value>/WEB-INF/springmvc-config.xml</param-value>
20.     </init-param>
21.     <load-on-startup>1</load-on-startup>
22.   </servlet>
23.   <servlet-mapping>
24.     <servlet-name>app</servlet-name>
25.     <url-pattern>*.do</url-pattern>
26.   </servlet-mapping>
27.   <welcome-file-list>
28.     <welcome-file>index.html</welcome-file>
29.     <welcome-file>index.htm</welcome-file>
30.     <welcome-file>index.jsp</welcome-file>
31.   </welcome-file-list>
32. </web-app>
```

- 第 8 行的 ContextLoaderListener 实现了 javax.servlet.ServletContextListener 接口，是 Servlet 监听器，其作用是可以在 Web 工程启动前和关闭时加入自定义的一些代码。如在 Web 启动前对 Spring IoC 容器的初始化，也可以在 Web 服务关闭时完成一些资源的释放。
- 第 11 行的 contextConfigLocation 是全局配置变量，用来配置 Spring IoC 配置文件的位置信息，这样 Spring 就会按照指定位置加载相应的配置文件。如果是多个配置文件，可以使用逗号将它们分隔开，并且它还支持正则表达式匹配。其默认值为/WEB-INF/applicationContext.xml。
- 第 14～22 行定义了一个名为 app 的 Servlet，其实现类是由 Spring 框架提供的 DispatcherServlet，它也是 Spring MVC 核心控制器，在此也可以由第 18 行的 contextConfigLocation 指定 Spring MVC 自己所需的配置文件的位置信息。在此指定的是/WEB-INF/目录下的 springmvc-config.xml 配置文件。
- 第 23 行的<servlet-mapping>中的标签<url-pattern>*.do/</url-pattern>定义了前面定义为 app 的 Servlet 对访问资源的控制范围和控制方式，这里的 *.do 表示拦截所有的以 *.do 结尾的请求。

3. 创建 Controller 控制器类

在 com.controller 包中创建 HelloWorldController 类，这个类要实现 Spring 框架中的 Controller 接口，如代码清单 2-2 所示。

代码清单 2-2：/chap2/src/com/controller/HelloController.java

```java
1.   package com.controller;
2.   import javax.servlet.http.HttpServletRequest;
3.   import javax.servlet.http.HttpServletResponse;
4.   import org.springframework.web.servlet.ModelAndView;
5.   import org.springframework.web.servlet.mvc.Controller;
6.   public class HelloController implements Controller{
7.       @Override
8.       public ModelAndView handleRequest(HttpServletRequest request,
         HttpServletResponse response) throws Exception {
9.           //创建 ModelAndView 对象
10.          ModelAndView mv=new ModelAndView();
11.          //将信息保存在 ModelAndView 对象中，这个对象中的值可传递到 JSP 页面中
12.          mv.addObject("msg","第一个 Hello World 程序");
13.          //设置要转发的页面
14.          mv.setViewName("/WEB-INF/jsp/helloWorld.jsp");
15.          return mv;
16.      }
```

第 8 行的 handleRequest()方法是实现 Controller 接口的方法，其作用是调用该方法处理请求，并返回一个包含视图名和模型数据的 ModelAndView 对象，在此返回的是 WEB-INF/jsp 目录中的 helloWorld.jsp 页面。

4. 创建 SpringMVC 的配置文件

在 WEB-INF 目录下创建配置文件 springmvc-config.xml，如代码清单 2-3 所示。

代码清单 2-3：/chap2/WebContent/WEB-INF/springmvc-config.xml

```xml
1.  <?xml version="1.0" encoding="UTF-8"?>
2.  <beans xmlns="http://www.springframework.org/schema/beans"
3.      xmlns:xsi="http://www.w3.org/2001/XMLSchema-instance"
4.      xsi:schemaLocation="http://www.springframework.org/schema/beans
5.      http://www.springframework.org/schema/beans/spring-beans-4.3.xsd">
6.      <!--配置处理器 Handle,映射"/helloController.do"请求 -->
7.      <bean name="/helloController.do"
8.          class="com.controller.HelloController" />
9.      <!--处理器映射器,将处理器 Handle 的 name 作为 url 进行查找 -->
10.     <bean class=
11.     "org.springframework.web.servlet.handler.BeanNameUrlHandlerMapping" />
12.     <!--处理器适配器,配置对处理器中 handleRequest()方法的调用-->
13.     <bean class=
14.     "org.springframework.web.servlet.mvc.SimpleControllerHandlerAdapter" />
15.     <!--视图解析器 -->
16.     <bean class=
17.     "org.springframework.web.servlet.view.InternalResourceViewResolver">
18.     </bean>
19. </beans>
```

配置文件中的第 7 行定义了 name="/helloController.do"的 Bean,表示定义了一个访问 URL 为/helloController.do 访问路径。

第 11 行配置了一个处理器映射器 BeanNameUrlHandlerMapping,作用是根据定义 Bean 的 name 值查找相应的处理类。如配置文件中定义的 URL 为 helloController.do 时,由映射器解析后将提交给 HelloController 类处理。

第 14 行配置了处理器适配器 SimpleControllerHandlerAdapter,作用是完成对处理类的方法的调用。如上面代码中的请求具体的处理方法是 HelloController 类中的 handleRequest()方法。

第 17 行配置了视图解析器 InternalResourceViewResolver,作用是完成视图渲染和解析,最终将 view 呈现给用户。

需要注意的是,在 Spring 4.0 以后,如果不配置处理器映射器、处理器适配器和视图解析器,Spring MVC 会配置默认的处理器,笔者在此写出配置处理过程,便于读者了解 Spring MVC 的工作过程,更好地理解 Spring 的工作原理。

5. 创建视图页面

在 WEB-INF/jsp 目录下创建 helloWorld.jsp 页面,页面中使用 EL 表达式获取 msg 中的信息,显示在页面上,如代码清单 2-4 所示。

代码清单 2-4：chap2/WebContent/WEB-INF/jsp/helloWorld.jsp

```jsp
1.  <%@page language="java" contentType="text/html; charset=UTF-8"
2.      pageEncoding="UTF-8"%>
```

```
3.    <!DOCTYPE html>
4.    <html>
5.    <head>
6.    <meta charset="UTF-8">
7.    <title>helloWorld</title>
8.    </head>
9.    <body>
10.   ${msg}
11.   </body>
12.   </html>
```

第 10 行的 ${msg} 就是 EL 表达式，将保存在 ModelAndView 中的数据取出显示在页面上。

6. 发布服务，测试应用

Spring MVC 示例的目录结构如图 2-5 所示。

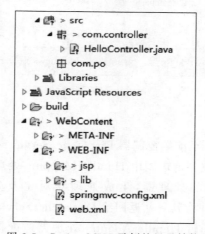

图 2-5　Spring MVC 示例的目录结构

将项目发布到 Tomcat 中并启动服务，在浏览器中访问地址 http://localhost/chap2/helloController.do，运行结果如图 2-6 所示。

图 2-6　Spring MVC 示例程序运行结果

从运行结果可以看到，页面中显示的信息就是保存在 ModelAndView 中 msg 变量的值,这是第一个 Spring MVC 框架的示例程序。

2.4 Spring MVC 的工作机制

下面以上面的示例为例讲解 Spring MVC 的工作过程。Spring MVC 运行原理如图 2-7 所示。

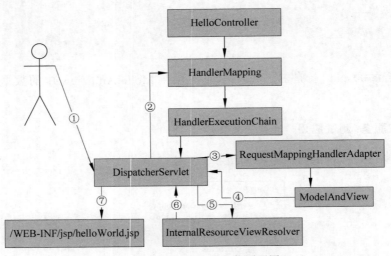

图 2-7 Spring MVC 运行原理图

其工作流程为:

① 用户发送请求 URL 至前端控制器 DispatcherServlet。

② DispatcherServlet 收到请求调用 HandlerMapping 处理器映射器。处理器映射器根据请求 URL 找到具体的处理器,生成处理器对象及处理器拦截器(两者组成 HandlerExecutionChain),并将其一并返回给 DispatcherServlet。

③ DispatcherServlet 通过 HandlerAdapter 处理器适配器调用处理器,执行处理器(HelloController,也叫后端控制器)。

④ HandlerAdapter 将 HelloController 执行结果 ModelAndView 返回给 DispatcherServlet。

⑤ DispatcherServlet 将 ModelAndView 传给 ViewReslover 视图解析器。

⑥ ViewReslover 解析后返回具体 View。

⑦ DispatcherServlet 对 View 进行渲染视图(即将模型数据填充至视图中),返回客户端。

以上 7 个步骤,DispatcherServlet、HandlerMapping、HandlerAdapter 和 ViewReslover 等对象协同工作,完成了 Spring MVC 请求-响应的整个流程,这些对象完成的工作对于开发者来说是不可见的,开发者并不需要关心这些对象是如何工作的,只在 Handler(Controller)中完成对请求的业务处理即可。

2.5 Spring MVC 基于注解的开发

目前开发的主流是采用注解开发,所以本书也采用注解为主的开发方式,使用注解开发 Spring MVC 非常简单高效。用到的注解主要是@Controller 和@RequestMapping 两个,@Controller 主要用于类上,@RequestMapping 可以用于类上,也可以用于方法上。当@Controller 主要用于类上,表示这个类用于控制器处理请求信息。

2.5.1 @Controller 注解

@Controller 注解用于标注类的实例是一个控制器,使用@Controller 注解的类不再需继承特定的父类或者实现特定的接口,相较之前的版本实现 Controller 接口变得更加简单。实现 Controller 接口的类只能处理一个单一的请求动作,而@Controller 注解的控制器可以同时支持处理多个请求动作,程序开发更加灵活、方便。

@Controller 用于标记一个类,使用它标记的类就是一个 Spring MVC Controller 对象,即一个控制器类。Spring 使用扫描机制查找应用程序中所有基于注解的控制器类。分发处理器会扫描使用了该注解的类的方法,并检测该方法是否使用了@RequestMapping 注解,而使用@RequestMapping 注解的方法才是真正处理请求的处理器。为了保证 Spring 能找到控制器,需要在 Spring MVC 的配置文件中做如下配置:

在 Spring MVC 的配置文件的头文件中引入 spring-context;

使用<context:component-scan/>元素启动包扫描功能,以便注册带有@Controller、@Service、@repository、@Component 等注解的类成为 Spring 的 Bean。配置文件如下所示:

```
<context:component-scan base-package="com.controller"/>
```

其中的 base-package 属性指定了要扫描的类包及其递归子包中所有的类都会被处理。所有的控制器类应该放在指定的包下,以方便 Spring IoC 容器生成相应的 Bean。

springmvc-config.xml 配置文件代码如代码清单 2-5 所示。

代码清单 2-5:/chapter12/src/springmvc-config.xml

```
1.  <?xml version="1.0" encoding="UTF-8"?>
2.  <beans xmlns="http://www.springframework.org/schema/beans"
3.     xmlns:xsi="http://www.w3.org/2001/XMLSchema-instance"
4.     xmlns:context="http://www.springframework.org/schema/context"
5.     xsi:schemaLocation="http://www.springframework.org/schema/beans
6.     http://www.springframework.org/schema/beans/spring-beans-4.3.xsd
7.     http://www.springframework.org/schema/context
8.     http://www.springframework.org/schema/context/spring-context-4.3.xsd">
9.     <!--指定需要扫描的包 -->
10.    <context:component-scan base-package="com.itheima.controller" />
11. </beans>
```

注意，代码中的黑体字部分是前面配置文件中没有的，这段代码的作用是Spring初始化时自动扫描包下的类，并根据注解生成相应的Bean，对应的控制器类的代码如代码清单2-6所示。

代码清单2-6：/chap2/src/com/controller/UserController.java

```
1.  package com.controller;
2.  import org.springframework.stereotype.Controller;
3.  @Controller
4.  public class UserController {
5.  }
```

2.5.2 @RequestMapping 注解

开发者需要控制器内部为每个请求动作开发相应的请求处理方法。用哪个类或方法处理请求可以通过@RequestMapping注解指示Spring，该注解可用于类或方法。当该注解作用于控制器类时，所有方法都将映射为相对于类级别的请求，表示该控制器处理的所有请求都被映射到value属性指示的路径下。

@RequestMapping注解的属性见表2-1。

表2-1 @RequestMapping 注解的属性

属性名	类型	是否必要	说明
consumes	String[]	否	用于指定处理请求的内容类型，如： consumes = "text/plain" consumes = {"text/plain", "application/*"}
headers	String[]	否	指定request中必须包含某些指定的header值，才能让该方法处理请求。 如 @RequestMapping(value = "/something", headers = "content-type=text/*")
name	String[]	否	指定映射地址的别名
params	String[]	否	指定request中必须包含某些参数值时，才让该方法处理
method	String[]	否	映射指定请求的方法类型，包括GET、POST、HEAD、OPTIONS、PUT、PATCH、DELETE、TRACE
produces	String[]	否	指定返回类型，返回类型必须是request请求头中包含的类型，如 produces = "text/plain" produces = {"text/plain", "application/*"} produces = MediaType.APPLICATION_JSON_UTF8_VALUE
value	String[]	否	用于映射一个请求的地址到方法上，可以标注在类和方法上，也是默认的属性。如 @RequestMapping("/login")相当于 @RequestMapping(value="/login")

1. @RequestMapping 注解标注在类上

@RequestMapping 是一个用来映射地址到请求方法上的注解，可以使用@RequestMapping 注释一个类，表示该控制器处理的所有请求都被映射到 value 属性值指定的路径下，示例代码如下所示。

```
1.  package com.controller;
2.  import org.springframework.stereotype.Controller;
3.  import org.springframework.web.bind.annotation.RequestMapping;
4.  @Controller
5.  @RequestMapping(value="/user")
6.  public class UserController {
7.  }
```

第 5 行的映射表示所有请求路径包含/user 的请求都由 UserController 处理，具体处理请求的方法根据方法映射的 value 值确定。

2. @RequestMapping 注解标注在方法上

将@RequestMapping 注解标注在方法上，表示该控制器处理的请求被映射到 value 属性值指定的方法。使用@RequestMapping 注解标注在方法上的示例代码如下。

```
1.  package com.controller;
2.  import org.springframework.stereotype.Controller;
3.  import org.springframework.web.bind.annotation.RequestMapping;
4.
5.  @Controller
6.  @RequestMapping(value="/user")
7.  public class UserController {
8.      @RequestMapping(value="/login")
9.      public String login(String name,String password ) {
10.         return null;
11.     }
12. }
```

示例代码中的@RequestMapping 分别作用到 UserController()和 login()方法上，此时可以通过下面的 url 向其发送请求处理：

http://localhost:8080/chap2/user/login

从客户端向服务器发送 HTTP 请求时，如果指定请求处理方法 POST()或 GET()，在服务端可通过@RequestMapping 的 method 属性是 get 或 post 进行相应的请求处理。示例代码如下。

```
1.  @Controller
2.  @RequestMapping(value="/user")
3.  public class UserController {
4.      @RequestMapping(value="/login",method=RequestMethod.POST)
5.      public String login(String name,String password) {
6.          return null;
7.      }
8.  }
```

第 4 行黑体字部分代码表示响应的是 POST()方法的请求,如果是 GET 请求,可以将 method 值设为 RequestMethod.GET。第 4 行的代码也可用@GetMapping 注解代替,如下所示。

```
1.  @Controller
2.  @RequestMapping(value="/user")
3.  public class UserController {
4.      @GetMapping(value="/login")
5.      public String login(String name,String password) {
6.          return null;
7.      }
8.  }
```

第 4 行黑体字部分的功能与下面的代码作用相同,

@RequestMapping(value="/login",method=RequestMethod.GET),

与请求方法 GET、POST、HEAD、OPTIONS、PUT、PATCH、DELETE、TRACE 都有对应的注解,如@PostMapping、@PutMapping 等。

2.5.3 @SessionAttribute 和@SessionAttributes 注解

这两个注解的作用是取代 HttpSession,当然也可以在 Controller 中用 HttpSession 完成这些会话功能,但这样就将 HttpSession 与控制器耦合在了一起,这不符合 Spring IoC 的设计原则。

@SessionAttributes 注解的作用是当某个类被注解后,Spring MVC 执行完控制器的逻辑后,将数据模型中对应的属性名称或者属性类型保存到 HTTP 的 Session 对象中。也就是说,配置这个注解时可以根据属性的名称或属性的类型匹配要保存的值或对象。具体的做法是将@SessionAttributes 注解添加到某一个 Controller 上,在这个 Controller 中满足@SessionAttributes 注解配置条件的值或对象就以键值对的方式被自动保存到 Session 作用域中。注意,@SessionAttributes 注解只能对类进行标注,示例代码如下。

```
1.  //处理用户请求的控制器
2.  @Controller
```

```
3.    @RequestMapping(value="/user")        //响应所有 user 路径下的处理请求
4.    @SessionAttributes(names={"regDate"},types={User.class})
5.    public class UserController {
6.
7.        @PostMapping(value="/reg")
                                  //这是处理用户注册的方法,响应 user/reg.do 的处理请求
8.        public String reg(User user,Model model) {
                                  //user 为响应请求的参数,接收用户的信息
9.            Date loginDate=new Date();        //注册时间
10.           model.addAttribute("user",user);//将 user 信息保存到 Model 对象中
11.           model.addAttribute("regDate",loginDate);        //保存注册时间
```

第 4 行配置了@SessionAttributes(names={"regDate"},types={User.class}),这个注解中的 names={"regDate"},表示在类中只要是与 regDate 这个名称匹配的变量,就会保存在 HttpSession 中,注解中的 types={User.class},表示在类中只要是与 User 这个类型的变量匹配,就会保存在 HttpSssion 中,其保存在 Session 中属性的名称就是类型名称,并且第一个字母小写,此处的名称是 user。

@SessionAttribute 注解的作用是获取保存在 HttpSession 中属性变量的值,一般作用于方法的形参上,示例代码如下。

```
1.    @RequestMapping(value="/login")        //这是处理用户登录的方法,响应 user/
                                              //reg.do 的处理请求
2.        public ModelAndView login (String loginname, String password,
      @SessionAttribute("user") User user1,ModelAndView mv ) {
3.        String msg;                        //登录是否成功信息
4.        if(loginname!=null&&loginname.equals(user1.getUsername())
5.            &&password!=null&&password.equals(user1.getPassword()))
6.        msg="登录成功";
7.        else
8.          msg="登录失败";
9.        mv.addObject("msg",msg);        //保存数据到 ModelAndView 对象中
10.       mv.setViewName("result");
                                  //将要转发的页面视图保存到 ModelAndView 对象中
11.       return mv;                        //返回 ModelAndView 对象
12.   }
```

第 2 行的代码@SessionAttribute("user") User user1,是将保存在 HttpSession 中的 user 变量的值取出后赋值给方法的形参 user1。

第 4 行代码是根据从页面传来的用户名和密码与保存在 session 中的 user 中的用户名和密码进行比对,如果相同,则将"登录成功"赋值给 msg 变量,否则将"登录失败"赋值给 msg 变量,最后将 msg 保存到 mv 对象中。

2.5.4 控制器处理请求方法的参数类型

控制器处理请求的方法可以有多个不同的参数,既可以是 Servlet 中的一些参数,如 HttpServletRequest、HttpSession 这样的参数,也可以是 Spring MVC 定义的一些数据类型,如 Model。下面列出的是可以在请求方法中使用的参数类型。

- javax.servlet.ServletRequest
- javax.servlet.ServletResponse
- javax.servlet.http.HttpSession
- javax.servlet.http.PushBuilder
- java.security.Principal
- java.util.Locale
- java.io.InputStream
- java.io.Reader
- java.io.OutputStream
- java.io.Writer
- @PathVariable
- @MatrixVariable
- @RequestParam
- @RequestHeader
- @CookieValue
- @RequestBody
- HttpEntity<B>
- @RequestPart
- RedirectAttributes
- @ModelAttribute
- java.util.Map
- org.springframework.ui.ModelMap
- org.springframework.ui.Model
- org.springframework.web.servlet.ModelAndView

上面列出的这些参数是将来有可能在请求方法中使用的参数,其中比较常用的是最后两个 Model 和 ModelAndView 类型参数。这两个不是 Servlet API 类型,而是 Spring MVC 中定义的类型,如果处理请求的方法中添加了这两个类型参数,则每次调用方法时,Spring 都会创建相应的对象,并将其作为参数传递给方法。利用此参数可以在不同的请求方法或 JSP 页面与控制器请求方法之间进行数据传递。Model 的源程序如下:

```
1.    package org.springframework.ui;
2.    import java.util.Collection;
3.    import java.util.Map;
```

```
4.    import org.springframework.lang.Nullable;
5.    public interface Model {
6.        Model addAttribute(String attributeName, @Nullable Object attributeValue);
7.        Model addAttribute(Object attributeValue);
8.        Model addAllAttributes(Collection<?>attributeValues);
9.        Model addAllAttributes(Map<String, ?>attributes);
10.       Model mergeAttributes(Map<String, ?>attributes);
11.       boolean containsAttribute(String attributeName);
12.       Map<String, Object>asMap();
13.   }
```

从源代码可以看出，Model 是一个接口，提供了一些保存属性值的方法，其中常用的方法是第 6 行的 addAttribute()方法，这个方法类似于 Servlet 中的 request.setAttribute()方法，其功能是将一个值或对象保存在 Model 对象中，在转发的 JSP 页面或请求的处理方法中可直接得到保存到 Model 对象中的值。ModelAndView 对象既可以保存键值对，也可以保存视图名。

2.6 请求处理方法的返回类型

在控制器类中，每个请求处理方法都可以有多个不同的参数，以及一个或多个类型的返回值。返回值可以是 String、Model 和 ModelAndView 等。下面列出了可以作为返回类型的参数列表。

- @ResponseBody
- @ModelAttribute
- HttpEntity<B>
- ResponseEntity<B>
- Java.lang.String
- java.util.Map
- org.springframework.ui.Model
- ModelAndView object
- Void
- DeferredResult<V>
- Callable<V>
- ListenableFuture<V>
- java.util.concurrent.CompletionStage<V>
- java.util.concurrent.CompletableFuture<V>
- ResponseBodyEmitter，SseEmitter
- StreamingResponseBody

2.6.1 Model 类型的使用

在请求处理方法中可出现和返回的参数类型中，最重要的是 Model 和 ModelAndView 两种类型。如果是只封装数据，可以用 Model；如果是数据和视图同时存在，可以用 ModelAndview 进行封装。在处理方法中，Model 对象常使用如下方法保存数据模型：

addAttribute(String attributeName,Object attributeValue)

示例代码如下：

```
1.    package com.controller;
2.    import org.springframework.stereotype.Controller;
3.    import org.springframework.ui.Model;
4.    import org.springframework.web.bind.annotation.GetMapping;
5.    import org.springframework.web.bind.annotation.RequestMapping;
6.    @Controller
7.    @RequestMapping(value="/user")
8.    public class UserController {
9.        @GetMapping(value="/login")
10.       public String login(String name,String password,Model model) {
11.           String msg;                              //登录是否成功信息
12.           if(name!=null&&name.equals("admin")&&password!=null&&password
              .equals("123456"))
13.               msg="登录成功";
14.           else
15.               msg="登录失败";
16.           model.addAttribute("msg",msg);   //保存数据到 Model 对象中
17.           return "result";                 //返回 result.jsp 页面
18.       }
19.   }
```

第 16 行将提示信息保存在 Model 对象中，在转发的 result.jsp 页面中可以利用 JSP 的 EL 表达式 ${msg}输出保存在 Model 对象中的值，这样就通过 Model 对象实现了数据在 Controller 和 View 之间的传递。当然，Model 对象中不只是能存入值对象，还可以保存任意类型的对象，包括集合类型的数据。

2.6.2 ModelAndView 类型的使用

在控制器的请求处理方法中，如果返回结果中既包含数据，也包含视图，则可以考虑使用 ModelAndView 返回类型，因为这个对象既可以保存数据，也可以保存视图。ModelAndView 的源代码如下：

```
1.   public class ModelAndView {
2.       /** 这里有很多构造方法重载 */
```

```
3.     public ModelAndView(View view, String modelName, Object modelObject) {
4.         this.view = view;
5.         addObject(modelName, modelObject);
6.     }
7.     //保存view
8.     public void setViewName(@Nullable String viewName) {
9.         this.view=viewName;
10.    }
11.    //以键值对的方式保存属性的值,实际上最终保存在ModelMap对象中
12.    public ModelAndView addObject (String attributeName, @Nullable
       Object attributeValue) {
13.        getModelMap().addAttribute(attributeName, attributeValue);
14.        return this;
15.    }
16. }
```

第8行代码表示当通过 ModelAndView 对象调用 setViewName() 方法时,实际上是将视图名保存在 View 对象中。第12行的 addObject() 表示当调用 addObject() 方法时,是将数据保存在 ModelMap 对象中,同时返回当前对象。下面给出使用 ModelAndView 对象的示例代码。

```
1.  @Controller
2.  @RequestMapping(value="/user")
3.  public class UserController {
4.      @GetMapping(value="/reg")
5.      public ModelAndView reg(User user, ModelAndView mv) {
6.          mv.addObject("user", user);
7.          mv.setViewName("success");
8.          return mv;
9.      }
10. }
```

这是一段控制器请求处理方法的代码,当 reg() 方法接收到请求后,首先将传递过来的 user 对象保存在 ModelAndView 的对象 mv 中,如第6行代码。其次,希望控制器处理完成后转发到 success.jsp 页面,将要转发的 JSP 页面的名称保存在 mv 中,如第7行代码所示。最后返回 mv 对象。这段代码处理完请求后会转发到 success.jsp 页面。

如果希望处理请求方法 reg() 处理请求完成后重定向到页面 success.jsp,可修改第7行代码为

```
mv.setViewName("redirect:/success.jsp");
```

如果希望处理请求方法 reg() 处理请求完成后重定向另外一个请求处理方法 login(),可修改第7行代码为

```
mv.setViewName("redirect:/login");
```

如果希望处理请求方法 reg() 处理请求完成后转发到另外一个请求处理方法，如 login() 方法，可修改第 7 行代码为

```
mv.setViewName("forward:/login");
```

2.6.3 返回类型为 String

在 Spring MVC 的控制类的请求处理方法中，可以将 String 作为方法的返回值。当 String 作为处理方法返回值时，会出现两种情况：一种是返回一个视图页面，如 JSP 页面；另一种是调用另一个处理方法。

现在假设有一个添加用户的处理请求方法 addUser()，处理完成后转发到一个表示处理成功的 success.jsp 页面，其示例代码如下所示。

```
1.  @Controller
2.  @RequestMapping(value="/user")
3.  public class UserController {
4.      @RequestMapping("/addUser")
5.      public String addUser(User user,Model model) {
6.          System.out.println("add user");
7.          model.addAttribute("user",user);
8.          return "success";
9.      }
10. }
```

这是一段控制器请求处理方法的代码，当 addUser() 方法接收到请求后，在控制台打印一个"add user"字符串，模拟添加用户操作，然后将传递过来的 user 对象保存在 model 对象中。第 8 行代码表示请求处理方法处理完成后转发到 success.jsp 页面。如果想将转发变为重定向到 success.jsp 页面，可将第 8 行代码改为

```
return "redirect:success.jsp";
```

这样，addUser() 方法处理完请求后将重定向到 success.jsp 页面。

在程序处理过程中经常需要处理完一个请求后，不是立即转发到一个页面，而是转发到另一个处理请求的方法，此时只在原来的字符串前面加上"forward:"即可。下面通过一个示例代码进行演示。

```
1.  @Controller
2.  @RequestMapping(value="/user")
3.  public class UserController {
4.      @RequestMapping("/addUser")
5.      public String addUser(User user,Model model) {
6.          System.out.println("add user");
```

```
7.            model.addAttribute("user",user);
8.            return "forward:list";      //转发到 list()方法
9.        }
10.       @RequestMapping("/list")
11.       public String list() {
12.           System.out.println("show all user");
13.           return "list";
14.       }
```

上面这段代码有两个方法：addUser()和 list()。addUser()方法的返回值如第 8 行代码，表示该方法处理完请求后，不是直接转发到 JSP 页面，而是调用 list()方法。list()方法处理完请求后才转发到 list.jsp 页面。

2.7 一个基于注解开发的示例

下面是一个示例，说明如何在 Controller 中使用注解进行开发。

例 2-2 项目的需求是基于 Spring MVC 注解的方式完成一个用户注册和登录的过程，具体要求是：

(1) 主页显示用户名和密码两个文本框，同时有一个"注册"按钮，页面为 reg.jsp。

(2) 单击"注册"按钮后提交给 UserController 的 reg()方法。

(3) 在 reg()方法中将从页面传递来的用户名和密码保存在 Model 对象中，同时转发到 login()方法，login()方法处理完成后转发到页面 success.jsp。

(4) 在 success.jsp 页面显示用户名和密码。

开发步骤：

(1) 在 Eclipse 创建一个 Dynamic Web Project 项目，名称为 chap21，在项目中添加所需的 JAR 包，具体操作与 2.3 节的示例步骤 1 相同。

(2) 配置前端控制器，在 web.xml 文件中配置 Spring MVC 的核心控制器 DispathcerServlet，具体操作与 2.3 节的示例步骤 2 相同。

(3) 创建 Controller 控制器类。

在 com.Controller 包中创建 UserController 类，在类中使用@Controller 注解，无须再实现 Controller 接口，如代码清单 2-7 所示。

代码清单 2-7：/chap21/src/com/controller/UserController.java

```
1.  /*添加@Controller 注解表示这是控制器,处理用户请求,不需再实现任何接口,
2.    Spring 通过包扫描机制会自动加载这个类,并创建 Bean */
3.  //处理用户请求的控制器
4.  @Controller
5.  @RequestMapping(value="/user")        //响应所有 user 路径下的处理请求
6.  @SessionAttributes(names={"regDate","user"})
```

```java
7.  public class UserController {
8.
9.      @PostMapping(value="/reg")
        //这是处理用户注册的方法,响应 user/reg.do 的处理请求
10.     public String reg(User user,Model model) {
        //user 为响应请求的参数,接收用户的信息
11.         Date loginDate=new Date();                          //注册时间
12.         model.addAttribute("user",user);
            //将 user 信息保存到 Model 对象中
13.         model.addAttribute("regDate",loginDate);            //保存注册时间
14.         return "forward:toLogin.do";
            //方法处理结束转发到 toLogin 方法
15.     }
16.     @RequestMapping(value="/toLogin")                       //响应 toLogin.do 请求
17.     public String toLogin() {
18.         return "login";                                     //转发到 login.jsp 页面
19.     }
20.     @RequestMapping(value="/login")
        //这是处理用户登录的方法,响应 user/reg.do 的处理请求
21.     public ModelAndView login (String loginname, String password,
        @SessionAttribute("user") User user,ModelAndView mv ) {
22.         String msg;                                         //登录是否成功信息
23.
24.         if(loginname!=null&&loginname.equals(user.getUsername())
25.             &&password!=null&&password.equals(user.getPassword()))
26.         msg="登录成功";
27.         else
28.             msg="登录失败";
29.         mv.addObject("msg",msg);
            //保存数据到 ModelAndView 对象中
30.         mv.setViewName("result");
            //将要转发的页面视图保存到 ModelAndView 对象中
31.
32.         return mv;                                          //返回 ModelAndView 对象
33.     }
34.
35. }
```

第 10 行响应注册请求处理的 reg()方法,在此方法中将注册时间和用户信息保存到 Model 类型变量 model 中,同时也保存到 HttpSession 中,然后转发到 toLogin()方法。在 reg()方法中有一个封装类 User 的对象 user,user 对象中的两个属性 username 和 password 可以接收从页面文本框传递过来的值,user 对象能正确接收变量值的前提是,文本框中的 name 的值应与 user 对象中的两个属性的值一致。

第17行的toLogin()方法是用来转发到login.jsp页面的,因为login.jsp页面在WEB-INF目录中,不能直接访问,只能通过这种方式访问。

第21行的login()方法,主要功能是处理登录请求,把登录页面传递过来的用户名和密码与保存在Session中的user进行对比,最后转发到视图result.jsp页面。

(4) 创建SpringMVC的配置文件。

在此创建的Spring MVC配置文件与2.3节的配置文件有所不同,这里要配置包扫描机制,对添加注解的类进行解析,同时配置视图解析器,如代码清单2-8所示。

代码清单2-8：/chap21/WebContent/WEB-INF/springmvc-config.xml

```xml
1.  <?xml version="1.0" encoding="UTF-8"?>
2.  <beans xmlns="http://www.springframework.org/schema/beans"
3.      xmlns:xsi="http://www.w3.org/2001/XMLSchema-instance"
4.      xmlns:context="http://www.springframework.org/schema/context"
5.      xsi:schemaLocation="http://www.springframework.org/schema/beans
6.      http://www.springframework.org/schema/beans/spring-beans-4.3.xsd
7.      http://www.springframework.org/schema/context
8.      http://www.springframework.org/schema/context/spring-context-4.3.xsd">
9.      <!--指定需要扫描的包 -->
10.     <context:component-scan base-package="com.controller" />
11.         <!--定义视图解析器 -->
12.     <bean id="viewResolver" class=
13.     "org.springframework.web.servlet.view.InternalResourceViewResolver">
14.         <!--设置前缀 -->
15.         <property name="prefix" value="/WEB-INF/jsp/" />
16.         <!--设置后缀 -->
17.         <property name="suffix" value=".jsp" />
18.     </bean>
19. </beans>
```

第10行代码配置包扫描机制,通过component-scan标签的base-package属性指定要扫描的包的路径,放在这个路径下的所有类都能被扫描到,Spring会根据标注的注解创建相应的Bean,以及根据它们之间的依赖关系完成属性的注入。

第12行配置的是视图解析器,在这里配置了id为viewResolver的视图解析器,这个类有prefix和suffix两个属性,分别表示前缀和后缀。配置这两个属性的目的是当控制器Controller中处理请求方法返回值类型为字符串时,只需指定返回的页面文件名称,无须指定文件所在文件夹和扩展文件名,这样省去了很多重复代码。如代码清单2-7中的login()方法的返回值为result,则其对应的页面的完整路径为Prefix的前缀＋result＋suffix的后缀,其结果为：/WEB-INF/jsp/result.jsp。

(5) 创建视图。

在WEB-INF/jsp目录下创建注册页面reg.jsp,此页面中有两个文本框,分别是用户名和密码，同时添加了不能为空的校验规则，处理请求的 action 为 user/reg.do，如代码清单 2-9 所示。

代码清单 2-9：/chap2/WebContent/WEB-INF/jsp/reg.jsp

```
1.   <script>
2.   // 判断登录账号和密码是否为空
3.   function check(){
4.       var username =$("#usename").val();
5.       var password =$("#password").val();
6.       if(username=="" || password==""){
7.           $("#message").text("账号或密码不能为空！");
8.           return false;
9.       }
10.      return true;
11.  }
12.  </script>
13.  </head>
14.  <body>
15.      <form action="${pageContext.request.contextPath}/user/reg.do"
16.            method="post" onsubmit="return check()">
17.          <br /><br />
18.          账号：<input id="username" type="text" name="username" />
19.          <br /><br />
20.          密码：<input id="password" type="password" name="password" />
21.          <br /><br />
22.          <center><input type="submit" value="注册" /></center>
23.      </form>
24.  </body>
```

第 3 行的 check() 方法是一个 JavaScript 函数，当 form 表单提交时，会调用这个函数进行表单的校验，验证两个文本框是否为空，如果为空，给出提示信息，同时返回 reg.jsp 页面；如果都不为空，才能提交给服务器。

第 15 行的 action 属性设置了表单提交服务器处理请求的地址，这里为 chap21/user/reg.do，经过 Spring MVC 的 DispatcherServlet 控制器处理后映射到 UserController 类的 reg() 方法处理这个具体的请求。

第 18 行和第 20 行是两个文本框，对应的是用户名和密码，注意这里的 name 属性的值要与 reg() 方法中的形参 user 对象的属性名称对应起来，user 对象有两个属性 username 和 password，所以两个文本框的 name 的值也必须与之一致。

在 WEB-INF/jsp 目录下还要创建一个 login.jsp 页面，这个页面用于接收用户登录的信息，如代码清单 2-10 所示。

代码清单 2-10：/chap21/WebContent/WEB-INF/jsp/login.jsp

```
1.   <%@page language="java" contentType="text/html; charset=UTF-8"
2.       pageEncoding="UTF-8"%>
3.   <!DOCTYPE HTML>
4.   <html>
5.   <body>
6.         <%--错误提示信息--%>
7.         <span id="message">${msg}</span>
8.   <form action="${pageContext.request.contextPath}/user/login.do"
9.   method="post" onsubmit="return check()">
10.    <br /><br />
11.  账号：<input id="usercode" type="text" name="loginname" />
12.    <br /><br />
13.  密码：<input id="password" type="password" name="password" />
14.    <br /><br />
15.
16.  <center><input type="submit" value="登录" /></center>
17.       </form>
18.    </body></html>
```

此页面用来接收用户登录的信息，并将用户名和密码提交给处理登录的方法 login()，此处要注意用户名 loginname、密码 password 要和 login() 方法中的形参一一对应。

在 WEB-INF/jsp 目录下还要创建一个 result.jsp 页面，在页面中使用 jsp 中的 EL 表达式 ${} 获取保存在 Model 中的数据信息，并显示在页面上，如代码清单 2-11 所示。

代码清单 2-11：/chap2/WebContent/WEB-INF/jsp/result.jsp

```
1.   <%@page language="java" contentType="text/html; charset=UTF-8"
2.       pageEncoding="UTF-8"%>
3.   <!DOCTYPE html>
4.   <html>
5.   <head>
6.   <meta charset="UTF-8">
7.   <title>helloWorld</title>
8.   </head>
9.   <body>
10.  ${user.username}
11.  ${msg}
12.  </body>
13.  </html>
```

由于注册页面 reg.jsp 存放在 WEB-INF/jsp 目录中，而从浏览器中不能访问 WEB-INF 目录下的资源，此时可在 WebContent 目录下创建一个 index.jsp 页面，在页面中添加一个 forward 转发语句，将请求转发到 reg.jsp 页面，如代码清单 2-12 所示。

代码清单 2-12：/chap2/WebContent/index.jsp

```
1.   <%@page language="java" contentType="text/html; charset=UTF-8"
2.       pageEncoding="UTF-8"%>
3.   <!--转发到注册页面 -->
4.   <jsp:forward page="/WEB-INF/jsp/reg.jsp"/>
```

index.jsp 页面存放在工程的根目录下，可以在 web.xml 中设置为主页，这样在访问这个网站时，可自动转发到 reg.jsp 页面。

（6）发布此 Web 工程，在浏览器中访问网站的主页，如图 2-8 所示。

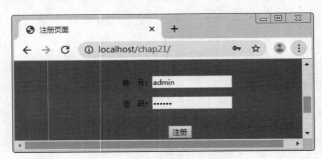

图 2-8　示例程序的注册页面

单击"注册"按钮后，提交给 Web 服务器的控制，在此处提交到了 UserController 中的 reg()方法，然后转发到 login()方法，该方法处理请求后转发到 result.jsp 页面，如图 2-9 所示。

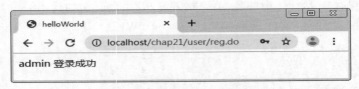

图 2-9　示例程序登录成功页面

至此，一个基于 Spring MVC 注解开发的示例基本完成，并运行成功。

习　　题

1. 下面有关 Spring MVC 的说法，错误的是(　　)。
    A. Spring MVC 是 MVC 模式的一种实现
    B. Spring MVC 是 Spring 的一个模块
    C. Spring MVC 支持多种视图，如 JSP 等
    D. Spring MVC 也是一种设计模式
2. Spring MVC 的核心控制器是一个(　　)。
    A. Filter　　　　　B. Servlet　　　　　C. Listener　　　　　D. Controller

3. 有关@RequestMapping注解的作用,正确的说法是(　　)。
   A. 这个注解只能作用于类上
   B. 这个注解的作用是用于映射处理请求的类或方法
   C. 这个注解只能作用于方法上
   D. 这个注解只能处理GET请求
4. 有关Spring MVC中@Controller注解的说法,错误的是(　　)。
   A. 可以标注一个类
   B. 这个注解可以与@RequestMapping结合使用
   C. 可以标注一个方法
   D. 可以通过包扫描机制加载@Controller标注的类
5. 在控制器的请求处理方法中,如果返回结果中既包含数据,也包含视图,则可以考虑使用返回类型(　　)。
   A. Model　　　　　　　　　　B. ModelAndView
   C. String　　　　　　　　　　D. ResponseBody

实验2　Spring MVC基于注解开发

1. 实验目的

掌握Spring MVC常用注解的使用方法,主要掌握@Controller、@RequestMapping注解。方法的返回类型主要是掌握Model和ModelAndView两种类型的使用。

2. 实验内容

现有Customer(客户)类,根据业务需求对客户信息进行登记,Customer类的代码如下:

```
1.   package com.po;
2.   public class Customer {
3.       private int id;                    //用户id
4.       private String username;           //客户名称
5.       private String telephone;          //电话号码
6.       private String sex;                //性别
7.       private int age;                   //年龄
8.       /**setter and getter **/
9.   }
10.  }
```

具体要求如下:
(1) 基于Spring MVC注解方式完成实验内容。
(2) 创建一个添加Customer信息的页面addCustomer.jsp,页面中的性别要用下拉列表框显示。

(3) 单击"添加"按钮后，在 showCustomer.jsp 页面显示 Customer 的信息。

(4) 所有页面都放在 WEB-INF/jsp 目录下。

3. 实现思路及步骤

(1) 创建 Web 工程，添加 Spring MVC 所需的 JAR 包，创建必要的包。

(2) 修改 web.xml，配置 DispatcherServlet 和 ContextLoaderListener。

(3) 创建 CustomerController 类和处理添加请求的方法 addCustomer()，同时为类和方法添加注解。在 addCustomer() 方法中，形参用 Customer 的对象接收页面 addCustomer.jsp 传递进来的数据。

(4) 创建 springmvc-config.xml 文件，并配置包扫描和视图解析器，视图解析器要配置前缀和后缀两个属性。

(5) 在 WEB-INF/jsp 目录下创建添加客户信息的页面 addCustomer.jsp 和显示信息的页面 showCustomer.jsp。

第 3 章 Spring MVC 的组件开发

本章目标
1. 了解 Spring MVC 框架中的拦截器,掌握拦截器的创建和配置方法。
2. 掌握文件的上传和下载的基本原理和使用方法。
3. 了解 Spring MVC 标签库,掌握几个常用标签的使用方法。

3.1 拦 截 器

Spring MVC 中的拦截器(interceptor)类似于 Servlet 中的过滤器(filter),主要用于拦截用户请求并作相应的处理。例如,通过拦截器可以进行权限验证、记录请求信息的日志、判断用户是否登录等。使用 Spring MVC 中的拦截器,须对拦截器类进行定义和配置。

3.1.1 Spring MVC 拦截器的设计

在开发拦截器之前,要先了解拦截器的设计结构。拦截器必须实现 HandlerInterceptor 接口。为增强功能,开发了多个拦截器。Spring MVC 拦截器接口与类设计结构如图 3-1 所示。

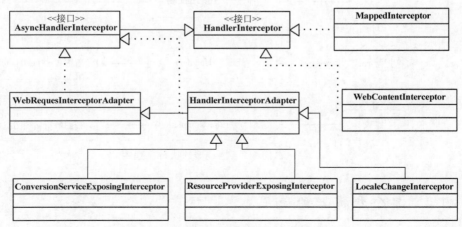

图 3-1 Spring MVC 拦截器接口与类设计结构

从图 3-1 中可以看出，所有的类都直接或间接实现了 HandlerInterceptor 接口，HandlerInterceptor 的源代码如下。

```
1.  package org.springframework.web.servlet;
2.  public interface HandlerInterceptor {
3.      default boolean preHandle(HttpServletRequest request, HttpServletResponse response, Object handler)
4.          throws Exception {
5.      return true;
6.      }
7.      default void postHandle(HttpServletRequest request, HttpServletResponse response, Object handler,
8.          @Nullable ModelAndView modelAndView) throws Exception {
9.      }
10.     default void afterCompletion(HttpServletRequest request, HttpServlet-Response response, Object handler,
11.         @Nullable Exception ex) throws Exception {
12.     }
13. }
```

HandlerInterceptor 接口中的 3 个方法描述如下。
- preHandle()方法：该方法在控制器方法被调用前执行，其返回值为一个 boolean 值，如果为 true，表示继续向下执行；如果为 false，会中断后面的所有操作。
- postHandle()方法：该方法在控制器方法被调用后执行，且在解析视图之前执行，可以通过此方法对请求域中的模型和视图做进一步修改。
- afterCompletion()方法：该方法在视图渲染结束之后执行，即该方法在整个请求完成后，无论是否产生异常都会执行。

拦截器的执行流程可用如图 3-2 所示的流程图表示。

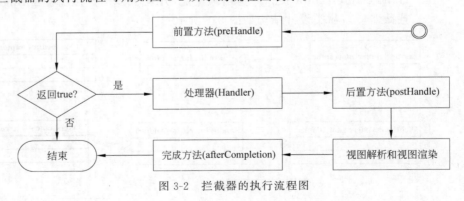

图 3-2　拦截器的执行流程图

3.1.2　单个拦截器的使用

要使用 Spring MVC 中的拦截器，需要对拦截器类进行定义和配置。通常，拦截器类可以通过两种方式定义。
- 通过实现 HandlerInterceptor 接口。
- 继承 HandlerInterceptor 接口的实现类(如 HandlerInterceptorAdapter)定义。

以继承 HandlerInterceptorAdapter 类为例,自定义拦截器,拦截对象是所有以 *.do 结尾的请求资源,类如代码清单 3-1 所示。

代码清单 3-1:/chap3/src/com/interceptor/UserInterceptor.java

```java
1.  package com.interceptor;
2.  import javax.servlet.http.HttpServletRequest;
3.  import javax.servlet.http.HttpServletResponse;
4.  import org.springframework.lang.Nullable;
5.  import org.springframework.web.servlet.ModelAndView;
6.  import org.springframework.web.servlet.handler.HandlerInterceptorAdapter;
7.  public class UserInterceptor extends HandlerInterceptorAdapter {
8.      public boolean preHandle(HttpServletRequest request, HttpServletResponse response, Object handler)
9.          throws Exception {
10.         System.out.println("preHandle of UserInterceptor ");
11.         return true;
12.     }
13.     public void postHandle(HttpServletRequest request, HttpServletResponse response, Object handler,
14.         @Nullable ModelAndView modelAndView) throws Exception {
15.         System.out.println("postHandle of UserInterceptor ");
16.     }
17.     public void afterCompletion ( HttpServletRequest request, HttpServletResponse response, Object handler,
18.         @Nullable Exception ex) throws Exception {
19.         System.out.println("afterCompletion of UserInterceptor ");
20.     }
21. }
```

在这个自定义的拦截器中,每个方法中只是在控制台上输出一个字符串,表示当前方法被调用。

下面是控制器的一个处理请求的方法,如代码清单 3-2 所示。

代码清单 3-2:com/controller/HelloController.java

```java
1.  @Controller
2.  public class HelloController {
3.      @RequestMapping("/hello")
4.      public ModelAndView handleRequest (HttpServletRequest request, HttpServletResponse response) throws Exception {
5.          //创建 ModelAndView 对象
```

```
6.        System.out.println("第一个 Hello World 程序");
7.        ModelAndView mv=new ModelAndView();
8.        //将信息保存在 ModelAndView 对象,这个对象中的值可传递到 JSP 页面中
9.        mv.addObject("msg","第一个 Hello World 程序");
10.       //设置要转发的页面
11.       mv.setViewName("helloWorld");
12.       return mv;
13.    }
14. }
```

在这个请求处理方法中,在控制台上打印了一个字符串,表示当前方法被调用,然后转发到 helloWorld.jsp 页面。

最后,在 Spring MVC 的配置文件中添加拦截器的配置信息,如代码清单 3-3 所示。

代码清单 3-3:/chap3/WebContent/WEB-INF/springmvc-config.xml

```xml
1.  <?xml version="1.0" encoding="UTF-8"?>
2.  <beans xmlns="http://www.springframework.org/schema/beans"
3.    xmlns:mvc="http://www.springframework.org/schema/mvc"
4.    xmlns:xsi="http://www.w3.org/2001/XMLSchema-instance"
5.    xmlns:p="http://www.springframework.org/schema/p"
6.    xmlns:context="http://www.springframework.org/schema/context"
7.    xsi:schemaLocation="
8.        http://www.springframework.org/schema/beans
9.        https://www.springframework.org/schema/beans/spring-beans.xsd
10.       http://www.springframework.org/schema/context
11.       https://www.springframework.org/schema/context/spring-context.xsd
12.       http://www.springframework.org/schema/mvc
13.       https://www.springframework.org/schema/mvc/spring-mvc.xsd">
14.   <!--指定需要扫描的包 -->
15.   <context:component-scan base-package="com.controller" />
16.   <!--定义视图解析器 -->
17.   <bean id="viewResolver" class=
18.   "org.springframework.web.servlet.view.InternalResourceViewResolver">
19.       <!--设置前缀 -->
20.       <property name="prefix" value="/WEB-INF/jsp/" />
21.       <!--设置后缀 -->
22.       <property name="suffix" value=".jsp" />
23.   </bean>
24.   <mvc:interceptors>
25.       <mvc:interceptor>
```

```
26.        <mvc:mapping path="/*.do"/>
27.        <bean class="com.interceptor.UserInterceptor"/>
28.    </mvc:interceptor>
29.  </mvc:interceptors>
30. </beans>
```

第 3、12、13 行黑体字部分是定义 mvc 命名空间和所需的 schema 文件位置信息。第 24 行的 <mvc:interceptors> 是一个容器标签，其子标签是 <mvc:interceptor>，用这个标签可定义多个拦截器。

在 <mvc:interceptor> 标签下有两个子标签，第 26 行的 <mvc:mapping path="/*.do">，表示拦截后缀为 *.do 的所有请求 uri。第 27 行定义实现这个拦截器的类。

首先，发布此服务，然后在浏览器中访问此服务，其运行结果如图 3-3 所示。

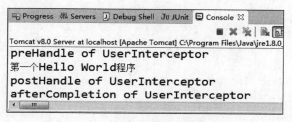

图 3-3　单个拦截器运行结果

从运行结果看，拦截器中的 3 个方法调用次序与前面分析的结果一致。

3.1.3　多个拦截器的使用

当一个项目比较大，业务逻辑复杂，需求也在不断变化的时候，可以添加多个拦截器满足实际工程的需求，此时多个拦截器中的方法执行次序也发生了变化。下面通过有多个拦截器存在时的示例说明拦截器中的方法与处理器的方法之间的先后执行次序。第一个拦截器的定义如代码清单 3-4 所示。

代码清单 3-4：/chap3/src/com/interceptor/Interceptor1.java

```
1.  public class Interceptor1 extends HandlerInterceptorAdapter {
2.    public boolean preHandle (HttpServletRequest request, HttpServletResponse response, Object handler)
3.        throws Exception {
4.      System.out.println("preHandle of UserInterceptor1 ");
5.      return true;
6.    }
7.    public void postHandle (HttpServletRequest request, HttpServletResponse response, Object handler,
8.        @Nullable ModelAndView modelAndView) throws Exception {
9.      System.out.println("postHandle of UserInterceptor1 ");
10.   }
```

```
11.    public void afterCompletion(HttpServletRequest request,
       HttpServletResponse response, Object handler,
12.        @Nullable Exception ex) throws Exception {
13.        System.out.println("afterCompletion of UserInterceptor1 ");
14.    }
15. }
```

第二个拦截器的定义如代码清单 3-5 所示。

代码清单 3-5：/chap3/src/com/interceptor/Interceptor2.java

```
1. public class Interceptor2 extends HandlerInterceptorAdapter {
2.     public boolean preHandle(HttpServletRequest request, HttpServletResponse
       response, Object handler)
3.         throws Exception {
4.     System.out.println("preHandle of UserInterceptor2 ");
5.     return true;
6.     }
7.     public void postHandle(HttpServletRequest request, HttpServletResponse
       response, Object handler,
8.         @Nullable ModelAndView modelAndView) throws Exception {
9.     System.out.println("postHandle of UserInterceptor2 ");
10.    }
11.    public void afterCompletion(HttpServletRequest request,
       HttpServletResponse response, Object handler,
12.        @Nullable Exception ex) throws Exception {
13.     System.out.println("afterCompletion of UserInterceptor2 ");
14.    }
15. }
```

在 Spring MVC 配置文件中添加拦截器配置，如代码清单 3-6 所示。

代码清单 3-6：/chap3/WebContent/WEB-INF/springmvc-config.xml

```xml
1.  <mvc:interceptors>
2.      <mvc:interceptor>
3.          <mvc:mapping path="/*.do"/>
4.          <bean class="com.interceptor.UserInterceptor1"/>
5.      </mvc:interceptor>
6.      <mvc:interceptor>
7.          <mvc:mapping path="/*.do"/>
8.          <bean class="com.interceptor.UserInterceptor2"/>
9.      </mvc:interceptor>
10. </mvc:interceptors>
```

控制器处理请求的方法还是用代码清单 3-2 所示代码，启动服务后，在浏览器中访问处理请求的映射地址，运行结果如图 3-4 所示。

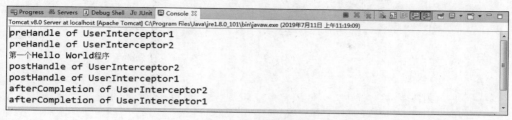

图 3-4　多个拦截器运行结果

从运行结果分析，在执行请求方法 handleRequest()之前，依次执行 Interceptor1 和 Interceptor2 中的 prdHandle()方法，在执行 handleRequest()之后，反序执行拦截器中的另外两个方法。多拦截器各方法之间执行次序如图 3-5 所示。

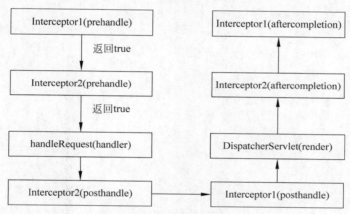

图 3-5　多拦截器各方法之间执行次序

3.1.4　拦截器应用——用户权限验证

拦截器的作用是在请求处理方法执行前，执行拦截器中的 prehandle()方法，这样可以预先处理一些事务，如判断一个用户是否已经登录一个网站，如果已经登录，可直接进入网站；如果没有登录，则转发到登录页面。下面是一个通过拦截器完成用户权限验证的实例。

例 3-1　当用户访问系统所有资源前进行权限的验证，如果是已登录的用户，可直接进入相应的请求处理方法；如果是未登录的用户，则重定向到登录页面。

（1）编写控制器类，如代码清单 3-7 所示。

代码清单 3-7：/chap3/src/com/controller/UserController1.java

```
1.   package com.controller;
2.   //处理用户请求的控制器
3.   @Controller
4.   //响应所有 userInterceptor 路径下的处理请求
```

```
5.    @RequestMapping(value="/userInterceptor")
6.    public class UserController1 {
7.        @GetMapping(value="/login")                    //响应login.do请求
8.        public String login() {
9.            return "/login";                            //转发到login.jsp页面
10.       }
11.   //这是处理用户登录的方法,响应userInterceptor/login.do的处理请求
12.       @PostMapping(value="/login")
13.       public String login(User user,Model model,HttpSession session ) {
14.           String msg;                                 //登录是否成功信息
15.           if(user.getUsername()!=null&&user.getUsername().equals("admin")
16.              &&user.getPassword()!=null&&user.getPassword().equals(
17.              "123456"))
18.           {
19.               msg="登录成功";
20.               model.addAttribute("user",user);        //保存数据到Model对象中
21.               session.setAttribute("user",user);      //保存数据到session对象中
22.               return "main";                          //转发到main.jsp
23.           }
24.           else
25.               msg="登录失败";
26.           model.addAttribute("msg",msg);              //保存数据到Model对象中
27.           return "login";                             //转发到login.jsp
28.       }
29.       @RequestMapping(value="/main")                  //响应main.do请求
30.       public String main() {
31.           return "main";                              //转发到main.jsp页面
32.       }
33.       @GetMapping(value="/logout")                    //响应logout.do请求
34.       public String logout(HttpSession session) {
35.           session.invalidate();                       //当前的session失效
36.   //重定向到Login.do请求
37.           return "redirect:/userInterceptor/login.do";
38.       }
39.   }
```

(2) 编写拦截器类,决定哪些请求是允许的,哪些请求是非法的并加以拒绝,如代码清单3-8所示。

代码清单3-8：/chap3/src/com/interceptor/LoginInterceptor.java

```
1.    public class LoginInterceptor extends HandlerInterceptorAdapter {
2.        //进入Handler()方法之前执行
3.        //可用于身份认证、身份授权。如果认证没有通过,表示用户没有登录,需要此方法拦
          //截不再往下执行,否则就放行
```

```
4.      @Override
5.      public boolean preHandle(HttpServletRequest request, HttpServletResponse
        response, Object handler)
6.              throws Exception {
7.          //获取请求的 url
8.          String url=request.getRequestURI();
9.          //判断 url 是否公开地址(实际使用时将公开地址配置到配置文件中)
10.         //这里假设公开地址是登录提交的地址
11.         if (url.indexOf("login.do") >0) {
12.             //如果进行登录提交,放行
13.             return true;
14.         }
15.         //判断 session
16.         HttpSession session=request.getSession();
17.         //从 session 中取出用户身份信息
18.         User user=(User) session.getAttribute("user");
19.         if (user !=null) { //如果用户存在,表示已经登录,放行
20.             return true;
21.         }
22.         //执行到这里表示用户身份需要验证,跳转到登录页面
23.         request.getRequestDispatcher("/WEB-INF/jsp/login.jsp").forward
            (request, response);
24.         return false;
25.     }
26. }
```

（3）在 springmvc-config.xml 中配置自定义的拦截器,如代码清单 3-9 所示。

代码清单 3-9：/chap3/WebContent/WEB-INF/springmvc-config.xml

```
1.  <mvc:interceptors>
2.  <mvc:interceptor>
3.  <!--只拦截/userInterceptor 路径下的请求-->
4.  <mvc:mapping path="/userInterceptor/*.do"/>
5.  <!--自定义的拦截器类-->
6.  <bean class="com.interceptor.LoginInterceptor"/>
7.  </mvc:interceptor>
8.  </mvc:interceptors>
```

配置文件中其他省略的代码与代码清单 3-3 相同。

（4）视图页面,main.jsp 作为主页,login.jsp 是登录页面,在没有登录前,只能显示 login.jsp 页面,其他页面是不允许访问的,只有登录后,才能访问其他页面,如 main.jsp。 main.jsp 页面如代码清单 3-10 所示。

代码清单 3-10：/chap3/WebContent/WEB-INF/jsp/main.jsp

```
1.  <body>
2.  当前用户：${user.username}
3.  <a href="${pageContext.request.contextPath}/userInterceptor/logout.do">退出</a>
4.  
5.  </body>
```

login.jsp 页面如代码清单 3-11 所示。

代码清单 3-11：/chap3/WebContent/WEB-INF/jsp/login.jsp

```
1.  <form action="${pageContext.request.contextPath}/userInterceptor/login.do" method="post" onsubmit="return check()">
2.      <br /><br />
3.  账号：<input id="usercode" type="text" name="username" />
4.      <br /><br />
5.  密码：<input id="password" type="password" name="password" />
6.      <br /><br />
7.  <center><input type="submit" value="登录" /></center>
8.  </form>
```

代码清单 3-11 中只给出了 form 部分的代码，其他代码与前面的相同。

（5）发布服务，启动 Tomcat，首先访问 main.do，出现如图 3-6 所示的页面。

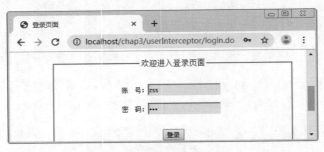

图 3-6　login.jsp 页面

由于用户没有登录，此时访问 main.do 时被拦截器拦截，跳转到 login.jsp 页面，此时输入正确的用户名和密码，单击"登录"按钮后进入 main.jsp 页面，如图 3-7 所示。

图 3-7　main.jsp 页面

登录成功后，可以随时进入主页面 main.jsp，但当单击"退出"链接后，当前的 session

失效,将不能再直接访问 main.jsp,而是跳转到 index.jsp 页面。

3.2 文件的上传与下载

在实际工程中,文件的上传与下载是必备的功能之一,Spring MVC 对此给出了很好的解决方案,可以方便地实现这个功能。

3.2.1 文件的上传

Spring MVC 为上传文件提供了良好的支持。首先,Spring MVC 的文件上传是通过 MultipartResolver(Multipart 处理器)处理的,对于 MultipartResolver 来说,它只是一个接口,有两个实现类。

- CommonsMultipartResolver 类,依赖于 Apache 下的 jakarta Common FileUpload 项目解析 Multipart 请求,它可以在 Spring 的各个版本中使用,只是它依赖于第三方 JAR 包才能实现。
- StandardServletMultipartResolver 类,是 Spring 3.1 版本后的产物,它依赖于 Servlet 3.0 或者更高版本的实现,但不依赖于第三方 JAR 包。

两个实现类与 MultipartResolver 接口的关系如图 3-8 所示。

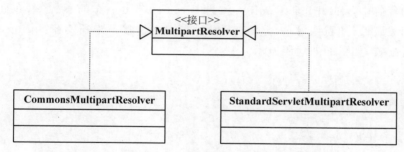

图 3-8 两个实现类与 MultipartResolver 接口的关系

下面以第二种方案实现为例,讲解 Spring MVC 如何实现文件的上传功能。

要实现文件上传,需要前端与后端服务器配合才能实现。首先,在前端页面上要有一个上传文件的表单,表单需要满足下面 3 个条件。

- form 表单的 method 属性设置为 post。
- form 表单的 enctype 属性设置为 multipart/form-data。
- 设置<input>标签的 type="file"。

示例代码如下:

```
1.  <form action="upload.do" enctype="multipart/form-data"
2.        method="post" >
3.  <input type="file" name="file" />
4.  <input type="submit" value="上传文件" multiple="multiple"/>
```

上述代码中,上传文件除了满足 3 个必备条件外,还增加了 multiple 属性,该属性是 HTML 5.0 中的新特性,如果使用了该属性,同时选中多个文件进行上传,即可实现多文件上传。

当客户端 form 表单的 enctype 属性为 multipart/form-data 时,浏览器会采用二进制流的形式处理表单的数据,服务器端就会对文件上传的请求进行解析处理。Spring MVC 为文件上传提供了直接的支持,这是通过 MultipartResolver 接口实现的。

MultipartResolver 接口有两个实现类,前面已经介绍过,这里用 Spring MVC 自己提供的 StandardServletMultipartResolver 类完成文件的上传。由于这个类是基于 Servlet 的,所以要在 Web.xml 中进行配置,示例代码如下。

```
1.   <multipart-config>
2.         <!--临时文件的目录-->
3.         <location>e:/mvc/upload</location>
4.         <!--上传文件最大 2MB -->
5.         <max-file-size>2097152</max-file-size>
6.         <!--上传文件整个请求不超过 4MB -->
7.         <max-request-size>4194304</max-request-size>
8.   </multipart-config>
```

上面的代码中,<multipart-config> 作为 DispatcherServlet 的子标签,在此是设置上传文件的相关属性,如默认目录,上传文件大小限制等。接下来还要在 Spring MVC 配置文件中进行配置,如下面的代码所示。

```
1.   <!--配置文件上传解析器-->
2.   <bean id="multipartResolver"
3.         class="org.springframework.web.multipart.support.StandardServletMultipartResolver">
```

因为 MultipartResolver 接口的实现类 StandardServletMultipartResolver,内部是引用 MultipartResolver 字符串获取该实现对象并完成文件解析的,所以在配置这个 Bean 的实例时,必须指定 Bean 的 id 为 multipartResolver。

当完成页面表单和文件上传解析器的配置后,需要在 Controller 中编写文件上传的方法实现文件上传的功能。在 Spring MVC,实现文件上传功能其实很简单,其代码如下。

```
1.   //处理上传文件请求的方法及方法的形参
2.   @RequestMapping(value ="/upload")
3.         public ModelAndView upload(
4.   @RequestParam("username") String username,
5.   HttpServletRequest request,
6.   MultipartFile file)
```

该方法是用来处理上传文件请求的,其中的 file 是一个 MultipartFile 引用变量,用来接收前端页面传递过来的文件,并将文件封装在这个对象中。MultipartFile 接口中提供了一些获取文件信息的方法，见表 3-1。

表 3-1　MultipartFile 接口中提供的获取文件信息的方法

方　　法	说　　明
Byte[] getBytes()	以字节数组的形式返回文件的内容
String getContentType()	返回文件的内容
InputStream getInputStream()	读取文件内容，返回一个 InputStream 流
String getName()	获取多部件 form 表单的参数名称
String getOriginalFilename()	获取上传文件的原始文件名
Boolean isEmpty()	判断上传文件的内容是否为空
Void transferTo(File file)	保存上传文件到目标目录中

例 3-2　文件上传示例，要求在一个 JSP 页面单击"浏览"按钮，打开一个文件管理窗口，选中某一个或多个文件后，可以上传到服务器指定的目录下。

具体实现步骤如下。

（1）创建 Web 工程，添加 Spring MVC 的 JAR 包，修改 web.xml 配置文件，添加 SpringMVC 必需的 listener 和 servlet。

（2）创建上传文件页面 fileUpload.jsp，如代码清单 3-12 所示。

代码清单 3-12：/chap3/WebContent/WEB-INF/jsp/fileUpload.jsp

```
1.    <form action="${pageContext.request.contextPath}/file/upload.do"
2.          enctype="multipart/form-data"
3.          method="post" onsubmit="return check()">
4.          <br /><br />
5.       上传人：<input id="username" type="text" name="username" />
6.          <br /><br />
7.       请选择文件：<input id="filename" type="file" name="file" />
8.          <br /><br />
9.          <center><input type="submit" value="上传" /></center>
10.   </form>
```

代码第 1 行中的 action 请求的地址是 file/upload.do，对应的是 FileUploadController 控制器类中的 upload() 方法，在此处理上传文件的请求。

（3）创建控制器类 FileUploadController，如代码清单 3-13 所示。

代码清单 3-13：/chap3/src/com/controller/FileUploadController.java

```
1.   @Controller
2.   @RequestMapping("/file")
3.   public class FileUploadController {
4.       @RequestMapping(value = "/toUpload")
5.       public String toUpload() {
6.           return "fileUpload";
```

```java
7.     }
8.     //处理上传文件请求的方法
9.     @RequestMapping(value="/upload")
10.    public ModelAndView upload(@RequestParam("username") String username,
            HttpServletRequest request,MultipartFile file) {
11.        ModelAndView mv=new ModelAndView();
12.        //指定要上传的文件所在路径
13.        String path =request.getServletContext().getRealPath("/upload/");
14.        //获取原始文件名
15.        String fileName=file.getOriginalFilename();
16.        fileName=path+File.separator+fileName;
17.        file.getContentType();
18.        //创建目标文件
19.        File dest=new File(fileName);
20.        String msg=null;
21.        try {
22.            //保存文件
23.            file.transferTo(dest);
24.            msg="上传文件成功";
25.        }catch(IllegalStateException|IOException e) {
26.            msg="上传文件失败";
27.            e.printStackTrace();
28.        }
29.        mv.addObject("msg", msg);        //提示信息保存在ModelAndView对象中
30.    //用户信息保存在ModelAndView对象中
31.        mv.addObject("username",username);
32.        mv.setViewName("result");
33.        return mv;
34.    }
35. }
```

第10行的MultipartFile file是MultipartFile接口的封装对象，其实现类是在Spring MVC中配置的StandardServletMultipartResolver类，此处的参数对应表单中的上传文件控件提交的文件信息。

第13行指定上传文件保存在服务中的路径，如果不指定，将保存在web.xml中指定的默认路径。

第15行得到上传文件原来的名字，在此也可以根据需要重新对文件进行命名。

第23行将上传文件保存到指定的服务器路径下。

（4）创建springmvc-config.xml配置文件，如代码清单3-14所示。

代码清单 3-14：/chap3/WebContent/WEB-INF/springmvc-config.xml

```xml
1.  <beans><!--schema 配置信息省略 -->
2.      <!--指定需要扫描的包 -->
3.      <context:component-scan base-package="com.controller" />
4.      <!--定义视图解析器 -->
5.      <bean id="viewResolver" class=
6.  "org.springframework.web.servlet.view.InternalResourceViewResolver">
7.          <!--设置前缀 -->
8.          <property name="prefix" value="/WEB-INF/jsp/" />
9.          <!--设置后缀 -->
10.         <property name="suffix" value=".jsp" />
11.     </bean>
12.     <!--配置文件上传解析器 -->
13.     <bean id="multipartResolver"
14.         class=" org. springframework. web. multipart. support. Standard-
            ServletMultipartResolver">
15.     </bean>
16. </beans>
```

第 14 行配置的 bean 是 MultipartFile 接口的实现类，此处通过 Spring IoC 创建类的实例。

（5）修改 web.xml 配置文件，修改后的代码如代码清单 3-15 所示。

代码清单 3-15：/chap3/WebContent/WEB-INF/web.xml

```xml
1.  <servlet-name>app</servlet-name>
2.  <servlet-class>org.springframework.web.servlet.DispatcherServlet
    </servlet-class>
3.  <init-param>
4.      <param-name>contextConfigLocation</param-name>
5.      <param-value>/WEB-INF/springmvc-config.xml</param-value>
6.  </init-param>
7.  <load-on-startup>1</load-on-startup>
8.  <multipart-config>
9.      <!--临时文件的目录-->
10.     <location>e:/mvc/upload</location>
11.     <!--上传文件最大 2MB -->
12.     <max-file-size>2097152</max-file-size>
13.     <!--上传文件整个请求不超过 4MB -->
14.     <max-request-size>4194304</max-request-size>
15. </multipart-config>
16.
17. </servlet>
```

第 8～15 行配置了 MultipartFile 接口实现类 StandardServletMultipartResolver 的几个属性，这几个属性是上传文件必需的。

（6）发布服务，启动 Tomcat，在浏览器中输入地址
http://localhost/chap3/file/toUpload.do，浏览器中出现如图 3-9 所示的页面。

图 3-9　上传文件页面

单击"选择文件"按钮，打开目录管理窗口，选择要上传的文件，并输入上传人的姓名，单击"上传"按钮后，出现上传文件成功的页面，此时文件被上传到项目工程所在目录的 upload 目录中。

如果工程的名字为 chap3，发布到 Eclipse 的 Tomcat 中，则发布后工程的实际目录为

F:\workspace\.metadata\.plugins\org.eclipse.wst.server.core\tmp0\wtpwebapps
\chap3

其中 F:\workspace 是 Eclipse 的工作空间所在目录。

3.2.2　文件的下载

文件下载看似简单，其实还是要分几种情况分别进行处理。最简单的情况是在页面给出一个超链接，该链接 href 的属性等于要下载文件的文件名，就可以实现文件下载了。这种情况只适用于非中文文件件名，而且文件不是很大。如果该文件的文件名为中文文件名，在某些早先的浏览器上就会导致下载失败；如果使用最新的 Firefox、Chrome、Opera、Safari，则都可以正常下载文件名为中文的文件。为了保证下载成功，此时要对文件名进行转码处理，避免出现中文乱码。如果要下载的文件很大，而且要随时掌握下载的状态，此时可通过服务器读取本地文件流，然后将文件流输出到客户端，在这个过程中可以轻易获取文件传输过程中的各个参数。下面给出的是利用 ResponseEntity 类型实现的一种文件下载方式。

例 3-3　创建一个下载文件的页面，单击下载链接可下载服务器端指定的文件，要求对汉字文件名也可以正常下载。

（1）创建一个下载页面，如代码清单 3-16 所示。

代码清单 3-16：/chap3/WebContent/WEB-INF/jsp/download.jsp

```
1.    <%@page language="java" contentType="text/html; charset=UTF-8"
2.      pageEncoding="UTF-8"%>
3.    <%@page import="java.net.URLEncoder"%>
```

```
4.    <!DOCTYPE html PUBLIC "-//W3C //DTD HTML 4.01 Transitional //EN"
5.       "http://www.w3.org/TR/html4/loose.dtd">
6.    <html>
7.    <head>
8.    <meta http-equiv="Content-Type" content="text/html; charset=UTF-8">
9.    <title>下载页面</title>
10.   </head>
11.   <body>
12.   <a href="${pageContext.request.contextPath}/file/download.do?
      filename=第1章 spring的基础知识.pptx">文件下载   </a>
13.   </body>
14.   </html>
15.   </html>
```

此处的下载文件名为"第1章 spring 的基础知识.pptx",是存放在服务器端指定下载目录中的文件,此处是发布工程目录下的 upload 目录。

（2）在控制器类 FileUploadController 中创建处理下载文件请求的方法 fileDownload()，如代码清单 3-17 所示。

代码清单 3-17：/chap3/src/com/controller/FileUploadController.java

```
1.    //处理文件下载请求的方法
2.    @RequestMapping("/download")
3.     public ResponseEntity< byte[ ]> fileDownload (HttpServletRequest request,
4.      @RequestParam("filename") String filename,
5.      @RequestHeader("User-Agent") String userAgent) throws Exception{
6.    //指定要下载的文件所在路径
7.     String path=request.getServletContext().getRealPath("/upload/");
8.     System.out.println(path);
9.    //创建该文件对象
10.    File file=new File(path+File.separator+filename);
11.   //该方法返回一个 static ResponseEntity.BodyBuilder 对象
12.    BodyBuilder builder=ResponseEntity.ok();
13.
14.    builder.contentType(MediaType.APPLICATION_OCTET_STREAM);
15.   //对文件名编码,防止中文文件乱码
16.    filename=URLEncoder.encode(filename,"UTF-8");
17.   //不同的浏览器,处理方式不同,要根据浏览器版本区别对待
18.   //如果是 IE,只用 UTF-8 字符集进行 URL 编码即可
19.    if(userAgent.indexOf("MSIE")>0) {
20.     builder.header("Content-Disposition","attachment;filename="+
       filename);}
```

```
21.      //而 FireFox、Chrome 等浏览器,则需要说明编码的字符集
22.      else
23.      {builder.header("Content-Disposition","attachment;filename*=UTF
         -8''"+filename);
24.      }
25.
26.      return builder.body(FileUtils.readFileToByteArray(file));
27.    }
```

ResponseEntity 继承了 HttpEntity,是 HttpEntity 的子类,第 12 行的 ResponseEntity.ok() 方法返回一个 BodyBuilde,可用于 RestTemplate 和 Controller 层方法。第 26 行的 builder.body() 方法返回一个 ResponseEntity 对象。

(3) 发布服务,并启动 Tomcat,分别在不同的浏览器中进行测试。下面是在 Chrome 浏览器中测试结果。在地址栏中输入网址 http://localhost/chap3/file/toDownload.do,在页面上单击"文件下载"超链接,出现如图 3-10 所示的页面。

图 3-10 下载文件页面

从运行结果看,中文文件名下载是正常的,对不同的浏览器进行文件下载测试,程序都能正常运行。

3.3 Spring 的表单标签库

从 Spring 2.0 版开始,Spring 提供了一组全面的可绑定数据的标签库,用于在使用 JSP 和 Spring Web MVC 时处理表单元素,每个标签都支持其相应 HTML 标签对应的属性集。与其他表单/输入标签库不同,Spring 的表单标签库与 Spring Web MVC 集成,使标签可以访问控制器处理的命令对象和引用数据。正如我们在以下示例中所示,表单标记使 JSP 更易于开发、读取和维护。

表单标签库实现类在 spring-webmvc.jar,标签库描述符文件是 spring-form.tld。要在页面上使用此标签库中的标签,需要在 JSP 页面的头部添加如下的 JSP 指令:

```
<%@taglib prefix="form" uri="http://www.springframework.org/tags/form" %>
```

3.3.1 form 标签

使用 Spring 的 form 标签的作用与 HTML 中的 form 功能类似,只是属性有所不同,其中一个重要的属性 modelAttribute 是 Spring 所特有的,它可以绑定模型属性名称,默认值是 command,可自定义修改。示例代码如下：

```
<form:form modelAttribute="user" method="post" action="">
...
</form:form>
```

包含这段代码的 JSP 页面要保证在 Model 中存在 user 这个属性变量名称。

3.3.2 input 标签

Spring MVC 中的 input 标签会被渲染成一个类型为 HTML input 的标签,使用 Spring 标签的好处是可以绑定表单数据,一般是通过 input 标签的 path 属性绑定 Model 中的值。示例代码如下：

```
1.  <form:form method="post" action="addProduct.do" >
2.    <table>
3.      <tr>
4.        <td>产品名称:</td>
5.        <td><form:input path="name"/></td>
6.      </tr>
7.      <tr>
8.        <td>生产日期:</td>
9.        <td><form:input path="pd"/></td>
10.     </tr>
11.     <tr>
12.       <td>价格:</td>
13.       <td><form:input path="price"/></td>
14.     </tr>
15.     <tr>
16.       <td><input type="submit" value="提交"/></td>
17.     </tr>
18.   </table>
19. </form:form>
```

这里的 path 属性相当于 HTML 中 input 标签的 name 属性,只是这里的 path 属性可以与 Model 中的值进行绑定。除此标签外,如 password、hidden、textarea 等标签都最终渲染为 HTML 对应的标签,主要区别是有一个 path 属性绑定与 Controller 中请求处理方法的 Model 属性对应的属性值。

3.3.3 checkboxes 标签

Spring MVC 的 checkboxes 标签会渲染多个类型为 checkbox 的普通 HTML 标签。checkboxes 常用的属性主要有两个：path 和 items，具体含义如下。
- path 属性：要绑定属性路径，与 Controller 请求方法的参数对应。
- items 属性：用于生成 checkbox 元素的对象的 Collection、Map 或者 Array。

使用 checkboxes 标签时，以上两个属性是必须指定的，items 表示当前要用来显示的项的集合数据，而 path 绑定的表单对象的属性表示当前表单对象拥有的项，即在 items 显示的所有项中，表单对象拥有的项会被设定为选中状态。

例 3-4 以用户的爱好为例演示 checkboxes 标签的使用方法。

（1）创建 TagController 控制器，生成填充复选框选项的值，如代码清单 3-18 所示。

代码清单 3-18：chap3/src/com/controller/TagController.java

```java
1.  package com.controller;
2.  import java.util.ArrayList;
3.  import java.util.List;
4.  import org.springframework.stereotype.Controller;
5.  import org.springframework.ui.Model;
6.  import org.springframework.web.bind.annotation.RequestMapping;
7.  import com.po.User;
8.  @Controller
9.  public class TagController {
10.     //处理选择爱好的请求方法
11.     @RequestMapping("/hobby")
12.     public String hobby(Model model) {
13.         //创建一个 user 对象
14.         User user=new User();
15.         //hobbys 用于存放当前用户的所有爱好
16.         List<String>hobbys=new ArrayList<String>();
17.         hobbys.add("篮球");
18.         hobbys.add("游泳");
19.         //将 hobbys 赋值给 user 对象
20.         user.setHobbys(hobbys);
21.         //用于在页面上显示所有爱好的选项
22.         List<String>hobbyList=new ArrayList<String>();
23.         hobbyList.add("篮球");
24.         hobbyList.add("围棋");
25.         hobbyList.add("游泳");
26.         hobbyList.add("画画");
27.         //将 user 保存在 Model 中
28.         model.addAttribute("user",user);
```

```
29.        //将 hobbyList 保存在 Model 中
30.        model.addAttribute("hobbyList",hobbyList);
31.        //转发到 checkboxesForm.jsp 页面
32.        return "checkboxesForm";
33.     }
34.     //处理显示已选择爱好的请求方法
35.     @RequestMapping("showHobbys")
36.     public String showHobby(User user,Model model) {
37.        //将页面传递过来的 user 保存在 Model 中,以便在 showHobbys 页面中显示
38.        model.addAttribute("user",user);
39.        //转发到 showHobbys.jsp
40.        return "showHobbys";
41.     }
42. }
```

这个 TagController 类中有两个方法：hobby()和 showHobby()。hobby()方法是生成了一个 user 对象和这个对象属性值 hobbys,同时生成一个 hobbyList,其值是页面上所需的所有爱好的选项。showHobby()方法用于将提交的爱好选项转发到 showHobbys.jsp 页面显示。

（2）创建 checkboxes.jsp 页面,在页面中显示爱好选项,并选择爱好,如代码清单 3-19 所示。

代码清单 3-19：/chap3/WebContent/WEB-INF/jsp/checkboxes.jsp

```
1.  <%@page language="java" contentType="text/html; charset=UTF-8"
2.    pageEncoding="UTF-8"%>
3.  <%@taglib prefix="form" uri="http://www.springframework.org/tags/form" %>
4.  <!DOCTYPE html PUBLIC "-//W3C//DTD HTML 4.01 Transitional//EN" "http://
    www.w3.org/TR/html4/loose.dtd">
5.  <html>
6.  <head>
7.  <meta http-equiv="Content-Type" content="text/html; charset=UTF-8">
8.  <title>测试 checkboxes 标签</title>
9.  </head>
10. <body>
11. <h3>form:checkboxes 测试</h3>
12. <form:form modelAttribute="user" method="post" action="showHobbys.do" >
13.    <table>
14.       <tr>
15.          <td>选择爱好:</td>
16.          <td>
17.             <form:checkboxes items="${hobbyList}" path="hobbys"/>
```

```
18.            </td>
19.            <td><input type="submit" value="提交"/></td>
20.        </tr>
21.    </table>
22. </form:form>
23. </body>
24. </html>
```

第 12 行 modelAttribute="user" 绑定了 Model 中的 user 属性值，此处的作用是在所有的爱好选项中，与 user 对象中的 hobbys 中的值对应的选项表示为选中。

第 17 行 items="${hobbyList}"将 Model 中的 hobbyList 属性绑定到 items 的属性上，用于显示所有的爱好选项，path="hobbys"表示提交服务器处理请求方法中的参数名称，此参数的类型是一个 String 数组。

（3）显示所选爱好信息的页面 showHobbys.jsp，如代码清单 3-20 所示。

代码清单 3-20：/chap3/WebContent/WEB-INF/jsp/showHobbys.jsp

```
1.  <%@page language="java" contentType="text/html; charset=UTF-8"
2.      pageEncoding="UTF-8"%>
3.      <%@taglib prefix="c" uri="http://java.sun.com/jsp/jstl/core"%>
4.  <!DOCTYPE html>
5.  <html>
6.  <head>
7.  <meta charset="UTF-8">
8.  <title>显示爱好</title>
9.  </head>
10. <body>
11.    <table>
12.    <tr><td>爱好:
13.    <c:forEach items="${user.hobbys}" var="hobby">
14.        ${hobby}
15.    </c:forEach></td>
16.    </tr>
17.    </table>
18. </body>
19. </html>
```

第 13 行用 JSTL 标签库中的循环标签 forEach 对 user 对象的 hobbys 集合属性进行遍历，第 14 行用 EL 表达式 ${hobby} 输出集合中的所有元素。

在浏览器地址栏中输入网址 http://localhost/chap3/hobby.do，显示如图 3-11 所示的页面。

选中所有的 4 个爱好，单击"提交"按钮，在 showHobbys.jsp 页面显示选中的爱好选项，如图 3-12 所示。

图 3-11　生成 checkboxes 标签

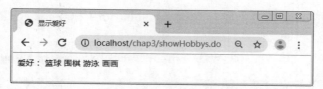

图 3-12　接收 checkboxes 标签选项并显示

3.3.4　radiobuttons 标签

radiobuttons 标签会渲染成多个类型为 radio 的普通 HTML input 标签。一个 radiobuttons 标签将根据其绑定的数据生成多个单选框按钮。radiobuttons 标签绑定的数据可以是数组、List 和 Map。radiobuttons 标签有两个重要的属性是必须设置的：一个是 items；另一个是 path。items 属性绑定一个集合，用来表示有哪些单选按钮可选。path 属性绑定的对象表示当前表单对象拥有的项，在 items 显示的所有项中，表单对象拥有的项会被设定为选中状态。

例 3-5　利用 radiobuttons 标签选择用户性别，并显示选择的性别。

（1）在 TagController 中添加 sex() 方法，用于处理选择性别的请求，如代码清单 3-21 所示。

代码清单 3-21：/chap3/src/com/controller/TagController.java

```
1.   //处理选择性别的请求方法
2.   @RequestMapping("/sex")
3.   public String sex(Model model) {
4.       //创建一个 user 对象
5.       User user=new User();
6.       user.setSex("男");
7.       //用于在页面上显示所有性别的单选按钮选项
8.       List<String>sexList=new ArrayList<String>();
9.       sexList.add("男");
10.      sexList.add("女");
11.      //将 user 保存在 Model 中
12.      model.addAttribute("user",user);
13.      //将 sexList 保存在 Model 中
```

```
14.        model.addAttribute("sexList",sexList);
15.        //转发到 radiobuttons.jsp 页面
16.        return "radiobuttons";
17.    }
```

第 6 行将 user 对象的 sex 属性赋值为"男",在页面上将 user 对象绑定到 radiobuttons 的 path 属性上,在生成 radio 时默认选中的选项是"男"。

第 8 行创建一个 List,用于存放可选的性别。List 元素的个数决定了页面上生成 radio 的可选项的个数。

(2) 创建 radiobuttons.jsp 页面,添加 radiobuttons 标签,修改后的代码如代码清单 3-22 所示。

代码清单 3-22:/chap3/WebContent/WEB-INF/jsp/radiobuttons.jsp

```
1.   <body>
2.   <h3>form:radiobuttons 测试</h3>
3.   <form:form modelAttribute="user" method="post" action="showHobbys.do" >
4.       <table>
5.           <tr>
6.               <td>选择性别:</td>
7.               <td>
8.                   <form:radiobuttons items="${sexList}" path="sex"/>
9.               </td></tr><tr>
10.              <td><input type="submit" value="提交"/></td>
11.          </tr>
12.      </table>
13.  </form:form>
14.  </body>
```

第 8 行语句 items="${sexList}",是将 Model 中的属性 sexList 值绑定到 radiobuttons 的 items 属性上。这个属性决定了 radio 的个数和内容。

(3) 发布服务,启动 Tomcat,在浏览器中输入网址 http://localhost/chap3/sex.do,可得到如图 3-13 所示的页面。

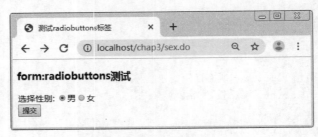

图 3-13 测试 radiobuttons 标签

在示例中用 List 绑定 radiobuttons 标签 items 的属性,也可以将 Map 作为 items 的

值,此时 Map 的 key 将提交给服务端处理请求的参数,Map 的值显示在页面的 radio 选项上,这样在页面上看到的值和提交给服务器的值是不一样的。读者可将示例中的 List 换为 Map 进行验证。

3.3.5 select 标签

Spring MVC 的 select 标签会渲染成一个 HTML 的 select 元素。其功能是提供一个下拉列表框,列表框中的数据可以由 select 标签的 items 元素的值决定。这个标签也有很多属性,使用最多的是 items 和 path 属性。

- items 属性:其值可以是 Collection、Map 或 Array,用于生成下拉列表框中的可选项。
- path 属性:绑定的属性变量,与处理请求的方法形参对应。

例 3-6 利用 select 标签为 user 对象的 addId 属性赋值。

(1) 在 TagController 控制中添加修改 user 对象 AddId 属性处理请求的方法 select(),如代码清单 3-23 所示。

代码清单 3-23:/chap3/src/com/controller/TagController.java

```
1.    //处理通过 select 选择 addId 的请求方法
2.    @RequestMapping("/select")
3.    public String select(Model model) {
4.        //创建一个 user 对象
5.        User user=new User();
6.        user.setAddId(3);
7.        //创建 map,用于生成 select 标签中地址列表框选项
8.        Map<Integer,String> addressIdMap=new HashMap<Integer,String>();
9.        addressIdMap.put(1,"南京");
10.       addressIdMap.put(2,"上海");
11.       addressIdMap.put(3,"南通");
12.       //将 user 保存在 Model 中
13.       model.addAttribute("user",user);
14.       //将 addressIdMap 保存在 Model 中
15.       model.addAttribute("addressIdMap",addressIdMap);
16.       //转发到 select.jsp 页面
17.       return "select";
18.   }
```

第 6 行将 user 对象的属性 AddId 赋值为 3,表示地址为"南通",这个值决定了在 select 标签中的默认选项。

第 8 行的 addressLdMap 是一个 Map,这个对象被保存在 Model 中,将赋值给 select 标签的 items 属性。

(2) 创建 select.jsp 页面,修改绑定的 user 对象的 addId 属性的值,如代码清单 3-24 所示。

代码清单 3-24：/chap3/WebContent/WEB-INF/jsp/select.jsp

```
1.   <h3>form:select 测试</h3>
2.   <form:form modelAttribute="user" method="post" action="showHobbys.do" >
3.      <table>
4.          <tr>
5.              <td>选择地址：</td>
6.              <td>
7.                  <form:select items="${addressIdMap}" path="addId"/>
8.              </td></tr><tr>
9.              <td><input type="submit" value="提交"/></td>
10.         </tr>
11.     </table>
12.  </form:form>
13.  </body>
```

第 7 行的 items="${addressIdMap}"表示与 Model 中的 addressIdMap 属性变量绑定，由于是一个 Map，在下拉列表框中显示的是 Map 集合的所有元素，当选中其中的某一项时，其键值 key 与 path 属性绑定，此处是 user 对象的 addId 属性变量。由于 user 对象的 addId 的值为 3，所以下拉列表框中当前的选项是值为 3 的地址，即"南通"。

（3）发布服务，启动 Tomcat，在浏览器输入地址 http://localhost/chap3/select.do，在浏览器中得到如图 3-14 所示的页面。

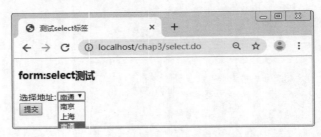

图 3-14　select 标签

从图 3-14 中可以看到，addressIdMap 的值已经显示在列表框中，当前选中的值是 user 对象中 addId 属性的值 3 对应的地址"南通"。

3.3.6　标签应用示例

下面以修改用户信息为例，应用前面学过的 Spring MVC 标签完成一个具有用户修改功能的应用程序。

例 3-7　利用 Spring MVC 标签在 JSP 页面上对 User 对象完成修改，并将修改结果显示出来。User 类中的 addId 属性是表示地址的 ID，要求从下拉列表框中选取。User 类中的 sex 属性表示性别，要求利用单选按钮选取。User 类中的 hobbys 属性是一个集合，要求利用复选框选取。显示修改页面时，要显示原来 user 对象的状态。

（1）创建 User 类，如代码清单 3-25 所示。

代码清单 3-25：/chap3/src/com/po/User.java

```
1.   package com.po;
2.   import java.util.List;
3.   public class User {
4.       private String username;        //用户名
5.       private String password;        //密码
6.       private List<String> hobbys;    //爱好
7.       private String sex;             //性别
8.       private Integer addId;          //地址的 ID
9.       //setter and getter 方法
10.  }
```

（2）在 UserController 类中添加 updateUser()方法和 showUser()方法，如代码清单 3-26 所示。

代码清单 3-26：/chap3/src/com/controller/UserController.java

```
1.   //处理修改用户请求的方法
2.   @RequestMapping("/updateUser")
3.   public String updateUser(Model model) {
4.       User user=new User();
5.       List<String>hobbys=new ArrayList<String>();
6.       hobbys.add("篮球");
7.       hobbys.add("围棋");
8.       user.setAddId(2);
9.       user.setHobbys(hobbys);
10.      user.setPassword("123456");
11.      user.setSex("男");
12.      user.setUsername("管理员");
13.      //存放所有性别选项,用于生成 radio 单选按钮
14.      List<String>sexList=new ArrayList<String>();
15.      sexList.add("男");
16.      sexList.add("女");
17.      //创建 map,用于生成 select 标签中的地址列表框选项
18.      Map<Integer,String>addressIdMap=new HashMap<Integer,String>();
19.      addressIdMap.put(1,"南京");
20.      addressIdMap.put(2,"上海");
21.      addressIdMap.put(3,"南通");
22.      //hobbyList 用于在页面上生成 checkboxes 标签显示所有爱好的选项
```

```
23.         List<String>hobbyList=new ArrayList<String>();
24.         hobbyList.add("篮球");
25.         hobbyList.add("围棋");
26.         hobbyList.add("游泳");
27.         hobbyList.add("画画");
28.         //将 user 保存在 Model 中
29.         model.addAttribute("user",user);
30.         //将 sexList 保存在 Model 中
31.         model.addAttribute("sexList",sexList);
32.         model.addAttribute("addressIdMap",addressIdMap);
33.         model.addAttribute("hobbyList",hobbyList);
34.
35.         return "updateUser";
36.     }
37. //显示修改后用户信息的请求处理方法
38.     @RequestMapping("showUser")
39.     public String showUser(User user,Model model) {
40.         model.addAttribute("user",user);
41.         return "showUser";
42.     }
```

在 updateUser()方法中分别生成了 checkboxes、radiobuttons 和 select 标签所需显示的集合数据，即 hobbyList、sexList 和 addressIdMap 3 个对象。这 3 个对象分别保存在 Model 中，这样可以在 JSP 页面上获取到它们的变量的值，并显示在相应的标签中。

（3）创建修改用户信息的 updateUser.jsp 页面，如代码清单 3-27 所示。

代码清单 3-27：/chap3/WebContent/WEB-INF/jsp/updateUser.jsp

```
1.  <%@page language="java" contentType="text/html; charset=UTF-8"
2.   pageEncoding="UTF-8"%>
3.  <%@ taglib prefix="form" uri="http://www.springframework.org/tags/form" %>
4.  <!DOCTYPE html PUBLIC "-//W3C//DTD HTML 4.01 Transitional//EN" "http://www.w3.org/TR/html4/loose.dtd">
5.  <html>
6.  <head>
7.  <meta http-equiv="Content-Type" content="text/html; charset=UTF-8">
8.  <title>Spring MVC 标签示例</title>
9.  </head>
10. <body>
11. <h3>Spring MVC 标签示例</h3>
12. <form:form modelAttribute="user" method="post" action="showUser.do">
13.     <table>
14.         <tr>
```

```
15.            <td>用户名:</td>
16.            <td>
17.                <form:input path="username"/>
18.            </td></tr>
19.            <tr>
20.            <td>密码:</td>
21.            <td>
22.                <form:input path="password"/>
23.            </td></tr>
24.            <tr>
25.            <td>选择地址:</td>
26.            <td>
27.                <form:select items="${addressIdMap}" path="addId"/>
28.            </td></tr>
29.            <tr>
30.            <td>选择爱好:</td>
31.            <td>
32.                <form:checkboxes items="${hobbyList}" path="hobbys"/>
33.            </td></tr>
34.            <tr>
35.            <td>选择性别:</td>
36.            <td>
37.                <form:radiobuttons items="${sexList}" path="sex"/>
38.            </td></tr><tr>
39.            <td><input type="submit" value="提交"/></td>
40.            </tr>
41.        </table>
42.    </form:form>
43.    </body>
44.    </html>
```

第 27 行代码生成了一个下拉列表框，列表框中的选项是由 addressIdMap 变量中的元素决定的，path 属性绑定了 user 对象中的 addId 属性。

第 32 行代码生成了多个复选框，复选框的个数和内容是由 hobbyList 变量中的元素决定的，path 属性绑定了 user 对象的 hobbys 属性。

第 37 行代码生成了多个单选按钮，单选按钮的个数和内容由 sexList 变量中的元素决定，path 属性绑定了 user 对象的 sex 属性。

（4）显示用户信息的页面 showUser.jsp，如代码清单 3-28 所示。

代码清单 3-28：/chap3/WebContent/WEB-INF/jsp/showUser.jsp

```
1.    <%@page language="java" contentType="text/html; charset=UTF-8"
2.        pageEncoding="UTF-8"%>
3.    <%@taglib prefix="c" uri="http://java.sun.com/jsp/jstl/core"%>
```

```
4.   <!DOCTYPE html>
5.   <html>
6.   <head>
7.   <meta charset="UTF-8">
8.   <title>显示爱好</title>
9.   </head>
10.  <body>
11.  <table>
12.  <tr><td>用户名：${user.username}</td></tr>
13.  <tr><td>密码：${user.password}</td></tr>
14.  <tr><td>爱好：
15.  <c:forEach items="${user.hobbys}" var="hobby">
16.      ${hobby}
17.  </c:forEach></td></tr>
18.  <tr>
19.  <td>性别：${user.sex}</td></tr>
20.  <tr><td>地址 Id：${user.addId}</td>
21.  </tr>
22.  </table>
23.  </body>
24.  </html>
```

第 15 行由于表示爱好的 hobbys 是一个 List，通过 forEach 循环标签显示这个集合中的每一个元素。

（5）发布服务，启动 Tomcat，输入网址 http：//localhost/chap3/user/updateUser.do，在浏览器中可看到如图 3-15 所示的页面。

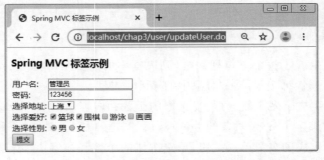

图 3-15　showUser.jsp 运行页面

从图 3-15 可以看到，所选的地址是上海，与在 updateUser()方法中 user 对象中 addId 的值为 2 是对应的。选择爱好的复合框中前两项被选中，这是与 updateUser()方法中 user 对象中 hobbys 属性值对应的。单击"提交"按钮，得到如图 3-16 所示的页面。

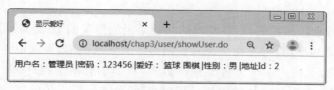

图 3-16 用户修改后的信息

习　题

1. 定义拦截器时,如果想在控制器方法被调用前执行一段代码,应该将代码写在如下的( )方法中。
   A. postHandle　　　　　　　　B. afterCompletion()
   C. preHandle()　　　　　　　　D. handler()
2. 在 Spring MVC 中实现文件的上传,在 JSP 页面中( )不是必需的。
   A. form 表单的 method 属性设置为 post
   B. form 表单的 enctype 属性设置为 multipart/form-data
   C. 设置<input>标签的 type="file"
   D. 设置<input>标签的 type="text"
3. @RequestHeader 注解的作用是获取请求头文件的信息,如果想获取浏览器的类型信息,选项( )是正确的。
   A. @RequestHeader("Accept-Language")
   B. @RequestHeader("Content-Length")
   C. @RequestHeader("Authorization")
   D. @RequestHeader("User-Agent")
4. 在 JSP 页面中使用 Spring MVC 标签 checkboxes 时,在控制器的请求处理方法中接收选中的标签选项变量的类型是( )。
   A. String　　　　B. String[]　　　　C. List<String>　　　　D. char[]

实验 3　组 件 开 发

1. 实验目的

掌握拦截器的定义和配置方法,根据需要编写上传和下载文件所需的 JSP 页面、控制器和 Spring MVC 配置文件,在 JSP 页面中要求使用 Spring MVC 标签。

2. 实验内容

创建一个 Spring MVC 的 Web 项目,要求项目具有以下功能。

(1) 根据用户名和密码进行登录,登录成功在页面显示"登录成功",否则显示"登录失败"。

（2）为登录请求添加拦截器，拦截器的功能是根据指定的 IP 地址判定是否是合法用户，只有合法的 IP 地址用户，才能提交到控制器中的登录请求，否则拒绝登录。

（3）登录成功后，进入一个上传文件的页面，页面中有两个组件，一个是添加用户的文本框，一个是选择要上传文件的按钮，通过此按钮可选择要上传的文件。上传的文件保存在工程的 upload 目录中，文件名为用户名＋文件原来的名字，文件类型不变。

3. 实现思路及步骤

（1）首先创建一个 Web 工程，并添加所需的 JAR 包，创建 Spring MVC 的配置文件，修改 web.xml，创建包的结构。

（2）创建一个用户的封装类 User，处理登录请求的控制器 loginController，登录页面 login.js，登录成功提示页面 success.jsp。

（3）创建拦截器类 loginInterceptor，并编程实现根据 IP 地址进行拦截的代码，在 Spring MVC 的配置文件中添加拦截器的配置信息。

提示：可通过 request.getRemoteAddr()方法得到发出请求客户端的 IP 地址。

（4）发布服务并启动服务，测试相应的功能。

第 4 章 Spring MVC 的数据转换与表单验证

本章目标

1. 掌握编写自定义数据类型转换的方法。
2. 掌握常用的用于格式化的注解，如 @DateTimeFormat、@NumberFormat。
3. 了解 JSON 数据格式，在 Spring MVC 中正确使用 JSON 进行数据传输。
4. 掌握 JSR 303 中常用的校验规则，并应用于 Web 开发中。

4.1 数据绑定过程

在 Spring MVC 中会将来自 Web 页面的请求和响应数据与 Controller 中对应的处理方法的参数进行绑定，即数据绑定。流程如下：

（1）Spring MVC 主框架将 ServletRequest 对象及目标方法的入参实例传递给 WebDataBinderFactory 实例，以创建 DataBinder 实例对象。

（2）DataBinder 对象调用装配在 Spring MVC 上下文中的 ConversionService 组件进行数据类型转换、数据格式化工作，将 Servlet 中的请求信息填充到入参对象中。

（3）调用 Validator 组件对已经绑定了请求消息的入参对象进行数据合法性校验，并最终生成数据绑定结果 BindingData 对象。

（4）Spring MVC 抽取 BindingRequest 中的入参对象和校验错误对象，将它们赋给处理方法的相应入参。

Spring MVC 数据绑定过程如图 4-1 所示。

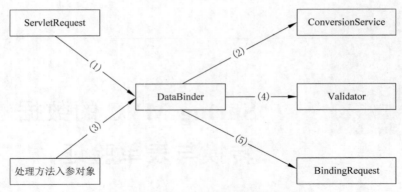

图 4-1 Spring MVC 数据绑定过程

4.2 数据类型转换

Java 标准的 PropertyEditor 的核心功能是将一个字符串转换为一个 Java 对象，以便根据界面的输入或配置文件中的配置字符串构造出一个 JVM 内部的 Java 对象。但是，Java 原生的 PropertyEditor 存在以下不足：

- 只能用于字符串到 Java 对象的转换，不能实现与任意两个 Java 类型之间的转换。
- 对源对象及目标对象所在的上下文信息，如注解、所在宿主类的结构等不敏感，在类型转换时不能利用这些上下文信息实施高级的转换逻辑。

鉴于此，Spring 在核心模型中添加了一个通用的类型转换模块，类型转换模块位于 org.springframework.core.convert 包中。Spring 希望用这个类型转换体系替换 Java 标准的 PropertyEditor。但由于历史原因，Spring 将同时支持二者，在 Bean 配置、Spring MVC 处理方法入参绑定中使用它们。

Spring MVC 上下文中内建了很多转换器，可以完成大多数 Java 类型转换工作。但是，如果要转换成我们自定义的类型，就要自定义类型转换器，并将其加入 ConversionService 中。ConversionService 中包含了很多 Spring MVC 内建的转换器。

ConversionService 是 Spring MVC 类型转换体系的核心接口，可以利用 ConversionServiceFactoryBean 在 Spring 的 IoC 容器中定义一个 ConversionService，Spring 将自动识别出 IoC 容器中的 ConversionService，并在 Bean 属性配置及 Spring MVC 处理方法入参绑定等场合使用它进行数据转换。

4.2.1 ConversionService

org.springframework.core.convert.ConversionService 是 Spring 数据类型转换的核心接口，其源代码如下：

```
1.    package org.springframework.core.convert;
2.    import org.springframework.lang.Nullable;
3.    public interface ConversionService {
```

```
4.    /** 判断是否可以将一个 Java 类转换为另一个 Java 类 */
5.    boolean canConvert(@Nullable Class<?> sourceType, Class<?> targetType);
6.    /** 需转换的类将以成员变量的方式出现在宿主类中。TypeDescriptor 不但描述
      了需转换类的信息,还描述了宿主类的上下文信息,如成员变量上的注解,成员变量
      是否以数组、集合或 Map 的方式呈现等。类型转换逻辑可以利用这些信息做出各种
      灵活控制
7.    */
8.    boolean canConvert(@Nullable TypeDescriptor sourceType, TypeDescriptor targetType);
9.    /** 将原类型对象转换为目标类型对象 */
10.   @Nullable
11.   <T>T convert(@Nullable Object source, Class<T> targetType);
12.
13.   /** 将对象从原类型对象转换为目标类型对象,此时往往会用到所在宿主类的上下文
      信息 */
14.   @Nullable
15.   Object convert(@Nullable Object source, @Nullable TypeDescriptor sourceType, TypeDescriptor targetType);
16.
17. }
```

可以利用 org.springframework.context.support.ConversionServiceFactoryBean 在 Spring 的上下文中定义一个 ConversionService。Spring 将自动识别出上下文中的 ConversionService,并在 Bean 属性配置及 Spring MVC 处理方法入参绑定等场合使用它进行数据转换。示例代码如下:

```
<bean id="ConversionService"
class="org.springframework.context.support.ConversionServiceFactoryBean"/>
```

在 ConversionServiceFactoryBean 中内建了很多转换器,使用它们可以完成大多数 Java 类型的转换工作。除了包括将 String 对象转换为各种基础类型的对象外,还包括 String、Number、Array、Collection、Map、Properties 及 Object 之间的转换器。示例代码如下:

```
1. <bean id="conversionService" class="org.springframework.context.support.ConversionServiceFactoryBean">
2.   <property name="converters">
3.     <list>
4.       <bean class="com.converter.CustomConverter1"/>
5.       <bean class="com.converter.CustomConverter2"/>
6.     </list>
7.   </property>
8. </bean>
```

4.2.2 Spring 支持的转换器

Spring 在 org.springframework.core.convert.converter 包中定义了 3 种类型的转换器接口,我们可以实现其中任意一种转换器接口,并将它作为自定义转换器注册到 ConversionServiceFactoryBean 中。这 3 种类型转换器接口如下所示。

(1) Converter<S,T>: 将 S 类型的对象转换为 T 类型对象,该接口只有一个方法:

```
T convert(S source)
```

(2) ConverterFactory: 如果将一种类型的对象转换为另一种类型及其子类型对象,就需一系列的 Converter(例如,将 String 转换为 Number 及 Number 的子类)。这个接口的作用就是将相同系列的多个 Converter 封装在一起,该接口只有一个方法:

```
<T extends R> Converter<s,T>getConverter(Class<T>targetType)
```

(3) GenericConverter: 会根据源类对象与目标类对象所在的宿主类中的上下文信息进行类型转换。该接口中定义了两个方法:

```
Set<GenericConverter.ConvertiblePair>getConvertibleTypes()
Object convert(Object source,TypeDscriptor sourceType,TypeDescriptor targetType)
```

4.2.3 自定义数据转换器

在实际工程中,经常需要日期类型的数据,我们从页面输入的是 String 类型的数据,在控制器中接收请求的参数是 Date 类型的,在 Spring MVC 中没有配置默认的从 String 转换为 Date 的数据类型转换器,这就需要我们自定义一个从 String 到 Date 的转换器。下面通过一个示例演示如何定义类型转换器。

例 4-1 自定义一个 String 到 Date 类型的类型转换器。

(1) 定义的转换器类如代码清单 4-1 所示。

代码清单 4-1: /chap4/src/com/converter/StringToDateConverter.java

```java
1.   package com.converter;
2.   import java.text.SimpleDateFormat;
3.   import java.util.Date;
4.   import org.springframework.core.convert.converter.Converter;
5.   //实现 Converter<S,T>接口
6.   public class StringToDateConverter implements Converter<String, Date>{
7.       //日期类型模板: 如 yyyy-MM-dd
8.       private String pattern;
9.       public void setPattern(String pattern) {
10.          this.pattern =pattern;
11.      }
12.      //Converter<S,T>接口的类型转换方法
13.      @Override
```

```
14.    public Date convert(String date) {
15.        try {
16.            SimpleDateFormat dateFormat=new SimpleDateFormat(this.pattern);
17.            //将日期字符串转换成 Date 类型返回
18.            return dateFormat.parse(date);
19.        } catch (Exception e) {
20.            e.printStackTrace();
21.            System.out.println("日期转换失败!");
22.            return null;
23.        }
24.    }
25. }
```

这个类型转换器实现了 Converter 接口,其中的 convert()方法实现了从 String 类型数据转换为 Date 类型数据。有了类型转换器,还要在 Spring MVC 配置文件中注册这个转换器。

(2) 在 springmvc-config.xml 文件中注册上面的转换器,如代码清单 4-2 所示。

代码清单 4-2:/chap4/WebContent/WEB-INF/springmvc-config.xml

```
1.  <!--指定需要扫描的包 -->
2.  <context:component-scan base-package="com.controller" />
3.  <!--装配自定义的类型转换器 -->
4.  <mvc:annotation-driven conversion-service="conversionService"/>
5.  <!--自定义的类型转换器 -->
6.  <bean id="conversionService"
7.      class="org.springframework.context.support.ConversionService-
        FactoryBean">
8.      <property name="converters">
9.          <list>
10.             <bean class="com.converter.StringToDateConverter"
11.                 p:pattern="yyyy-MM-dd"></bean>
12.         </list>
13.     </property>
14. </bean>
```

第 6~14 行定义了一个 id 为 conversionService 的类型转换器,第 4 行将这个转换器注册到 Spring IoC 容器中。第 10 行表示将前面定义的转换器赋值给 converters 属性,如果还有其他自定义类型转换器,可以添加到这个<list>中。

(3) 为了验证转换器能否正常工作,可创建一个如代码清单 4-3 所示的 UserController。

代码清单 4-3:/chap4/src/com/controller/UserController.java

```
1.  package com.controller;
2.  import org.springframework.stereotype.Controller;
3.  import org.springframework.ui.Model;
4.  import org.springframework.web.bind.annotation.RequestMapping;
```

```
5.    import com.po.User;
      //处理用户请求的控制器
6.    @RequestMapping("/user")
7.    @Controller
8.    public class UserController {
9.        @RequestMapping(value="/reg")
                                    //这是处理用户注册的方法,响应 reg.do 的处理请求
10.       public String reg(User user,Model model) {
11.           System.out.println(user);
12.           model.addAttribute("user",user);
13.           return "/success.jsp";
14.       }
15.   }
```

第 10 行的 reg() 方法中的 user 对象中有一个 birthday 属性是 Date 类型,可以从 reg.jsp 页面将参数传递给 user 对象。

(4) reg.jsp 页面代码如代码清单 4-4 所示。

代码清单 4-4:/chap4/WebContent/reg.jsp

```
1.    <body>
2.        <form action="${pageContext.request.contextPath}/user/reg.do" method="post">
3.        用户名:<input type="text" name="username" id="username"><br />
4.        密码:
5.            <input type="password" name="password" id="password"><br />
6.        生日:<input type="text" name="birthday" id="birthday"><br />
7.            <input type="submit" value="注册" />
8.        </form>
9.    </body>
```

第 2 行的 action 属性指向了 red.do,即提交给了 UserController 中的 reg() 方法,此方法处理完请求后转发到 success.jsp 页面输出 user 对象的信息。

(5) success.jsp 页面源码如代码清单 4-5 所示。

代码清单 4-5:/chap4/WebContent/success.jsp

```
1.    <%@page language="java" contentType="text/html; charset=UTF-8"
2.        pageEncoding="UTF-8"%>
3.    <%@taglib uri="http://java.sun.com/jsp/jstl/fmt" prefix="fmt"%>
4.    <!DOCTYPE html PUBLIC "-//W3C//DTD HTML 4.01 Transitional//EN" "http://www.w3.org/TR/html4/loose.dtd">
5.    <html>
6.    <head>
7.    <meta http-equiv="Content-Type" content="text/html; charset=UTF-8">
```

```
8.    <title>测试数据类型转换ConversionService</title>
9.    </head>
10.   <body>
11.   登录名：${user.username}<br>
12.   生日：<fmt:formatDate value="${user.birthday}"
13.        pattern="yyyy年MM月dd日"/><br>
14.   </body>
15.   </html>
```

第12行引用了 Spring MVC 的标签库的 formatDate 标签，按 pattern 属性给定格式输出日期型数据。

发布服务，启动 Tocmat，程序可以正常运行。为了验证自定义类型转换器的必要性，可先将 springmvc-config.xml 文件中的类型转换注册部分去掉，启动服务，并运行程序，程序会抛出异常，再重新在配置文件中加入自定义类型转换器，程序又重新恢复正常。

如果在实际工程中需要其他类型的转换器，可以参照以上步骤进行类型转换器的设计和配置。

4.3 基于注解格式化数据

在 Spring 中，除非有特殊需求需要自定义格式转换器或格式化器，否则可以使用 Spring 中提供的一些可对数据格式化的注解。Spring 内部用得比较多的两个注解是 @DateTimeFormat 和 @NumberFormat。

4.3.1 @DateTimeFormat 注解

@DateTimeFormat 注解可以对 java.util.Date、java.util.Calendar 等时间类型的属性进行标注。它支持以下几个互斥的属性。具体说明如下：

- Iso，类型为 DateTimeFormat.ISO。以下是几个常用的可选值。
  DateTimeFormat.ISO.DATE：格式为 yyyy-MM-dd。
  DateTimeFormat.ISO.DATE_TIME：格式为 yyyy-MM-dd hh:mm:ss .SSSZ。
  DateTimeFormat.ISO.TIME：格式为 hh:mm:ss .SSSZ
  DateTimeFormat.ISO.NONE：表示不使用 ISO 格式的时间。
- pattern，类型为 String，使用自定义的时间格式化字符串，如"yyyy-MM-dd hh:mm:ss"。
- style，类型为 String，通过样式指定日期时间的格式，由两位字符组成，第一位表示日期的样式，第2位表示时间的格式。以下是几个常用的可选值。
  S：短日期/时间的样式。
  M：中日期/时间的样式。
  L：长日期/时间的样式。
  F：完整日期时间的样式。

-：忽略日期时间的样式。

4.3.2 @NumberFormat 注解

@NumberFormat 注解可对类似数字类型的属性进行标注,它拥有两个互斥的属性,具体说明如下：

pattern,类型为 String,使用定义的数字格式化串,如"＃＃.＃＃＃.＃＃"。
style,类型为 NumberFormat.Style,以下是几个常用的可选值。
NumberFormat.CURRENCY：货币类型。
NumberFormat.NUMBER：正常数字类型。
NumberFormat.PERCENT：百分数类型。

4.3.3 基于注解格式化数据示例

例 4-2 利用@DateTimeFormat 和@NumberFormat 注解格式化 Product 类中的属性。

(1) Product 类的源码如代码清单 4-6 所示。

代码清单 4-6：/chap4/src/com/po/Product.java

```java
1.  package com.po;
2.  import java.util.Date;
3.  import org.springframework.format.annotation.DateTimeFormat;
4.  import org.springframework.format.annotation.NumberFormat;
5.  public class Product {
6.      private String name;          //商品名称
7.      @DateTimeFormat(pattern="yyyy-MM-dd")
8.      private Date pd;              //生产日期
9.      @NumberFormat(pattern="##,###.##")
10.     private float price;          //价格
11.     public String getName() {
12.         return name;
13.     }
14.  /* setter and getter*/
15.     @Override
16.     public String toString() {
17.         return "Product [name=" +name +", pd=" +pd +", price=" +price
            +"]";
18.     }
19.
20.  }
```

这是一个 Product 封装类,第 7 行在生产日期 date 属性上添加@DateTimeFormat 标注,并通过 pattern 属性指定日期格式。

第 9 行在产品价格 price 属性上添加 @NumberFormat 标注,并通过 pattern 属性指定数值的格式。

（2）编写 ProductController，添加处理请求的 addProduct()方法，如代码清单 4-7 所示。

代码清单 4-7：/chap4/src/com/controller/ProductController.java

```
1.   package com.controller;
2.   import org.springframework.stereotype.Controller;
3.   import org.springframework.ui.Model;
4.   import org.springframework.web.bind.annotation.GetMapping;
5.   import org.springframework.web.bind.annotation.RequestMapping;
6.   import org.springframework.web.bind.annotation.RequestParam;
7.   import com.po.Product;
8.   //处理用户请求的控制器
9.   @Controller
10.  public class ProductController {
11.      //处理 toAddProduct.do 请求
12.      @GetMapping(value="toAddProduct")
13.      public String toAddProduct(Model model) {
14.          Product product=new Product();        //创建一个 product 对象，
15.          //在 addProduct.jsp 页面要引用 product 属性
16.          model.addAttribute("product",product);
17.          //转发到 addProduct.jsp 页面
18.          return "addProduct.jsp";
19.      }
20.      //这是处理产品注册的方法,响应 addProduct.do 的处理请求
21.      @RequestMapping(value="/addProduct")
22.      public String addProduct(Product product,Model model) {
23.          System.out.println(product);
24.          //将 product 添加到 model 中,以便在 JSP 页面中访问此变量的值
25.          model.addAttribute("product",product);
26.          return "showProduct.jsp";
27.      }
28.  }
```

第 22 行的 product 是请求方法的参数，其参数中的属性被格式化注解标注，可对入参进行数据转换并格式化。

（3）输入商品信息的页面 addProduct.jsp 如代码清单 4-8 所示。

代码清单 4-8：/chap4/WebContent/addProduct.jsp

```
1.   <body>
2.   <form:form modelAttribute="product" method="post" action="addProduct.do" >
3.       <table>
```

```
4.           <tr>
5.               <td>产品名称:</td>
6.               <td><form:input path="name"/></td>
7.           </tr>
8.           <tr>
9.               <td>生产日期:</td>
10.              <td><form:input path="pd"/></td>
11.          </tr>
12.          <tr>
13.              <td>价格:</td>
14.              <td><form:input path="price"/></td>
15.          </tr>
16.          <tr>
17.              <td><input type="submit" value="提交"/></td>
18.          </tr>
19.      </table>
20.  </form:form>
21.  </body>
```

第 2 行中的 modelAttribute="product" 是绑定的 Model 属性的变量名称,如果有这个属性,就不能直接进入这个 JSP 页面,而是要从一个 Controller 的 RequestMapping 转发过来,在请求处理的方法中要添加 Model 类型的形参,并在这个形参中添加 product 属性,如代码清单 4-7 的第 16 行,否则会抛出异常。

(4) 为了验证数据格式化是否成功,编写 showProduct.jsp 页面显示 product 对象的信息,如代码清单 4-9 所示。

代码清单 4-9:/chap4/WebContent/showProduct.jsp

```
1.  <title>基于注解格式化输出</title>
2.  </head>
3.  <body>
4.  <form:form modelAttribute="product" method="post" action="" >
5.      <table>
6.          <tr>
7.              <td>产品名称:</td>
8.              <td><form:input path="name" readonly="true"/></td>
9.          </tr>
10.         <tr>
11.             <td>生产日期:</td>
12.             <td><form:input path="pd" readonly="true"/></td>
13.         </tr>
14.         <tr>
15.             <td>价格:</td>
16.             <td><form:input path="price" readonly="true"/></td>
```

```
17.            </tr>
18.         </table>
19.     </form:form>
20. </body>
```

(5) sprintmvc-config.xml 配置文件与例 4-1 相同,发布服务,启动 Tomcat,在浏览器地址栏中输入 http://localhost/chap4/toAddProduct.do,出现添加商品页面,如图 4-2 所示。

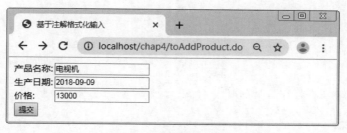

图 4-2　添加商品页面

单击"提交"按钮后,得到如图 4-3 所示的页面。可以看出,页面上已输出格式化后的生产日期和价格。

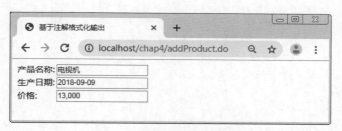

图 4-3　显示格式化后的数据

4.4　JSON 数据格式的转换

JSON(JavaScript Object Notation,JS 对象简谱)是一种轻量级的数据交换格式。它基于 ECMAScript(欧洲计算机制造商协会制定的 JS 规范)的一个子集,采用完全独立于编程语言的文本格式存储和表示数据。简洁和清晰的层次结构使得 JSON 成为理想的数据交换语言。JSON 易于人阅读和编写,同时也易于机器解析和生成,并有效地提升网络传输效率。

Douglas Crockford 是 Web 开发领域知名的技术权威之一,ECMA JavaScript 2.0 标准化委员会委员,被 JavaScript 之父 Brendan Eich 称为 JavaScript 的大宗师(Yoda),曾任 Yahoo! 资深 JavaScript 架构师,现任 PayPal 高级 JavaScript 架构师。JSON 是 Douglas Crockford 在 2001 年开始推广使用的数据格式。

4.4.1 JSON 格式简介

JSON 格式分为两类：数组格式和对象格式。

1. 数组格式

数组是值（value）的有序集合。一个数组以"["开始，以"]"结束，值之间使用","分隔。

格式：

```
[
    {"key1": "value1"},
    {"key2": "value2"}
]
```

例如：["南京","上海","杭州"]

2. 对象格式

对象是一个无序的"键值对"集合。一个对象以"{"开始，以"}"结束。每个"键"后跟一个":"，后面是一个"值"，"键值对"之间使用","分隔。

格式：

```
{
    "key1: "value1",
    "key2": "value2",
    "key3": [
        {"key31": "value31"},
        {"key32": "value32"}
    ]
}
```

例如：{"id":"1","username":"张红","age":"18"}

上面两种数据结构可以根据需要组合成更复杂的 JSON 数据结构。例如，一个 User 对象包含 name、sex、age 3 个属性，如果用 JSON 格式表示，其格式如下：

{"name":"lili","sex":["男","女"],"age":20}

4.4.2 JSON 数据格式转换

Spring MVC 默认用 MappingJackson2HttpMessageConverter 类对 JSON 数据进行转换，MappingJackson2HttpMessageConverter 类实现了 HttpMessageConverter 接口。该类利用 Jackson 开源包读写数据，将 Java 对象转换为 JSON 对象和 XML 对象，同时也可以将 JSON 对象和 XML 文档转换为 Java 对象。要想正常进行 JSON 数据转换，需要加入 Jackson 的 JAR 包，如下所示。

- Jackson-annotations-2.9.2.jar
- Jackson-core-2.9.2.jar
- Jackson-databind-2.9.2.jar

在 Spring MVC 中，可以使用@RequestBody 和@ResponseBody 两个注解，分别完成请求报文到对象和对象到响应报文的转换，底层这种灵活的消息转换机制是 Spring 3.x 中新引入的 HttpMessageConverter，即消息转换器机制。

@RequestBody 注解用于读取 http 请求的内容（字符串），通过 Spring MVC 提供的 HttpMessageConverter 接口将读到的内容转换为 json、xml 等格式的数据，并绑定到 Controller 类方法的参数上。

@ResponseBody 注解用于将 Controller 的请求处理方法返回的对象，通过适当的 HttpMessageConverter 转换为指定格式后，写入到 Response 对象的 body 数据区。当返回的数据不是 HTML 标签的页面，而是其他某种格式的数据时（如 JSON、XML 等），可以使用@ResponseBody 注解。下面是一个 JSON 数据与 Java 对象互相转换的示例。

例 4-3 从 JSP 页面输入用户名和密码，以 JSON 格式向服务器发送请求，在 Controller 中用@RequestBody 注解标注 User 对象参数接收 JSON 数据，用@ResponseBody 注解标注请求方法，实现将 Customer 对象转换为 JSON 对象，并输出到页面上。

（1）发送请求的页面 index.jsp，如代码清单 4-10 所示。

代码清单 4-10：/chap4/WebContent/index.jsp

```jsp
1.  <%@page language="java" contentType="text/html; charset=UTF-8"
2.      pageEncoding="UTF-8"%>
3.  <!DOCTYPE html PUBLIC "-//W3C //DTD HTML 4.01 Transitional //EN"
4.  "http://www.w3.org/TR/html4/loose.dtd">
5.  <html>
6.  <head>
7.  <title>测试 JSON 交互</title>
8.  <meta http-equiv="Content-Type" content="text/html; charset=UTF-8">
9.  <script type="text/javascript"
10.     src="${pageContext.request.contextPath}/js/jquery-1.11.3.min.js">
11. </script>
12. <script type="text/javascript">
13. function testJson(){
14.     //获取输入的用户名和密码
15.     var username =$("#username").val();
16.     var password =$("#password").val();
17.     $.ajax({
18.         url : "${pageContext.request.contextPath}/login.do",
19.         type : "post",
20.         //data 表示发送的数据
21.         data :JSON.stringify({username:username,password:password}),
```

```
22.              //定义发送请求的数据格式为JSON字符串
23.              contentType : "application/json;charset=UTF-8",
24.              //定义回调响应的数据格式为JSON字符串,该属性可以省略
25.              dataType : "json",
26.              //将响应的结果写入table中
27.              success : function(data){
28.                  $.each(data,function(){
29.                      var tr=$("<tr align='center'/>");
30.                      $("<td/>").html(this.id).appendTo(tr);
31.                      $("<td/>").html(this.username).appendTo(tr);
32.                      $("<td/>").html(this.telephone).appendTo(tr);
33.                      $("<td/>").html(this.sex).appendTo(tr);
34.                      $("<td/>").html(this.age).appendTo(tr);
35.                      $("#booktable").append(tr);
36.                  })
37.              }
38.          });
39.      }
40.      </script>
41.      </head>
42.      <body>
43.          <form>
44.              用户名:<input type="text" name="username" id="username"><br />
45.              密  码:
46.              <input type="password" name="password" id="password"><br />
47.      <input type="button" value="测试JSON交互" onclick="testJson()" />
48.          </form>
49.          < table id =" booktable" border =" 1" style =" border - collapse: collapse;">
50.          <tr align="center">
51.              <th>id</th>
52.              <th>用户名</th>
53.              <th>电话</th>
54.              <th>性别</th>
55.              <th>年龄</th>
56.          </tr>
57.      </table>
58.      </body>
59.      </html>
```

第10行在页面的<head>部分引入jQuery的js文件,页面使用jQuery发送请求。

第 15、16 行获得两个文本框的值,分别赋给 username 和 password 两个变量。

第 18 行以 ajax 方式发送请求到 login.do,也就是 CustomerController 类的 login()方法。

第 21 行以 JSON 字符串格式发送数据。

第 23 行通过 contentType 在请求头里设定浏览器请求格式,这里设为 JSON 格式。

第 27～35 行表示请求数据处理完成并成功返回后,对返回数据做处理,这里是将返回的 JSON 字符串写入表格中,在页面显示。

第 47 行单击"测试 JSON 交互"按钮后,首先运行的是 testJson()函数,此函数中完成了数据的提交和返回数据的处理。

第 50～56 行设置了输出表格的表头信息。

(2)处理请求的控制器 CustomerController 如代码清单 4-11 所示。

代码清单 4-11:/chap4/src/com/controller/CustomerController.java

```java
1.   package com.controller;
2.   import java.util.ArrayList;
3.   import java.util.Date;
4.   import java.util.List;
5.   import org.springframework.stereotype.Controller;
6.   import org.springframework.ui.Model;
7.   import org.springframework.web.bind.annotation.PostMapping;
8.   import org.springframework.web.bind.annotation.RequestBody;
9.   import org.springframework.web.bind.annotation.RequestMapping;
10.  import org.springframework.web.bind.annotation.ResponseBody;
11.  import com.po.Customer;
12.  import com.po.User;
13.  //处理用户请求的控制器
14.  @Controller
15.  public class CustomerController {
16.      //这是处理用户登录的方法,响应 login.do 的处理请求
17.      @RequestMapping(value="/login")
18.      //@ResponseBody 注解的方法表示将方法返回类型转换为 JSON 格式数据
19.      @ResponseBody
20.      //通过@RequestBody 注解,user 对象可以将 JSON 格式数据转换为 Java 对象
21.      public Object testJson(@RequestBody User user) {
22.          System.out.println(user);
23.          //创建一个集合 list
24.          List<Customer>list=new ArrayList<Customer>();
25.          list.add(new Customer(1,"刘洋 1","13678652341","男",30));
26.          list.add(new Customer(1,"刘洋 2","13678652342","女",40));
27.          list.add(new Customer(1,"刘洋 3","13678652343","男",20));
```

```
28.        return list;
29.    }
30. }
```

在 CustomerController 控制器中,由于 user 对象被@RequestBody 标注,所以可以将请求中的 JSON 数据转换为 Java 对象 user,这一点可以通过第 22 行从控制台打印出的 user 对象的内容得到验证。从第 24~27 行构造了一个集合对象 list,list 中的元素是 user 对象,当将这个对象作为方法的返回值时,由于方法被 @ResponseBody 所标注,list 对象就被转换为 JSON 数据返回到客户端。在控制器中,通过 @RequestBody 和 @ResponseBody 两个注解完成了 Java 对象与 JSON 数据的相互转换。

(3) Spring MVC 配置文件 springmvc-config.xml 与第 3 章中的例 3-1 相同,只是这里不需要定义拦截器。

(4) 发布服务,启动 Tomcat,在浏览器中输入网址 http://localhost/chap4/index.jsp,得到如图 4-4 所示的页面。

图 4-4　运行 JSON 提交页面 index.jsp

在文本框中分别输入 admin 和 11,单击"测试 JSON 交互"按钮,控制台上输出了用户名 admin 和密码 11,说明从页面发送的 JSON 数据在控制器中转换成了 Java 对象。同时可看到如图 4-5 所示的页面。

图 4-5　Java 对象转换为 JSON 对象页面

从图 4-5 中可以看出,控制器中的 Java 对象 list 在页面上完整地呈现出来,说明从控制器到页面实现了从 Java 对象到 JSON 数据的转换。

(5) 实现 JSON 数据格式转换所需的 JAR 包如图 4-6 所示。

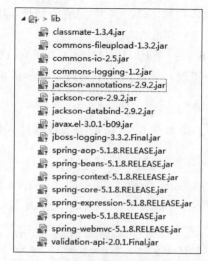

图 4-6 实现 JSON 数据格式转换所需的 JAR 包

4.5 表单验证

数据的校验是 Web 网站一个不可或缺的功能，前端的 JS 校验可以涵盖大部分的校验职责，如用户名唯一性、生日格式、邮箱格式校验等常用的校验，但这需要编写大量的 JS 函数。Spring MVC 中也提供了数据校验功能，使用方法更加简单高效。其分为两种方式：一种是利用 JSR 303/JSR 380（Java 验证规范）实现校验功能；另一种是利用 Spring 自带的 Validation 校验框架。由于 Validation 框架通过硬编码完成数据校验，在实际开发中显得比较麻烦，因此现在推荐使用 JSR 303/JSR 380 完成数据校验。

4.5.1 JSR 303 校验规则

JSR 303 是一项标准，JSR 380 是它的升级版本，添加了一些新特性，它们规定一些校验规范，即校验注解，如 @Null、@NotNull、@Pattern，它们位于 javax.validation.constraints 包下，只提供规范，不提供实现。hibernate validation 是对这个规范的实践（不要将 hibernate 和数据库 orm 框架联系在一起），它提供了相应的实现，并增加了一些其他校验注解，如 @Email、@Length、@Range 等，它们位于 org.hibernate.validator.constraints 包下。Spring 为了给开发者提供便捷，对 Hibernate Validation 进行了二次封装，显示校验 Validated Bean 时，可以使用 Spring Validation 或者 Hibernate Validation，而 Spring Validation 的另一个特性是其在 Spring MVC 模块中添加了自动校验，并将校验信息封装进了特定的类中。这无疑便捷了 Web 开发。本文主要介绍在 Spring MVC 中自动校验的机制。

JSR 380 是一个规范，它的核心接口是 javax.validation.Validator，该接口根据目标对象中标注的校验注解进行数据校验，并得到校验结果。JSR 380 规范是 bean 验证的 Java API 规范，使用的注解如 @NotNull、@Min、@Max，确保 bean 属性符合一定条件。JSR

380 需要 Java 8.0 或以上版本,利用 Java 8.0 中新增的特性(如注解类型)支持新的类(如 Optional 和 LocalDate)。

JSR 380 目前有两个实现:一个是 Hibernate Validator;另一个是 Apache。这里只介绍 Hibernate Validator 的实现。可以从以下网站下载所需的 JAR 包:

https://sourceforge.net/projects/hibernate/files/hibernate-validator/

进入网站后可看到最新的版本是 6.0.17,单击下载链接下载 JAR 包为

hibernate-validator-6.0.17.Final-dist.zip

解压文件后,在目录中找到 dist 目录,将其中的 3 个 JAR 包和 lib\required 目录中的 4 个 JAR 包一起复制到工程的 lib 目录中。一共 7 个 JAR 包,如图 4-7 所示。

文件名	日期	类型	大小
classmate-1.3.4.jar	2019/6/13 13:25	JAR 文件	64 KB
hibernate-validator-6.0.17.Final.jar	2019/6/13 13:25	JAR 文件	1,129 KB
hibernate-validator-annotation-proce...	2019/6/13 13:25	JAR 文件	133 KB
hibernate-validator-cdi-6.0.17.Final.jar	2019/6/13 13:25	JAR 文件	37 KB
javax.el-3.0.1-b09.jar	2019/6/13 13:25	JAR 文件	233 KB
jboss-logging-3.3.2.Final.jar	2019/6/13 13:25	JAR 文件	65 KB
validation-api-2.0.1.Final.jar	2019/6/13 13:25	JAR 文件	91 KB

图 4-7 实现 JSR 380 Hibernate Validator 校验框架 JAR 包

JSR 380 中定义了一套可标注在类成员变量、属性方法上的校验注解,见表 4-1。

表 4-1 JSR 380 注解

注 解	功 能 描 述	示 例
@NotNull	验证注解的属性值不能为 null	@NotNull String username
@AssertTrue	验证注解的属性值为 true	@AssertTrue Boolean isEmpty
@Size	验证注解的属性值大小在 max 和 min 之间,也可以应用在 String、Collection、Map、Array 类型的属性上	@Size(min=20,max=90) Int age
@Min	验证注解的属性值不小于 value 属性指定的值	@Min(30) Int age
@Max	验证注解的属性值不大于 value 属性指定的值	@Max Int age
@Email	验证注解的属性值是一个有效的 Email 地址	@Email String email
@NotEmpty	验证注解的属性值不为 null 或 empty;可以应用在 String、Collection、Map 或 Array 类型的属性上	@NotEmpty String username
@NotBlank	验证注解的属性值不为 null 或 whitespace,应用在字符串值验证	@NotBlank String username

续表

注　解	功能描述	示　例
@URL	验证是否是合法的 URL	@URL String url
@Range（min,max,message）	验证属性值必须在合适的范围内	@Range（min＝20,max＝70,message＝"职工年龄必须在25到60之间"） Int age

4.5.2 校验规则示例

例 4-4 现有一添加客户信息的表单 addCustomer.jsp，要求对表单添加 JSR 校验功能，具体要求可查看下面类的设计。

（1）创建 Customer1 类，在类的属性中添加校验注解，如代码清单 4-12 所示。

代码清单 4-12：/chap4/src/com/po/Customer1.java

```
1.   package com.po;
2.   import javax.validation.constraints.NotEmpty;
3.   import javax.validation.constraints.NotNull;
4.   import javax.validation.constraints.Pattern;
5.   import org.hibernate.validator.constraints.Range;
6.   public class Customer1 {
7.       @Range(min=0,message="ID不能小于0")
8.       private int id;
9.       @NotEmpty(message="用户名不能为空")
10.      private String username;
11.      @Pattern(regexp="[1][3,8][3,6,9][0-9]{8}",
12.      message="电话号码格式不对,第1位为1,第2位为3或8,第3位为3、6、9,长度为11位")
13.      private String telephone;
14.      private String sex;
15.      @Range(min=20,max=60,message="职工年龄在20与60之间")
16.      private int age;
17.      //setter and getter 方法}
18.  }
```

（2）创建控制器类 CustomerController，在类中添加注册方法 reg()和跳转到 addCustomer.jsp 的方法 toRed()，如代码清单 4-13 所示。

代码清单 4-13：/chap4/src/com/controller/CustomerController.java

```
1.   package com.controller;
2.   import javax.validation.Valid;
3.   import org.springframework.stereotype.Controller;
4.   import org.springframework.ui.Model;
```

```
5.  import org.springframework.validation.Errors;
6.  import org.springframework.web.bind.annotation.GetMapping;
7.  import org.springframework.web.bind.annotation.PostMapping;
8.  import com.po.Customer1;
9.
10. @Controller
11. public class CustomerController {
12.     @GetMapping(value="toReg")
13.     public String toReg(Model model) {
14.         Customer1 customer1=new Customer1();
15.         model.addAttribute("customer1",customer1);
16.         return "addCustomer";
17.     }
18.     @PostMapping(value="reg")
19. //@Valid 注解表示 customer 对象将会被检验,Errors 的对象 errors 用于保存错误
    //信息,也就是当采用 JSR 规范进行校验后,它会将错误信息保存到这个参数中
20.     public String reg(@Valid Customer1 customer1, Errors errors, Model model) {
21.         if(errors.hasErrors()){
22.             return "addCustomer";
23.         }
24.         model.addAttribute("customer1",customer1);
25.         return "showCustomer";
26.     }
27. }
```

第 13 行的 toReg()方法中创建了一个 customer 对象,并将对象保存在 Model 中。这样,在转发的 addCustomer.jsp 页面中可以获得这个对象。

第 20 行的 reg()方法中的参数 customer1 添加了@Valid 注解,表示提交表单时对属性进行校验。

(3) 创建客户注册页面 addCustomer.jsp,如代码清单 4-14 所示。

代码清单 4-14：/chap4/WebContent/addCustomer.jsp

```
1. <%@page language="java" contentType="text/html; charset=UTF-8"
2.     pageEncoding="UTF-8"%>
3. <%@taglib prefix="form"
4.   uri="http://www.springframework.org/tags/form" %>
5. <!DOCTYPE html>
6. <html>
7. <head>
8. <meta charset="UTF-8">
9. <title>添加客户信息测试 JSR</title>
```

```
10.    </head>
11.    <body>
12.    <form:form modelAttribute="customer1" method="post" action="reg.do">
13.       <table>
14.         <tr>
15.            <td>客户 ID:</td>
16.            <td><form:input path="id"/></td>
17.            <td><form:errors path="id" cssStyle="color:red"/></td>
18.         </tr>
19.         <tr>
20.            <td>用户名:</td>
21.            <td><form:input path="username"/></td>
22.            <td><form:errors path="username"
23.    cssStyle="color:red"/></td>
24.         </tr>
25.         <tr>
26.            <td>电话号码:</td>
27.            <td><form:input path="telephone"/></td>
28.            <td><form:errors path="telephone"
29.    cssStyle="color:red"/></td>
30.         </tr>
31.         <tr>
32.            <td>年龄:</td>
33.            <td><form:input path="age"/></td>
34.            <td><form:errors path="age" cssStyle="color:red"/></td>
35.         </tr>
36.         <tr>
37.            <td><input type="submit" value="提交"/></td>
38.         </tr>
39.       </table>
40.    </form:form>
41.    </body>
42.    </html>
```

addCustomer.jsp 页面中使用了 Spring MVC 的标签。使用 Spring MVC 的标签,要在页面代码前面添加引用标签库的语句:

```
<%@taglib prefix="form" uri="http://www.springframework.org/tags/form" %>
```

其中 prefix="form"定义了一个前缀,在 JSP 页面中可以通过这个前缀引用 Spring MVC 的标签。第 21 行的代码<form:input path="username"/>标签相当于 HTML 中的<input type="text" name="username"/>标签,其中的"path"属性表示绑定的属性名称。

`<form:errors path="username" cssStyle="color:red"/>`标签用于输出出错信息,其中"path"属性值与"input"标签对应的属性值一致。

(4) springmvc-congif.xml 配置文件如代码清单 4-15 所示。

代码清单 4-15:/chap4/WebContent/WEB-INF/springmvc-config.xml

```xml
1.  <?xml version="1.0" encoding="UTF-8"?>
2.  <beans schema 这部分省略>
3.      <!--指定需要扫描的包-->
4.      <context:component-scan base-package="com.controller" />
5.          <!--默认装配方案-->
6.      <mvc:annotation-driven/>
7.      <!--静态资源处理-->
8.      <mvc:default-servlet-handler/>
9.      <!--定义视图解析器-->
10.     <bean id="viewResolver" class=
11.     "org.springframework.web.servlet.view.InternalResourceViewResolver">
12.         <!--设置前缀-->
13.         <property name="prefix" value="/WEB-INF/jsp/" />
14.         <!--设置后缀-->
15.         <property name="suffix" value=".jsp" />
16.     </bean>
17. </beans>
```

(5) 注册成功后,转发到 showCustomer.jsp 页面,如代码清单 4-16 所示。

代码清单 4-16:/chap4/WebContent/showCustomer.jsp

```jsp
1.  <%@page language="java" contentType="text/html; charset=UTF-8"
2.      pageEncoding="UTF-8"%>
3.  <!DOCTYPE html>
4.  <html>
5.  <head>
6.  <meta charset="UTF-8">
7.  <title>helloWorld</title>
8.  </head>
9.  <body>
10. 客户 ID:${customer1.id}<br>
11. 用户名:${customer1.username}<br>
12. 年龄:${customer1.age}<br>
13. 电话号码:${customer1.telephone}
14. </body>
15. </html>
```

第 10~13 行代码利用 EL 表达式显示控制器中保存在 Model 中的数据,实际上是保存在 request 对象中的。

（6）发布服务，启动 Tomcat 服务器，在浏览器中输入网址 http://localhost/chap4/toReg.do，出现如图 4-8 所示的页面。

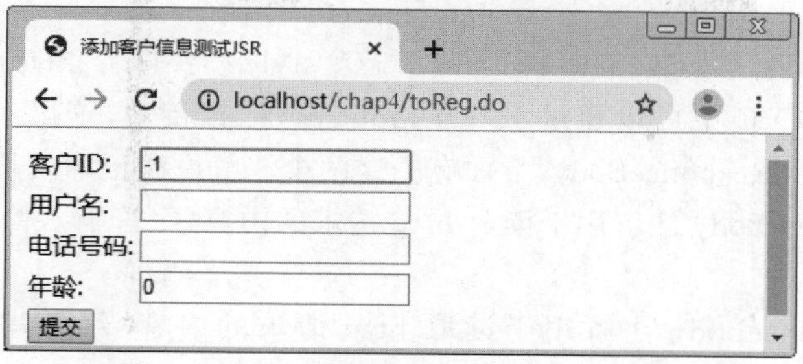

图 4-8　注册客户信息页面

此时在客户 ID 中添加−1，其他不添加数据，单击"提交"按钮，则出现如图 4-9 所示的页面。

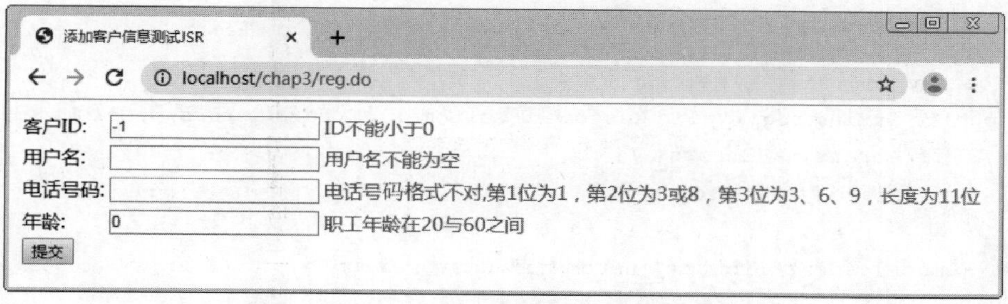

图 4-9　注册客户信息校验出错页面

从运行结果看，校验规则已经有效，由于输入的数据不符合验证规则，因此页面上会显示提示信息。按格式要求输入正确的数据，提交表单，得到如图 4-10 所示的页面。

图 4-10　注册客户信息成功页面

习　　题

1. 在 Spring MVC 中用于数据类型转换的接口有 3 个，其中(　　)不属于类型转换接口。
   A. GenericConverter　　　　　　　　B. ConverterFactory
   C. Converter　　　　　　　　　　　　D. DataConverter

2. @DateTimeFormat 注解可以对 java.util.Date、java.util.Calendar 等时间类型的属性进行标注，其作用是对日期进行格式化。下面选项中的（　　）不是@DateTimeFormat 注解的属性。

    A. iso   B. pattern   C. style   D. patter

3. 在 Spring MVC 中可以通过 JSON 格式进行数据传递，这里用到的两个注解是@RequestBody 和@ResponseBody，下面有关这两个注解说法正确的是（多选）（　　）。

    A. @RequestBody 注解用于读取 http 请求的内容（字符串），将内容转换为 JSON 格式

    B. @ResponseBody 注解用于读取 http 请求的内容（字符串），将内容转换为 JSON 格式

    C. @ResponseBody 注解的方法表示将方法返回类型转换为 JSON 格式数据

    D. @ResponseBody 注解用于读取 http 请求的内容（字符串），将内容转换为 JSON 格式

4. 在控制器类中有如下一段代码：

```
@PostMapping(value="reg")
public String reg(@Valid Customer1 customer1,Errors errors,Model model) {
    if(errors.hasErrors()){
      return "addCustomer";
    }
    model.addAttribute("customer1",customer1);
    return "showCustomer";
}
```

有关这段代码描述错误的是（　　）。

    A. @Valid 注解表示 customer 对象将被校验

    B. Errors 的对象 errors 用于保存错误信息

    C. 如果有校验错误信息，则返回 addCustomer.jsp 页面

    D. 如果没有核验错误信息，则返回 addCustomer.jsp 页面

实验 4　数据转换与表单验证

1. 实验目的

了解 Sping 的数据转换规则，掌握自定义数据转换器的方法，特别是掌握 JSON 数据格式的转换。掌握利用 JSR 303 校验规则对表单进行校验的方法。

2. 实验内容

现有用户类 User，代码如下。

```
1.    package com.po;
2.    import java.util.Date;
3.    public class User1 {
4.        private String username;        //用户名
5.        private int age;                //年龄
6.        private String telephone;       //手机号
7.        private String email;           //电子邮箱
8.        private Date birthday;          //生日
9.        //setter and getter
10.   }
```

要求如下:

(1) 自定义一个日期类型转换器,并在配置文件进行配置,完成字符到日期类型数据的自动转换。

(2) 为 User 类中的所有属性添加 JSP 303 校验,具体要求是:
- 属性 username 不能为空,长度为 6~8 个字符串;
- 属性 age 的范围为 20~50;
- 属性 telephone 必须是 0~9 的数字,长度为 11 个;
- 属性 email 为电子邮件格式;
- 属性 birthday 日期必须是以前的日期。

(3) 创建一个 UserController,并添加两个方法 toAddUser()和 addUser(),toAddUser()方法用于转发到 addUser.jsp 页面,addUser()方法用于处理添加的请求,并对 user 对象进行校验。若校验通过,则在 showUser.jsp 页面显示 user 对象的信息;若校验没有通过,则返回 addUser.jsp 页面,并在页面上显示错误信息。

3. 实现思路及步骤

(1) 首先创建一个 Web 工程,并添加所需的 JAR 包,创建 Spring MVC 的配置文件,修改 web.xml,创建包的结构。

(2) 创建一个用户的封装类 User,对每个属性添加校验规则。

(3) 创建处理用户请求的控制器 UserController,在类中添加用户的请求方法 addUser()中,添加 JSR 校验所需的注解,根据 errors 对象是否为空决定转发的页面。

(4) 创建添加用户信息的页面 addUser.jsp 和显示用户信息的页面 showUser.jsp。

(5) 发布服务并启动服务,测试相应的功能。

第 5 章 MyBatis 基础知识

本章目标
1. 了解 ORM 的概念及与 MyBatis 的关系。
2. 掌握 MyBatis 中常用的 API 的使用方法。
3. 正确编写 MyBatis 的配置文件。
4. 正确编写 MyBatis 中的映射文件。

MyBatis 本是 Apache 的一个开源项目 iBatis，2010 年这个项目由 Apache Software Foundation 迁移到 Google Code，并且改名为 MyBatis，2013 年 11 月迁移到 GitHub。

MyBatis 是一款优秀的持久层框架，它支持定制化 SQL、存储过程以及高级映射。MyBatis 将 JDBC 进行了封装，避免了直接使用 JDBC 操作数据库。MyBatis 可以使用简单的 XML 或注解配置映射类和表之间的关系，将接口和 Java 的 POJO（Plain Ordinary Java Object，普通的 Java 对象）映射成数据库中的记录。

5.1 ORM 与 MyBatis

MyBatis 框架是 ORM 的一种实现。所谓的 ORM（Object-Relational Mapping，对象-关系映射）就是为了解决面向对象程序设计与数据库之间不匹配的技术，它通过描述 Java 对象与数据库表之间的映射关系，自动将 Java 应用程序中的对象持久化到关系数据库的表中。

对象-关系映射是随着面向对象的软件开发方法发展产生的。面向对象的开发方法是当今企业级应用开发环境中的主流开发方法，关系数据库是企业级应用环境中永久存放数据的主流数据存储系统。对象和关系是业务实体的两种表现形式，业务实体在内存中表现为对象，在数据库中表现为关系数据。内存中的对象之间存在关联和继承关系，而在数据库中，关系数据无法直接表达多对多关联和继承关系。因此，对象-关系映射系统一般以中间件的形式存在，主要实现程序对象到关系数据库数据的映射。

当开发一个应用程序的时候,如果不使用 ORM,则可能会写不少数据访问层的代码,用来从数据库保存、删除、读取对象信息等。通常在数据库编程中写很多方法以完成读取对象数据、改变状态对象等任务,这些代码不可避免会有重复。

ORM 解决的主要问题是对象关系的映射。域模型和关系模型都建立在概念模型基础上。域模型是面向对象的,而关系模型是面向关系的。一般情况下,一个持久化类和一个表对应,类的每个实例对应表中的一条记录,类的每个属性对应表的每个字段。

ORM 技术的特点如下:

(1)提高了开发效率。由于 ORM 可以自动对实体对象与数据库中的表进行字段与属性的映射,所以我们已经不需要一个专用的、庞大的数据访问层。

(2)ORM 提供了对数据库的映射,不用 SQL 直接编码,就能像操作对象一样从数据库获取数据。

从系统结构上看,采用 ORM 的系统一般都是多层系统,系统的层次多了,效率就会降低。ORM 是一种完全的面向对象的做法,而面向对象的做法也会对性能产生一定的影响。

当前的 ORM 框架产品有很多,常见的主要有 Hibernate 和 MyBatis 两个框架,这两个框架的对比见表 5-1。

表 5-1 Hibernate 框架和 MyBatis 框架的对比

对比项目	Hibernate 框架	MyBatis 框架
自动化程度	Hibernate 是一个开放源代码的对象关系映射框架,它对 JDBC 进行了非常轻量级的对象封装,建立对象与数据库表的映射,是一个全自动的、完全面向对象的持久层框架	MyBatis 是一个开源对象关系映射框架,是一个半自动化的持久层框架
性能	Hibernate 自动生成 SQL,有些语句较为烦琐,会多降低性能	MyBatis 手动编写 SQL,可以避免不需要的查询,提高系统性能
对象管理	Hibernate 是完整的对象-关系映射的框架,在开发工程中无须过多关注底层实现,只要管理对象即可	MyBatis 需要自行管理对象与表的映射关系
缓存管理	Hibernate 的二级缓存配置在 SessionFactory 生成的配置文件中进行详细配置,然后再在具体的表-对象映射中配置是哪种缓存	MyBatis 的二级缓存配置都是在每个具体的表-对象映射中进行详细配置,这样针对不同的表可以自定义不同的缓存机制
学习难易	学习门槛较高	容易学习和掌握

5.2 MyBatis 的开发环境

5.2.1 MyBatis 框架的 JAR 包下载

想在 Web 开发项目中添加 MyBatis 的支持非常简单,只将 MyBatis 的 JAR 包加入构建路径即可。编写本书时,MyBatis 的最新版本是 mybatis-3.5.2,本书中的所有

MyBatis 案例都基于这个版本。MyBatis 的 JAR 包下载地址为 https://github.com/mybatis/mybatis-3/releases。单击网址,可打开如图 5-1 所示的页面。

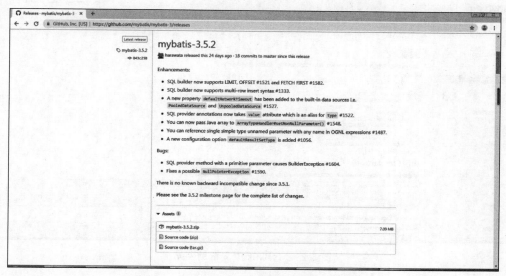

图 5-1　MyBatis 框架 JAR 包下载页面

在网页中单击 mybatis-3.5.2.zip 超链接即可下载 JAR 包。下载后将其中的 mybatis-3.5.2.jar 文件复制到工程的 lib 目录中即可,同时不要忘记添加 MySQL 的数据库驱动程序。此处,我们添加的驱动程序是 mysql-connector-java-5.1.40-bin.jar。除了上面的网站还有一个网站,对开发者会有帮助,网址为 http://www.mybatis.org/mybatis-3/zh/getting-started.html。

这个网站为开发者提供了最新的 MyBatis 相关技术内容,同时提供了 MyBatis 框架所需的一些基本知识,网站提供了 5 种语言,包括简体中文,非常便于查阅。MyBatis 中文帮助网站如图 5-2 所示。

5.2.2　日志信息配置

使用 MyBatis 的时候,有时需要查看 SQL 语句执行的详细信息,为此 MyBatis 提供了非常简单有效的方法——日志。

MyBatis 的内置日志工厂提供日志功能。内置日志工厂将日志交给以下其中一种工具作代理:

- SLF4J;
- Apache Commons Logging;
- Log4j 2;
- Log4j;
- JDK logging。

MyBatis 内置日志工厂基于运行时自省机制选择合适的日志工具。它会使用第一个查找得到的工具(按上文列举的顺序查找)。如果一个都未找到,日志功能就会被

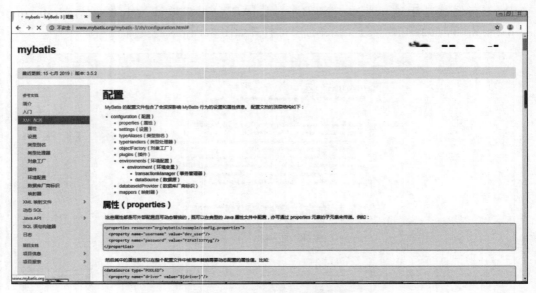

图 5-2　MyBatis 中文帮助网站

禁用。

不少应用服务器（如 Tomcat 和 WebShpere）的类路径中已经包含 Commons Logging，所以在这种配置环境下的 MyBatis 会把它作为日志工具。这将意味着，在诸如 WebSphere 的环境中，它提供了 Commons Logging 的私有实现，你的 Log4J 配置将被忽略。如果你的应用部署在一个类路径已经包含 Commons Logging 的环境中，而你又想使用其他日志工具，可以通过在 MyBatis 配置文件 mybatis-config.xml 里添加一项 setting 选择别的日志工具，如下面的代码所示。

```
<configuration>
  <settings>
    ...
    <setting name="logImpl" value="LOG4J"/>
    ...
  </settings>
</configuration>
```

logImpl 可选的值有 SLF4J、LOG4J、LOG4J2、JDK_LOGGING、COMMONS_LOGGING、STDOUT_LOGGING、NO_LOGGING，或者是实现了接口 org.apache.ibatis.logging.Log 且构造方法是以字符串为参数的类的完全限定名（可以参考 org.apache.ibatis.logging.slf4j.Slf4jImpl.java 的实现）。

下面的例子将使用 Log4J 配置完整的日志服务，共两个步骤。

步骤 1：添加 Log4J 的 JAR 包。

因为我们使用的是 Log4J，就要确保它的 JAR 包在应用中是可用的。要启用 Log4J，只要将 JAR 包添加到应用的类路径中即可。对于 Web 应用或企业级应用，需要将 log4j.jar 添加到 WEB-INF/lib 目录下；对于独立应用，可以将它添加到 JVM 的 -classpath 启动参数中。

步骤 2：配置 Log4J。

配置 Log4J 比较简单，假如需要记录下面这个映射器接口的日志：

```
Package com.mapper;
public interface CustomerMapper {
    @Select("SELECT * FROM customer WHERE id=#{id}")
    Customer selectCustomer(int id);
}
```

在应用的类路径中创建一个名称为 log4j.properties 的文件，文件的具体内容如下：

```
#Global logging configuration
log4j.rootLogger=ERROR, stdout
#MyBatis logging configuration...
log4j.logger.com.mapper.CustomerMapper=TRACE
#Console output...
log4j.appender.stdout=org.apache.log4j.ConsoleAppender
log4j.appender.stdout.layout=org.apache.log4j.PatternLayout
log4j.appender.stdout.layout.ConversionPattern=%5p [%t] - %m%n
```

添加以上配置后，Log4J 就会记录 com.mapper.Customer 的详细执行操作，且仅记录应用中其他类的错误信息。

也可以将日志的记录方式从接口级别切换到语句级别，从而实现更细粒度的控制。下面的配置只对 selectCustomer 语句记录日志：

```
log4j.logger.com.mapper.CustomerMapper.selectCustomer=TRACE
```

与此相对，可以对一组映射器接口记录日志，只要对映射器接口所在的包开启日志功能即可。

```
log4j.logger.com.mapper=TRACE
```

某些查询可能会返回庞大的结果，此时只想记录其执行的 SQL 语句，而不想记录结果该怎么办？MyBatis 中 SQL 语句的日志级别被设为 DEBUG（JDK 日志设为 FINE），结果的日志级别为 TRACE（JDK 日志设为 FINER）。所以，只要将日志级别调整为 DEBUG 即可达到目的：

```
log4j.logger.com.mapper=DEBUG
```

要记录日志的是类似下面的映射器文件，而不是映射器接口又该怎么做呢？

```
1.  <?xml version="1.0" encoding="UTF-8" ?>
2.  <!DOCTYPE mapper
3.      PUBLIC "-//mybatis.org//DTD Mapper 3.0//EN"
```

```
4.         "http://mybatis.org/dtd/mybatis-3-mapper.dtd">
5.    <mapper namespace="com.mapper.ClazzMapper">
6.      <select id="selectClazzById" resultType="Clazz">
7.         select * from tb_clazz where id = #{id}
8.      </select>
9.    </mapper>
```

如需对 XML 文件记录日志,只对命名空间增加日志记录功能即可。

`log4j.logger.com.mapper.ClazzMapper=TRACE`

要记录具体语句的日志可以这样做:

`log4j.logger.com.mapper.ClazzMapper.selectClazzById=TRACE`

从上面的配置可以看到,为映射器接口和 XML 文件添加日志功能的语句没有差别。

5.3 MyBatis 中的 API

使用 MyBatis 的主要 Java 接口就是 SqlSession。可以通过这个接口执行命令,获取映射器和管理事务。我们会概括讨论一下 SqlSession 本身,但是首先还是要了解如何获取一个 SqlSession 实例。SqlSessions 是由 SqlSessionFactory 实例创建的。SqlSessionFactory 对象包含创建 SqlSession 实例的所有方法。而 SqlSessionFactory 本身是由 SqlSessionFactoryBuilder 创建的,它可以从 XML、注解或手动配置 Java 代码创建 SqlSessionFactory。

5.3.1 SqlSessionFactoryBuilder

SqlSessionFactoryBuilder 有 9 个 build() 方法,每种都允许从不同的资源中创建一个 SqlSession 实例。SqlSessionFactoryBuilder 的部分源码如下。

```
1.  public class SqlSessionFactoryBuilder {
2.
3.    public SqlSessionFactory build(InputStream inputStream) {
4.      return build(inputStream, null, null);
5.    }
6.    public SqlSessionFactory build(Configuration config) {
7.      return new DefaultSqlSessionFactory(config);
8.    }
9.    public SqlSessionFactory build ( InputStream inputStream, String
      environment, Properties properties) {
10.     try {
11.       XMLConfigBuilder parser = new XMLConfigBuilder ( inputStream,
          environment, properties);
```

```
12.        return build(parser.parse());
13.    } catch (Exception e) {
14.      throw ExceptionFactory.wrapException("Error building SqlSession.", e);
15.    } finally {
16.        ErrorContext.instance().reset();
17.        try {
18.            inputStream.close();
19.        } catch (IOException e) {
20.            //Intentionally ignore. Prefer previous error.
21.        }
22.    }
23.    }
24. }
```

在上面的代码中，XMLConfigBuilder 是一个解析 XML 的工具类，如果调用有参数 environment 的 build() 方法，MyBatis 将会使用 configuration 对象配置这个 environment。当然，如果指定了一个不合法的 environment，就会得到错误提示。如果调用了不带 environment 参数的 build() 方法，就使用默认的 environment（在上面的示例中指定为 default="development" 的代码）。

如果调用了有参数 properties 实例的方法，MyBatis 就会加载那些 properties（属性配置文件）。那些属性可以用 ${propName} 语法形式多次用在配置文件中。

5.3.2 SqlSessionFactory

这里给出的是 3 个比较常用的生成 SqlSessionFactory 的 build() 方法，下面的代码演示了第 3 行的 build() 方法生成 SqlSessionFactory 示例的应用。

```
1. String resource="mybatis-config.xml";
2. InputStream inputStream=Resources.getResourceAsStream(resource);
3. SqlSessionFactoryBuilder builder=new SqlSessionFactoryBuilder();
4. SqlSessionFactory factory=builder.build(inputStream);
```

以上代码是读取 mybatis-config.xml 配置文件构建一个 InputStream 输入流，然后调用 SqlSessionFactoryBuilder 的 build() 方法创建一个 SqlSessionFactory 工厂类。典型的 mybatis-config.xml 配置文件如下所示。

```
1. <?xml version="1.0" encoding="UTF-8" ?>
2. <!DOCTYPE configuration PUBLIC "-//mybatis.org//DTD Config 3.0//EN"
3.     "http://mybatis.org/dtd/mybatis-3-config.dtd">
4. <configuration>
5.     <!--1.配置环境,默认的环境 id 为 mysql-->
6.     <environments default="mysql">
7.         <!--1.2.配置 id 为 mysql 的数据库环境 -->
8.         <environment id="mysql">
```

```
9.         <!--使用JDBC的事务管理-->
10.        <transactionManager type="JDBC" />
11.        <!--数据库连接池-->
12.        <dataSource type="POOLED">
13.            <property name="driver" value="com.mysql.jdbc.Driver" />
14.            <property name ="url"
15.                      value="jdbc:mysql://localhost:3306/mybatis" />
16.            <property name="username" value="root" />
17.            <property name="password" value="root" />
18.        </dataSource>
19.      </environment>
20.    </environments>
21.    <!--2.配置mapper的位置-->
22.    <mappers>
23.        <mapper resource="com/mapper/CustomerMapper.xml" />
24.    </mappers>
25. </configuration>
```

配置文件中的<configuration>标签描述了数据库连接的信息,<mappers>标签描述了映射文件的位置信息。SqlSessionFactory 是一个接口,它有两个实现类,分别是 SqlSessionFactoryManager 和 DefaultSqlSessionFactory。前面讲到的几个类之间的关系可用如图 5-3 所示的 UML 类图表示。

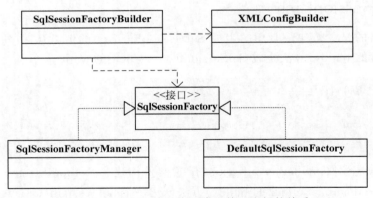

图 5-3　MyBatis 中的主要类和接口之间的关系

在基于 MyBatis 框架的应用程序中,都是以 SqlSessionFactory 为中心,它的主要作用是生产 MyBatis 的核心接口对象 SqlSession。

5.3.3　SqlSession

SqlSession 是 MyBatis 的关键对象,是执行持久化操作的对象,类似于 JDBC 中的 Connection。它是应用程序与持久层之间执行交互操作的一个单线程对象,也是 MyBatis 执行持久化操作的关键对象。SqlSession 对象完全包含对数据库所有执行 SQL

操作的方法，对 JDBC 操作数据库进行了封装，可以用 SqlSession 实例直接执行被映射的 SQL 语句。每个线程都应该有它自己的 SqlSession 实例，SqlSession 的实例不能被共享，同时也是线程不安全的。绝不能将 SqlSession 实例的引用放在一个类的静态字段，甚至是实例字段中。也绝不能将 SqlSession 实例的引用放在任何类型的管理范围中，如放在 Servlet 中的 HttpSession 对象中。使用完 SqlSession 后必须关闭，应该确保使用 finally 块来关闭它。

SqlSession 类中有 20 多个方法，将它们按功能进行分组，便于理解。

1. 对数据库进行增、删、改、查的方法

- <T>T selectOne(String statement，Object parameter)

查询方法，查询结果是一个泛型对象或为 null。

- <E>List<E>selectList(String statement，Object parameter)

查询方法，查询结果可以是 0 到多个泛型对象的集合，也可以为 null。

- <K，V>Map<K，V>selectMap(String statement，Object parameter，String mapKey)

查询方法，查询结果是一个 Map 集合。

- int insert(String statement，Object parameter)

更新方法，执行插入该表中记录的操作。

- int update(String statement，Object parameter)

更新方法，执行更新该表中记录的操作。

- int delete(String statement，Object parameter)

更新方法，执行删除该表中记录的操作。

- seqSession.selectCursor(String statement，Object parameter);

方法中的参数 statement 是映射文件中对应的元素的 id，参数 parameter 是执行 SQL 语句所需绑定的动态参数。

selectOne 和 selectList 的区别是，selectOne 必须返回一个对象或 null 值。如果返回值多于一个，就会抛出异常。如果不能确定返回对象的数量，请使用 selectList。如果需要查看返回对象是否存在，一个好的办法是返回一个值即可（0 或 1）。selectMap 稍微特殊一点，因为它会将返回的对象的其中一个属性作为 key 值，将对象作为 value 值，从而将多结果集转为 Map 类型值。

与上面 7 个方法对应的还有 7 个重载的方法，除了 selectMap() 方法有两个参数外，其他的都只有一个参数 statement。

- <T>T selectOne(String statement);
- <E>List<E>selectList(String statement);
- <T>Cursor<T>selectCursor(String statement);
- <K，V>Map<K，V>selectMap(String statement，String mapKey);
- int insert(String statement);
- int update(String statement);

- int delete(String statement)。

这 7 个方法中没有动态绑定参数 parameter，所以一般是将参数写在映射文件中，这种应用场景不是很多。

如果查询结果是一个集合，要对返回结果的数量进行控制，可以使用以下的重载方法。

- < E > List < E > selectList（String statement，Object parameter，RowBounds rowBounds）
- < K，V > Map< K，V > selectMap（String statement，Object parameter，String mapKey，RowBounds rowbounds）

这两个方法是前面 selectList 和 selectMap 两个方法的重载，其中的 rowbounds 对象参数可以设置返回结果集合的起始位置和数量。

2. 事务控制方法

控制事务作用域有 3 个方法。如果已经设置了自动提交或正在使用外部事务管理器，这 3 个方法不会起作用。然而，如果正在使用 JDBC 事务管理器，由 Connection 实例控制，那么这 3 个方法就会派上用场。

- void commit()：事务提交。
- void rollback()：回滚事务。
- void close()：关闭 SqlSession 对象。

3. 缓存管理方法

MyBatis 使用到两种缓存：本地缓存（local cache）和二级缓存（second level cache）。

- void clearCache()：清空本地缓存。

4. 获取 Mapper 接口

上述 SqlSession 的 insert、update、delete 和 select 方法都很强大，但也有些烦琐，可能会产生类型安全问题，并且对于 IDE 和单元测试也很不方便。因此，一个更通用的方式是通过使用映射器类执行映射语句。一个映射器类就是一个仅需声明与 SqlSession 方法相匹配的方法的接口类。

<T>T getMapper(Class<T>type)：获取映射器类接口的方法。

Mapper 接口的使用后面会详细介绍。

5.4　MyBatis 的配置文件

MyBatis 的配置文件是 MyBatis 框架的必备文件，包含了影响 MyBatis 行为的设置和属性信息。配置文档的顶层结构如下。

```
1.    <?xml version="1.0" encoding="UTF-8" ?>
2.    <!DOCTYPE configuration PUBLIC "-//mybatis.org//DTD Config 3.0//EN"
```

```
3.         "http://mybatis.org/dtd/mybatis-3-config.dtd">
4.     <configuration>
5.         <properties />
6.         <settings />
7.         <typeAliases />
8.         <typeHandlers />
9.         <plugins />
10.        <environments>
11.            <environment>
12.                <transactionManager />
13.                <dataSource />
14.            </environment>
15.        </environments>
16.        <databaseIdProvider />
17.        <mappers />
18.    </configuration>
```

这是配置文件的基本结构，文档中的各个元素的次序不能颠倒，如果次序不对，在 MyBatis 的启动阶段就会发生异常，导致程序无法运行。下面会对配置文件中常用的元素进行详细介绍。

5.4.1 \<properties\>元素

\<properties\>是一个配置属性的元素，该元素通常用于将内部的配置外在化，即通过外部的配置文件动态地替换内部定义的属性。常用的如数据库的连接等属性，就可以通过典型的 Java 属性文件中的配置来替换。下面给出的是一个数据库配置的属性文件，具体步骤如下。

（1）在 src 目录下创建名为 db.properties 的配置文件，文件内容如下。

```
1.  jdbc.driver=com.mysql.jdbc.Driver
2.  jdbc.url=jdbc:mysql://localhost:3306/ssm
3.  jdbc.username=root
4.  jdbc.password=root
```

（2）在 MyBatis 配置文件 mybatis-config.xml 中通过\<properties\>元素的 resource 属性引入 db.properties 文件。

```
<properties resource="db.properties"/>
```

（3）由于有了 db.properties 属性文件的导入，因此可修改配置文件中连接数据库的信息，代码如下。

```
1.  <dataSource type="POOLED">
2.      <!--数据库驱动 -->
3.      <property name="driver" value="${jdbc.driver}" />
```

```
4.          <!--连接数据库的 url -->
5.          <property name="url" value="${jdbc.url}" />
6.          <!--连接数据库的用户名 -->
7.          <property name="username" value="${jdbc.username}" />
8.          <!--连接数据库的密码 -->
9.          <property name="password" value="${jdbc.password}" />
10.     </dataSource>
```

配置数据源的 dataSource 元素有 4 个属性，分别是 driver、url、username 和 password，原来是将这些与数据库连接的相关信息直接写在配置文件中，现在是由属性文件 db.properties 替换，这样将数据库配置信息与 MyBatis 的配置信息分离开，修改数据库信息时不会影响其他配置信息，便于系统维护。

5.4.2 <settings>元素

这是 MyBatis 中极为重要的调整设置，它们会改变 MyBatis 的运行时行为。表 5-2 描述了设置中各项的含义、默认值等。

表 5-2 <settings>元素的 name 属性的值的含义

name 属性的值	描述	有效值	默认值
cacheEnabled	开启或关闭配置文件中的所有映射器已经配置的任何缓存	true|false	true
lazyLoadingEnabled	延迟加载的全局开关。当开启时，所有关联对象都会延迟加载。特定关联关系中可通过设置 fetchType 属性覆盖该项的开关状态	true|false	false
aggressiveLazyLoading	当开启时，任何方法的调用都会加载该对象的所有属性，否则每个属性会按需加载（参考 lazyLoadTriggerMethods）	true|false	false（在 3.4.1 及之前的版本默认值为 true）
multipleResultSetsEnabled	是否允许单一语句返回多结果集（需要驱动支持）	true|false	true
useGeneratedKeys	允许 JDBC 支持自动生成主键，需要驱动支持。如果设置为 true，则这个设置强制使用自动生成主键，尽管一些驱动不能支持，但仍可正常工作，如 Derby	true|false	false
autoMappingBehavior	指定 MyBatis 应如何自动映射列到字段或属性。NONE 表示取消自动映射；PARTIAL 只会自动映射没有定义嵌套结果集映射的结果集。FULL 会自动映射任意复杂的结果集（无论是否嵌套）	NONE, PARTIAL, FULL	PARTIAL
defaultStatementTimeout	设置超时时间，它决定驱动等待数据库响应的秒数	任意正整数	未设置（null）

续表

name 属性的值	描述	有效值	默认值
defaultFetchSize	为驱动的结果集获取数量（fetchSize）设置一个提示值。此参数只可以在查询设置中被覆盖	任意正整数	未设置（null）
mapUnderscoreToCamelCase	是否开启自动驼峰命名规则（camel case）映射，即从经典数据库列名 A_COLUMN 到经典 Java 属性名 aColumn 的类似映射	true|false	false

表 5-2 给出的是<settings>元素的部分 name 属性值，这个属性的值决定了 MyBatis 的一些运行时的状态和行为，这里比较重要的是 lazyLoadingEnabled 和 aggressiveLazyLoading 两个值。当有多表关联操作时，lazyLoadingEnabled 决定了是否加载关联对象，而在查询时，对列的选择是由 aggressiveLazyLoading 值决定的。

一个配置完整的 settings 元素的示例如下。

```
1.    <settings>
2.    <setting name="cacheEnabled" value="true"/>
3.    <setting name="lazyLoadingEnabled" value="true"/>
4.    <setting name="multipleResultSetsEnabled" value="true"/>
5.    <setting name="useColumnLabel" value="true"/>
6.    <setting name="useGeneratedKeys" value="false"/>
7.    <setting name="autoMappingBehavior" value="PARTIAL"/>
8.    <setting name="autoMappingUnknownColumnBehavior" value="WARNING"/>
9.    <setting name="defaultExecutorType" value="SIMPLE"/>
10.   <setting name="defaultStatementTimeout" value="25"/>
11.   <setting name="defaultFetchSize" value="100"/>
12.   <setting name="safeRowBoundsEnabled" value="false"/>
13.   <setting name="mapUnderscoreToCamelCase" value="false"/>
14.   <setting name="localCacheScope" value="SESSION"/>
15.   <setting name="jdbcTypeForNull" value="OTHER"/>
16.   <setting name="lazyLoadTriggerMethods" value="equals,clone,hashCode,toString"/>
17.   </settings>
```

5.4.3 <typeAliases>元素

在 MyBatis 配置文件中经常用到有限定名的类，而限定名有时很长，写起来不方便，在 MyBatis 配置文件中允许定义一个简写代表这个类，这就是别名。别名分为系统定义别名和自定义别名。在 MyBatis 中别名不区分大小写。

MyBatis 框架中已经定义了一些别名，见表 5-3。

表 5-3　系统自定义的别名

别　　名	映射的 Java 类型	别　　名	映射的 Java 类型
_byte	Byte	string	String
_long	Long	byte	Byte
_short	Short	long	Long
_int	int	short	Short
_integer	Int	Int	Integer
_double	Double	integer	Integer
_float	Float	double	Double
_boolean	boolean	float	Float
boolean	Boolean	date	Date
decimal	BigDecimal	bigdecimal	BigDecimal
object	Object	map	Map
hashmap	HashMap	list	List
arraylist	ArrayList	collection	Collection
iterator	Iterator		

使用<typeAliases>元素配置别名的格式如下。

```
<!--定义别名-->
<typeAliases>
    <typeAlias alias="customer" type="com.po.Customer">
</typeAliases>
```

<typeAliases>可以有多个<typeAlias>子元素，定义多个别名。上面的代码是为限定类 com.po.Customer 指定了别名 customer，这个别名可以在 MyBatis 的配置文件和后面的映射文件中使用。

实际项目中会有很多这样的限定类，如果每个类都这样定义，显得很烦琐，MyBatis 提供了另外一种定义别名的方式，就是通过包名定义包中所有类的别名。其定义格式如下。

```
<typeAliases>
    <pakage name="com.po">
</typeAliases>
```

每个在包 com.po 中的 Java Bean 在没有注解的情况下，都会使用 Bean 的首字母小写的非限定类名作为它的别名。例如，com.po.User 的别名为 user，若有注解，则别名为其注解值。注意，在 MyBatis 中别名不区分大小写。

5.4.4　<typeHandlers>元素

在 MyBatis 的 SQL 映射配置文件中，为 SQL 配置的输入参数最终要从 javaType 类型转换成数据库能识别的 jdbcType 类型，而从 SQL 的查询结果集中获取的数据也要从数据库的 jdbcType 数据类型转换为对应的 javaType 类型。typeHandler 的作用是承担 jdbcType 和 javaType 之间的相互转换。

和别名一样，在 MyBatis 中存在系统定义 typeHandler 和自定义 typeHandler。多数情况下，MyBatis 会根据 javaType 和数据库的 jdbcType 决定采用哪个 typeHandler 处理这些转换规则。系统提供的 typeHandler 能覆盖大部分应用场景，但是有些情况下是不够的，当项目有特殊的数据类型转换需求，如枚举类型，此时要自定义 typeHandler，同时在 MyBatis 配置文件中需用 <typeHandlers> 元素配置自定义的 typeHandler。系统已定义的 typeHandler 见表 5-4。

表 5-4 系统已定义的 typeHandler

类型处理器	Java 类型	JDBC 类型
BooleanTypeHandler	java.lang.Boolean，boolean	数据库兼容的 BOOLEAN
ByteTypeHandler	java.lang.Byte，byte	数据库兼容的 NUMERIC 或 BYTE
ShortTypeHandler	java.lang.Short，short	数据库兼容的 NUMERIC 或 SMALLINT
IntegerTypeHandler	java.lang.Integer，int	数据库兼容的 NUMERIC 或 INTEGER
LongTypeHandler	java.lang.Long，long	数据库兼容的 NUMERIC 或 BIGINT
FloatTypeHandler	java.lang.Float，float	数据库兼容的 NUMERIC 或 FLOAT
DoubleTypeHandler	java.lang.Double，double	数据库兼容的 NUMERIC 或 DOUBLE
BigDecimalTypeHandler	java.math.BigDecimal	数据库兼容的 NUMERIC 或 DECIMAL
StringTypeHandler	java.lang.String	CHAR，VARCHAR
LocalDateTypeHandler	java.time.LocalDate	DATE

这些就是 MyBatis 框架已经创建好的部分 typeHandler。当 MyBatis 框架提供的这些 typeHandler 不能满足需求时，如使用枚举类型时，可以通过自定义的方式对 typeHandler 进行扩展。自定义的 typeHandler 要实现 org.apache.ibatis.type.TypeHandler 这个接口或继承 org.apache.ibatis.type.BaseTypeHandler 类。<typeHandlers> 元素就是用于在配置文件中注册自定义的 typeHandler。它的注册方式有两种，具体如下。

1. 注册一个类的 typeHandler

```
<typeHandlers>
    <typeHandler handler="com.type.CustomertypeHandler"/>
</typeHandlers>
```

上述代码中，<typeHandler> 子元素的 handler 属性指向自定义的 typeHandler。

2. 注册一个包中的 typeHandler

如果在工程中需要定义多个 typeHandler，可以将这些类型处理器放在一个包中，然后以包的形式统一进行注册，代码如下所示。

```
<typeHandlers>
    <package name="com.type"/>
<typeHandlers>
```

上述代码中，<typeHandler>子元素的 name 属性用于指定自定义的 typeHandler 所在的包，系统会在启动时自动扫描 com.type 包下的所有文件，并将它们作类型处理器。

5.4.5 <environments>元素

在 MyBatis 配置文件中，<environments>元素用于配置程序运行环境，其实主要是数据源的配置。MyBatis 可以配置成适应多种数据源，这种机制有助于将 SQL 映射应用于多种数据库中。现实情况中有多种需求，例如，开发、测试和生产环境需要有不同的配置，或者想在具有相同 Schema 的多个生产数据库中使用相同的 SQL 映射等。

尽管可以配置多个运行环境，但每个 SqlSessionFactory 实例只能对应一种环境。所以，如果想连接两个数据库，就需要创建两个 SqlSessionFactory 实例，每个数据库对应一个运行环境，而如果是 3 个数据库，就需要 3 个实例，以此类推。下面是用环境配置元素<environments>配置数据源的示例。

```
1.  <environments default="development">
2.    <environment id="development">
3.      <transactionManager type="JDBC">
4.        <property name="..." value="..."/>
5.      </transactionManager>
6.      <dataSource type="POOLED">
7.        <property name="driver" value="${driver}"/>
8.        <property name="url" value="${url}"/>
9.        <property name="username" value="${username}"/>
10.       <property name="password" value="${password}"/>
11.     </dataSource>
12.   </environment>
13. </environments>
```

在第 1 行的代码中，<environments>是根元素，它包含一个属性 default，该属性指定了默认的环境 ID 为 development。<environments>是一个容器，它可以有多个<environment>子元素，也就可以定义多个数据源。每个<environment>都有一个 ID，代表一个数据源的定义。

第 3 行的<transactionManager>元素用于配置数据库的事务管理器，它有一个 type 属性用于指定事务管理方式。MyBatis 中有两种类型的事务管理器（也就是 type＝"[JDBC|MANAGED]"）。Type 的这两个值的含义如下。

- JDBC：这个配置就是直接使用了 JDBC 的提交和回滚设置，它依赖于从数据源得到的连接来管理事务作用域。
- MANAGED：这个配置几乎没做什么。它从来不提交或回滚一个连接，而是让容器管理事务的整个生命周期（例如，JavaEE 应用服务器的上下文）。默认情况下，它会关闭连接，然而一些容器并不希望这样，因此需要将 closeConnection 属性设置为 false 阻止它默认的关闭行为，这可通过如下代码实现。

```
1.    <transactionManager type="MANAGED">
2.        <property name="closeConnection" value="false"/>
3.    </transactionManager>
```

在开发中如果使用 Spring + MyBatis，则没有必要配置事务管理器，因为 Spring 框架会使用自带的管理器覆盖前面的配置。

第 6 行<dataSource>元素用于配置数据源的具体信息，其属性 type 的值表示配置数据源的类型，它有 3 种内建的数据源类型，对应的 type 有 3 个值，具体含义如下。

UNPOOLED：这个数据源的实现只是每次被请求时打开和关闭连接。虽然有点慢，但对于在数据库连接可用性方面没有太高要求的简单应用程序来说，是一个很好的选择。不同的数据库在性能方面的表现也不一样，对于某些数据库来说，使用连接池并不重要，这个配置就很适合这种情形。UNPOOLED 类型的数据源具有以下属性。

- driver：这是 JDBC 驱动的 Java 类的完全限定名（并不是 JDBC 驱动中可能包含的数据源类）。
- url：这是数据库的 JDBC URL 地址。
- Username：登录数据库的用户名。
- Password：登录数据库的密码。
- defaultTransactionIsolationLevel：默认的连接事务隔离级别。作为可选项，也可以传递属性给数据库驱动。只需在属性名加上"driver."前缀即可，如 driver.encoding=UTF8。

这将通过 DriverManager.getConnection(url, driverProperties) 方法传递值为 UTF8 的 encoding 属性给数据库驱动。

POOLED：这种数据源的实现利用"连接池"的概念将 JDBC 连接对象组织起来，避免了创建新的连接实例时所需的初始化和认证时间。这是一种使得并发 Web 应用快速响应请求的流行处理方式。配置这种数据源时，除了 UNPOOLED 具有的 5 个属性外，还有下面一些属性。

- poolMaximumActiveConnections：在任意时间可以存在的活动（也就是正在使用）连接数量，默认值为 10。
- poolMaximumIdleConnections：任意时间可能存在的空闲连接数。
- poolMaximumCheckoutTime：在被强制返回之前，池中连接被检出（checked out）时间，默认值为 20000ms（即 20s）。
- poolTimeToWait：这是一个底层设置，如果获取连接花费了相当长的时间，连接池会打印状态日志并重新尝试获取一个连接（避免在误配置的情况下没有信息显示），默认值为 20000ms（即 20s）。
- poolMaximumLocalBadConnectionTolerance：这是一个关于坏连接容忍度的底层设置，作用于每个尝试从缓存池获取连接的线程。如果这个线程获取到的是一个坏的连接，那么这个数据源允许这个线程尝试重新获取一个新的连接，但是这个重新尝试的次数不应该超过 poolMaximumIdleConnections 与 poolMaximumLocalBadConnectionTolerance 之和，默认值为 3。

- poolPingQuery：发送到数据库的侦测查询，用来检验连接是否正常工作并准备接受请求。默认是 NO PING QUERY SET，这会导致多数数据库驱动失败时带有一个恰当的错误消息。
- poolPingEnabled：是否启用侦测查询。若开启，则需要设置 poolPingQuery 属性为一个可执行的 SQL 语句（最好是一个速度非常快的 SQL 语句），默认值为 false。
- poolPingConnectionsNotUsedFor：配置 poolPingQuery 的频率。可以被设置为和数据库连接超时时间一样，避免不必要的侦测，默认值为 0（即所有连接每一时刻都被侦测，当然仅当 poolPingEnabled 为 true 时适用）。

JNDI：这个数据源的实现是为了能在如 EJB 或应用服务器这类容器中使用，容器可以集中或在外部配置数据源，然后放置一个 JNDI 上下文的引用。这种数据源配置只需要以下两个属性。

- initial_context：这个属性用来在 InitialContext 中寻找上下文，即 initialContext.lookup(initial_context)。这是一个可选属性，如果忽略，将会直接从 InitialContext 中寻找 data_source 属性。
- data_source：这是引用数据源实例位置的上下文的路径，提供了 initial_context 配置时会在其返回的上下文中进行查找，没有提供时则直接在 InitialContext 中查找。

5.4.6 \<mappers\>元素

MyBatis 需要开发者自己编写 SQL 语句，这些 SQL 语句保存在映射文件中。在配置文件中，<mappers>元素用于指定映射文件的位置。MyBatis 提供了 4 种引入映射文件的方式。

1. 使用类路径方式引入

```
<mappers>
    <mapper resource="com/mapper/Customer.xml"/>
</mappers>
```

2. 使用本地文件绝对路径引入

```
<mappers>
    <mapper url="file:///e:/com/mapper/Customer.xml"/>
</mappers>
```

3. 使用接口类引入

```
<mappers>
    <mapper class="com/mapper/CustomerMapper"/>
</mappers>
```

4. 使用包名引入

```
<mappers>
    <package name="com.mapper"/>
</mappers>
```

在配置文件中有了这些配置信息，MyBatis 在启动时就能自动加载这些映射文件。下面主要讲解映射文件的内容。

5.5 MyBatis 映射器

映射器是 MyBatis 最复杂且最重要的部分，它由一个接口加上 XML 文件（或者注解）组成。在映射器中可以配置各种参数、SQL 语句、存储过程、缓存和级联等复杂的内容。在映射文件中通过映射规则映射到指定的 POJO 或者其他对象上，映射器能有效减少 JDBC 底层烦琐的代码。

映射器有两种实现方式：XML 映射文件和注解映射。这两种映射器各有所长，读者都应该掌握，后面对这两种技术会有详细的讲解。

5.5.1 XML 映射文件的主要元素

在映射文件中，<mapper> 是映射文件的根元素，其他元素都是它的子元素，这些子元素见表 5-5。

表 5-5　XML 映射文件中的元素

元素名称	描　　述	备　　注
select	SQL 查询语句	可自定义参数，返回结果集
insert	插入语句	执行后返回一个整数，代表插入记录数
update	更新语句	执行后返回一个整数，代表更新记录数
delete	删除语句	执行后返回一个整数，代表删除记录数
sql	定义一个 SQL 片段	这个 SQL 片段可以在映射文件的其他地方引用
resultMap	定义查询结果集	这个结果集一般用于关联操作中
cache	给定命名空间的缓存配置	
Cache-ref	引用已定义的缓存配置	

5.5.2 <select> 元素

<select> 查询语句是 MyBatis 中最常用的元素之一，它可以帮助我们从数据库中读取数据，多数应用也都是查询比修改要频繁和复杂。对每个插入、更新或删除操作，通常也需要多个查询操作，这是 MyBatis 的基本原则之一。<select> 元素的常用属性见表 5-6。

表 5-6　select 元素的常用属性

元素	描述
id	命名空间中唯一的标识符，在其他地方可以通过 id 引用或执行这条语句
parameterType	将传入这条语句的参数类的完全限定名或别名。这个属性是可选的，因为 MyBatis 可以通过类型处理器（TypeHandler）推断出具体传入语句的参数，默认值为未设置（unset）
resultType	从这条语句中返回的期望类型的类的完全限定名或别名。注意，如果返回的是集合，应该设置为集合包含的类型，而不是集合本身。可以使用 resultType 或 resultMap，但不能同时使用
resultMap	外部 resultMap 的命名引用，可以使用 resultMap 或 resultType，但不能同时使用
flushCache	将其设置为 true 后，清空本地缓存和二级缓存，默认值为 false
userCache	将其设置为 true 后，本条查询语句的结果被二级缓存缓存起来，默认值：对 select 元素为 true
timeout	设置等待数据库返回请求结果的时间，单位为秒
fetchSize	设定查询结果返回的行数
statementType	可以取值为 STATEMENT、PREPARED 或 CALLABLE 中的一个。这会让 MyBatis 分别使用 Statement、PreparedStatement 或 CallableStatement，默认值为 PREPARED
resultSetType	可以取值为 FORWARD_ONLY（游标允许向前访问）、SCROLL_SENSITIVE（双向流动，但不及时更新）、SCROLL_INSENSITIVE（双向流动，并及时跟踪数据库的更新），默认值为 unset（依赖驱动）
databaseId	如果配置了数据库厂商标识（databaseIdProvider），MyBatis 会加载所有不带 databaseId 或匹配当前 databaseId 的语句；如果带或者不带的语句都有，则不带的会被忽略
resultOrdered	这个设置仅针对嵌套结果 select 语句适用：如果为 true，就假设包含了嵌套结果集或是分组，这样，当返回一个主结果行的时候，就不会有对前面结果集的引用的情况。这就使得在获取嵌套的结果集的时候不至于导致内存不够用。默认值为 false
resultSets	适合多个结果集的情况，它将列出执行后返回的结果集每个结果集的名称，名称之间用逗号分隔

下面通过一个查询示例说明 select 语句的使用方法。

例 5-1　现有数据库 homework_db，有一个班级表 clazz(id(int),cname(varchar))。要求利用 MyBatis 的<select>元素根据 id 查询班级信息。

（1）创建一个 Web 工程 chap5，构建 MyBatis 开发环境。在 lib 中添加 mybatis-3.5.2.jar 包和数据库驱动程序 mysql-connector-java-5.1.40-bin.jar，为了使用日志，再添加 log4j 相应的 JAR 包和 log4j.properties 文件。在 src 下规划包结构，创建 com.po、com.mapper、com.test 和 com.utils 包。

（2）在 src 下创建属性文件 db.properties，存放连接数据的信息。

```
1. jdbc.driver=com.mysql.jdbc.Driver
2. jdbc.url=jdbc:mysql://localhost:3306/homework_db
```

3. jdbc.username=root
4. jdbc.password=root

（3）在 src 下创建 MyBatis 配置文件 mybatis-config.xml。

```xml
1.  <?xml version="1.0" encoding="UTF-8" ?>
2.  <!DOCTYPE configuration
3.    PUBLIC "-//mybatis.org //DTD Config 3.0//EN"
4.    "http://mybatis.org/dtd/mybatis-3-config.dtd">
5.  <configuration>
6.    <properties resource="db.properties" />
7.    <!--定义别名 -->
8.    <typeAliases>
9.      <package name="com.po" />
10.   </typeAliases>
11.   <!--配置环境,默认的环境 id 为 mysql -->
12.   <environments default="mysql">
13.     <!--配置 id 为 mysql 的数据库环境 -->
14.     <environment id="mysql">
15.       <!--使用 JDBC 的事务管理 -->
16.       <transactionManager type="JDBC" />
17.       <!--数据库连接池 -->
18.       <dataSource type="POOLED">
19.         <!--数据库驱动 -->
20.         <property name="driver" value="${jdbc.driver}" />
21.         <!--连接数据库的 url -->
22.         <property name="url" value="${jdbc.url}" />
23.         <!--连接数据库的用户名 -->
24.         <property name="username" value="${jdbc.username}" />
25.         <!--连接数据库的密码 -->
26.         <property name="password" value="${jdbc.password}" />
27.       </dataSource>
28.     </environment>
29.   </environments>
30.   <!--配置 mapper 的位置 -->
31.   <mappers>
32.     <mapper resource="com/mapper/ClazzMapper.xml" />
33.   </mappers>
34. </configuration>
```

配置文件中第 2~4 行是针对当前 xml 的文档类型定义，这部分内容可以从帮助文档中复制。第 18 行数据源 dataSource 的配置用到前面 db.properties 文件中包含的数据库信息。第 31 行<mappers>元素指定了映射文件的位置信息。

（4）在 com.po 包中创建与 clazz 表对应的类 Clazz.java。

```
1.   package com.po;
2.   public class Clazz {
3.       private Integer id;           //班级 id
4.       private String cname;         //班级名称
5.       @Override
6.       public String toString() {
7.           return "Clazz [id=" + id +", cname=" + cname +"]";
8.       }
9.       //setter and getter 方法
10.  }
```

（5）在 com.mapper 包中创建映射文件 ClazzMapper.xml。

```
1.   <?xml version="1.0" encoding="UTF-8" ?>
2.   <!DOCTYPE mapper
3.       PUBLIC "-//mybatis.org //DTD Mapper 3.0//EN"
4.       "http://mybatis.org/dtd/mybatis-3-mapper.dtd">
5.   <mapper namespace="com.mapper.ClazzMapper">
6.       <select id="selectClazzById" resultType="Clazz">
7.           select * from tb_clazz where id =#{id}
8.       </select>
9.   </mapper>
```

映射文件中的第 2～4 行是 DTD 文档类型的定义，每个 xml 的 DTD 文档类型的定义都是不一样的，这部分内容可从 MyBatis 帮助文档中复制过来。第 5 行的<mapper>元素的 namespace 属性定义了命名空间，这个命名空间在所有的映射文件中是唯一的，它与下面的<select>元素中的 id 属性的值一起构成唯一标识。这个标识就是

com.mapper.ClazzMapper.selectClazzById

<select>元素的属性 resultType 表示查询结果的返回值的类型，此处是用别名 Clazz 代替全称 com.po.Clazz。第 7 行是<select>元素或<select>标签的标签体，是标准的 SQL 语句，作用是根据 clazz 表的 id 查询某一个班级的信息，其返回值是一个 Clazz 类的对象。

（6）在 com.utils 包中创建一个工具类，作用是加载 mybatis-config.xml 配置文件，生成 SqlSessionFactory 类的实例，调用 sqlSessionFactory 的 openSession（）方法得到 SqlSession 的一个实例。

```
1.   package com.utils;
2.   import java.io.Reader;
3.   import org.apache.ibatis.io.Resources;
```

```
4.    import org.apache.ibatis.session.SqlSession;
5.    import org.apache.ibatis.session.SqlSessionFactory;
6.    import org.apache.ibatis.session.SqlSessionFactoryBuilder;
7.    /**
8.     * 工具类
9.     */
10.   public class MybatisUtils {
11.       private static SqlSessionFactory sqlSessionFactory=null;
12.       //初始化 SqlSessionFactory 对象
13.       static {
14.           try {
15.               //使用 MyBatis 提供的 Resources 类加载 MyBatis 的配置文件
16.               Reader reader=
17.                   Resources.getResourceAsReader("mybatis-config.xml");
18.               //构建 SqlSessionFactory 工厂
19.               sqlSessionFactory=
20.                   new SqlSessionFactoryBuilder().build(reader);
21.           } catch (Exception e) {
22.               e.printStackTrace();
23.           }
24.       }
25.       //获取 SqlSession 对象的静态方法
26.       public static SqlSession getSession() {
27.           return sqlSessionFactory.openSession();
28.       }
29.   }
```

（7）在 com.test 包中创建一个 jUnit 测试用例。

```
1.    package com.test;
2.    import org.apache.ibatis.session.SqlSession;
3.    import org.junit.jupiter.api.Test;
4.    import com.po.Clazz;
5.    import com.utils.MybatisUtils;
6.    class ClazzTest {
7.        @Test
8.        void findCalzzByIdTest() {
9.            SqlSession sqlSession=MybatisUtils.getSession();
10.           Clazz clazz = sqlSession.selectOne("com.mapper.ClazzMapper.
                  selectClazzById",1);
11.           System.out.println(clazz);
12.       }
13.   }
```

在测试用例中调用 MybatisUtils 的静态方法 getSession()，得到一个 SqlSession 的实例 sqlSession，相当于得到一个数据库的连接，通过 sqlSession 的 selectOne()方法执行在映射文件中指定的 SQL 语句，此处是根据 clazz 表的 id 查询 id=1 的记录，如果存在这条记录，则返回一个 Clazz 的对象，对应的是 clazz 表的一条记录，如果不存在，则返回 null。

（8）运行测试用例，在控制台上可得到如图 5-4 所示的结果。

```
DEBUG [main] - ==>  Preparing: select * from tb_clazz where id = ?
DEBUG [main] - ==> Parameters: 1(Integer)
DEBUG [main] - <==      Total: 1
Clazz [id=1, cname=软件201601]
```

图 5-4　运行 jUnit 测试用例 ClazzTest 的结果

从运行结果分析，实际执行了标准的 SQL 语句，映射文件中的 id=#{id}，相当于 id=?，实际运行时，id=1。

5.5.3　<insert>元素

<insert>元素用于映射插入语句，执行完映射语句后会返回一个整数，表示插入记录数。<insert>元素的属性见表 5-7。

表 5-7　<insert>元素的属性

属　　性	描　　述
id	命名空间中的唯一标识符，在其他地方可以通过 id 引用或执行这条语句
parameterType	将要传入语句的参数的完全限定类名或别名。这个属性是可选的，因为 MyBatis 可以通过类型处理器推断出具体传入语句的参数，默认值为未设置（unset）
flushCache	将其设置为 true 后，只要语句被调用，都会导致本地缓存和二级缓存被清空，默认值为 true（对于 insert、update 和 delete 语句）
timeout	这个设置是在抛出异常之前，驱动程序等待数据库返回请求结果的时间，单位为秒，默认值为未设置（unset）（依赖驱动）
statementType	STATEMENT、PREPARED 或 CALLABLE 中的一个。这会让 MyBatis 分别使用 Statement、PreparedStatement 或 CallableStatement，默认值为 PREPARED
userGeneratedKeys	（仅对 insert 和 update 有用）这会令 MyBatis 使用 JDBC 的 getGeneratedKeys() 方法取出由数据库内部生成的主键（例如，像 MySQL 和 SQL Server 这样的关系数据库管理系统的自动递增字段），默认值为 false
keyProperty	（仅对 insert 和 update 有用）唯一标记一个属性，MyBatis 会通过 getGeneratedKeys 的返回值或者通过 insert 语句的<selectKey>子元素设置它的键值，默认值为未设置（unset）。如果希望得到多个生成的列，也可以是逗号分隔的属性名称列表
keyColumn	（仅对 insert 和 update 有用）通过生成的键值设置表中的列名，这个设置仅在某些数据库（像 PostgreSQL）是必需的，当主键列不是表中的第一列时需要设置。如果希望使用多个生成的列，也可以设置为逗号分隔的属性名称列表

表 5-7 的属性中最常用的属性是 parameterType 和 userGeneratedKeys。parameterType 用于执行插入操作时传入的参数,一般是要插入表对应的 PO 类的实例。userGeneratedKeys 用于执行插入操作时,如果主键由数据库自动生成,并且希望立即得到插入记录后的主键的值,此时可以将此属性设置为 true。下面是映射文件中使用<insert>元素映射插入语句的片段。

```xml
<insert id="insertClazz" parameterType="clazz" keyProperty="id"
    useGeneratedKeys="true">
    insert into tb_clazz(cname) values(#{cname})
</insert>
```

分析上述代码,这段代码用<insert>元素定义了一个插入表的 SQL 语句片段,此处 parameterType 属性指定了执行这个片段时传入的参数的类型为 clazz,这个 clazz 是在配置文件中定义的别名,实际对应的是 com.po.Clazz 类。用 keyProperty 设定了 id 为主键,useGenerateKeys 设置为 true 表示在插入记录成功后返回生成的主键的值。在 insert 语句中的#{cname}是动态绑定的参数,它与传入的参数 Clazz 类中的属性名一致。为了验证映射文件的正确性,可创建一个测试用例,代码如下。

```java
@Test
void insertClazzTest() {
    SqlSession sqlSession=null;
    try {
        sqlSession=MybatisUtils.getSession();
        Clazz clazz=new Clazz();
        clazz.setCname("软件 201602");
        sqlSession.insert("com.mapper.ClazzMapper.insertClazz",clazz);
        sqlSession.commit();
        System.out.println(clazz);
    }catch(Exception e) {
        e.printStackTrace();
    }finally {
        sqlSession.close();
    }
}
```

分析上述代码,第 8 行调用 sqlSession 的 insert()方法,执行映射文件中 id 为 insertClazz 的 SQL 片段,在这个 SQL 片段中需要传入一个参数,这个参数是一个 Clazz 类的实例,完成插入 tb_clazz 表中的记录。运行插入测试用例结果如图 5-5 所示。

图 5-5 运行插入测试用例结果图

从运行结果分析，实际运行了 SQL 的 insert 语句。

对于不支持自动生成类型的数据库或可能不支持自动生成主键的 JDBC 驱动，MyBatis 有另外一种方法来生成主键。下面是调用 UUID()函数生成一个 36 位的唯一字符串作为表的主键的例子。

```xml
1.  <mapper namespace="com.mapper.StuMapper">
2.     <insert id="insertByStringId" parameterType="com.po.Stu">
3.       <selectKey keyProperty="id" order="BEFORE"
4.            resultType="java.lang.String">
5.            SELECT UUID()
6.       </selectKey>
7.       insert into tb_stu(id,name) values(#{id},#{name})
8.     </insert>
9.  </mapper>
```

tb_stu 表的主键 id 设为 varchar(36)，在上面的 SQL 片段执行时，会先运行<selectKey>元素，它会调用 UUID()生成一个唯一的 36 位长度的字符串，然后执行插入操作。测试这段代码的测试类如下。

```java
1.  public class StuTest {
2.     @Test
3.     public void StuInsertTest(){
4.         SqlSession sqlSession=null;
5.         try {
6.             sqlSession=MybatisUtils.getSession();
7.             Stu stu=new Stu();
8.             stu.setName("李红");
9.             sqlSession.insert("com.mapper.StuMapper.insertByStringId",stu);
10.            sqlSession.commit();
11.            System.out.println(stu);
12.        }catch(Exception e) {
13.            e.printStackTrace();
14.        }finally {
15.            sqlSession.close();
16.        }
17.    }
18. }
```

分析上述代码，在给 insert()方法传入参数时，第二个参数是 Stu 类的一个实例，Stu 类有两个属性 id 和 name，由于主键 id 是在<selectKey>元素中生成的，只需对 name 属性赋值即可。运行测试类的 StuInsertTest()方法，运行结果显示在控制台上，如图 5-6 所示。

图 5-6 测试方法 StuInsertTest() 运行结果

<selectKey>元素的属性见表 5-8。

表 5-8 <selectKey>元素的属性

属　性	描　述
keyProperty	selectKey 语句结果应该被设置的目标属性。如果希望得到多个生成的列，也可以是逗号分隔的属性名称列表
keyColumn	匹配属性的返回结果集中的列名称。如果希望得到多个生成的列，也可以是逗号分隔的属性名称列表
resultType	结果的类型。MyBatis 通常可以推断出来，但是为了更加精确，写上也不会有什么问题。MyBatis 允许将任何简单类型用作主键的类型，包括字符串。如果希望作用于多个生成的列，可以使用一个包含期望属性的 Object 或一个 Map
order	这可以被设置为 BEFORE 或 AFTER。如果设置为 BEFORE，它会首先生成主键，设置 keyProperty 然后执行插入语句。如果设置为 AFTER，那么先执行插入语句，然后是 selectKey 中的语句，这和 Oracle 数据库的行为相似，在插入语句内部可能有嵌入索引调用
statementType	与前面相同，MyBatis 支持 STATEMENT、PREPARED 和 CALLABLE 语句的映射类型，分别代表 PreparedStatement 和 CallableStatement 类型

5.5.4 <update>和<delete>元素

<update>和<delete>元素的使用相对比较简单，它们的属性配置也基本相同，下面是在映射文件中使用这两个元素完成更新和删除的示例。

```
1.  <update id="updateClazz" parameterType="clazz">
2.  update tb_clazz set cname=#{cname} where id=#{id}
3.  </update>
4.  <delete id="deleteClazzById" parameterType="Integer">
5.  delete from tb_clazz where id=#{id}
6.  </delete>
```

第 1 行 parameterType="clazz"表示传递的参数是 com.po.Clazz 类的对象，其中的 SQL 语句是根据 clazz 对象中的 id 更新 tb_clazz 表中的记录。第 4 行 parameterType="Integer"表示传递的参数是一个整数，其中的 SQL 语句表示根据传递的 id 值删除指定的记录。这两个 SQL 片段的测试类代码如下：

```java
1.   @Test
2.   void updateClazzTest() {
3.       SqlSession sqlSession=null;
4.       try {
5.           sqlSession=MybatisUtils.getSession();
6.           Clazz clazz=new Clazz();  //构造一个Clazz类的对象,用于更新表中的记录
7.           clazz.setId(1);
8.           clazz.setCname("软件2016022");
9.           //调用sqlSession.update()方法执行更新操作,
10.          //其参数是映射文件中id为updateClazz的SQL片段
11.          sqlSession.update("com.mapper.ClazzMapper.updateClazz",clazz);
12.          sqlSession.commit();
13.          System.out.println(clazz);
14.      }catch(Exception e) {
15.          e.printStackTrace();
16.      }finally {
17.          sqlSession.close();
18.      }
19.  }
20.  @Test
21.  void deleteClazzTest() {
22.      SqlSession sqlSession=null;
23.      try {
24.          sqlSession=MybatisUtils.getSession();
25.  //调用sqlSession的delete()方法,其参数是映射文件中id为deleteClazzById
    //的SQL片段
26.          sqlSession.delete("com.mapper.ClazzMapper.deleteClazzById",10);
27.          sqlSession.commit();
28.      }catch(Exception e) {
29.          e.printStackTrace();
30.      }finally {
31.          sqlSession.close();
32.      }
33.  }
```

在jUnit测试类中添加两个测试方法,void updateClazzTest()用于测试映射文件中的<update>元素中的SQL片段,此处要求传递一个Clazz对象参数,才能完成更新操作。void deleteClazzTest()用于测试映射文件中<delete>元素中的SQL片段,此处要求传递一个Integer类型的参数,才能完成删除操作。

5.5.5 <sql>元素

<sql>元素可以用来定义可重用的SQL代码片段,可以包含在其他语句中,也可以(在加载的时候)被静态地设置参数。在不同的包含语句中可以设置不同的值到参数占位

符上。例如，下面的语句：

```
<sql id="userColumns">${alias}.id,${alias}.username,${alias}.password </sql>
```

这个 SQL 片段可以包含在其他语句中，例如：

```
1.  <select id="selectUsers" resultType="map">
2.      select
3.          <include refid="userColumns"><property name="alias" value="t1"/>
4.  </include>,
5.      <include refid="userColumns"><property name="alias" value="t2"/>
6.  </include>
7.      from some_table t1
8.          cross join some_table t2
9.  </select>
```

第 3～6 行引用了前面定义的 userColumns 这个 SQL 片段，完成了数据库查询中的列的映射操作。这段代码翻译过来相当于下面的代码：

```
t1.id,t1.username,t1.password,t2.id,t2.username,t2.password
```

如果在映射文件中经常有重复的 SQL 片段，采用<sql>元素可避免编写重复的代码，提高代码的可重用率。

属性值也可以被用在<include>元素的 refid 属性里或<include>元素的内部语句中，例如：

```
1.  <sql id="sometable">
2.    ${prefix}Table
3.  </sql>
4.  <sql id="someinclude">
5.    from
6.      <include refid="${include_target}"/>
7.  </sql>
8.  <select id="select" resultType="map">
9.    select
10.     field1, field2, field3
11.   <include refid="someinclude">
12.     <property name="prefix" value="Some"/>
13.     <property name="include_target" value="sometable"/>
14.   </include>
15. </select>
```

第 9～14 行翻译过来相当于下面的 SQL 片段：

```
Select field1,field2,field3 from Sometable
```

习　题

1. 什么是 MyBatis?
2. MyBatis 框架适用于哪些场合?
3. MyBatis 与 Hibernate 有哪些不同?
4. 在 MyBatis 配置文件中如何设置全局延迟加载开关?
5. 在 MyBatis 配置文件中如何注册映射文件的信息?
6. MyBatis 映射文件都有哪些标签?
7. #{} 和 ${} 的区别是什么?
8. 如何获取自动生成的(主)键值? 请给出一个具体的示例。

实验 5　用 MyBatis 完成单表的增、删、改、查操作

1. 实验目的

了解 MyBatis 框架的基本工作原理,掌握在映射文件中 <select>、<insert>、<update>、<delete> 4 个元素的使用方法,并编写 jUnit 测试用例对映射文件中的 SQL 片段进行测试。

2. 实验内容

(1) 创建数据库 ssm,创建 Student 表,表的结构为主键(id,Integer,要求设置为自动增长)、学号(loginname,varchar(10))、姓名(username,varchar(20))、年龄(age,Integer)。

(2) 创建针对 Student 表的 PO 类和映射文件,在映射文件中利用 <insert>、<delete>、<update> 和 <select> 4 个元素实现表的增、删、改、查操作。

(3) 创建 jUnit 测试类,分别测试映射文件中 4 个元素对应的 SQL 片段,完成表的增、删、改、查操作。

3. 实现思路及步骤

(1) 创建一个 Web 工程,添加所需的 JAR 包,规划包的结构。
(2) 创建日志文件 log4j.properties。
(3) 创建属性文件 db.properties,存放数据连接信息。
(4) 在 src 下创建 MyBatis 配置文件 mybatis-config.xml。
(5) 在 PO 包中创建 Student 类。
(6) 在 mapper 包中创建 StudentMapper.xml 映射文件。
(7) 在 utils 包中创建工具类 MybatisUtils。
(8) 在 test 包中创建 jUnit 测试类 StudentText,并添加相应的测试方法。

第 6 章 结果映射与动态 SQL

本章目标

1. 了解结果映射（<resultMap> 元素）的用途和作用，掌握<resultMap>的编写方法。

2. 掌握动态 SQL 中几个常用标签 if、choose、where、set、foreach、bind、trim 的正确使用方法。

6.1 结果映射（<resultMap>元素）

我们在编写实体类的时候，一般实体类的属性和数据表的列名都是一一对应的，但难免有不能对应的情况，这时候就需要配置 resultMap，使得实体类的属性和数据表的列名对应，让 MyBatis 完成封装。还有一种情况需要 resultMap，就是多表关联操作。此时由于是多表关联，查询的结果集会非常复杂，包含多个表的字段和表达式，此时只有用 resultMap 在外部重新定义一个结果映射，才是最好的选择。

在 MyBatis 中，对查询结果具备一定的自动识别功能，如果查询的结果是一个集合，它会将结果自动映射为一个 List 或一个 Map，例如下面的 SQL 查询片段。

```
1.  <select id="selectAllClazz" resultType="clazz">
2.  select * from tb_clazz
3.  </select>
```

这个查询的结果可能为空，也可能是多个结果的集合，虽然在此我们给出的是 rsultType="clazz"，但实际上返回的结果是一个 List<Clazz>的泛型。这可以通过运行测试类的结果得到验证。下面是测试类的代码。

```
1.  @Test
2.     void selectAllClazzTest() {
3.        SqlSession sqlSession=MybatisUtils.getSession();
4.        List<Clazz> clazzes=sqlSession.selectList("com.mapper.
          ClazzMapper.selectAllClazz");
5.        clazzes.forEach(cla->System.out.println(cla));
6.     }
```

在测试方法中，通过 sqlSession 的 selectList（）方法运行映射文件中 id 为 selectAllClazz 对应的查询片段，运行测试类中的测试方法 selectAllClazzTest（），在控制台上得到的运行结果如图 6-1 所示。

```
Markers  Properties  Servers  Data Source Explorer  Snippets  Problems  Console ⊠  JUnit
<terminated> ClazzTest.selectAllClazzTest [JUnit] C:\Program Files\Java\jre1.8.0_101\bin\javaw.exe (2019年8月14日 下午4:49:
DEBUG [main] - ==>  Preparing: select * from tb_clazz
DEBUG [main] - ==> Parameters:
DEBUG [main] - <==      Total: 5
Clazz [id=1, cname=软件2016022]
Clazz [id=2, cname=计算机201601]
Clazz [id=3, cname=软件201602]
Clazz [id=4, cname=计算机201602]
Clazz [id=5, cname=软件201601]
```

图 6-1　测试方法 selectAllClazzTest 运行结果

从运行结果可知，MyBatis 已经成功将查询结果映射成了 List 集合，其元素为 Clazz 类的对象。

在实际项目中经常会遇见表的字段与 PO 类的属性不一致，或者查询结果非常复杂，特别是多表关联查询，这时需要通过<resultMap>定义一个结果集，这样可以简化查询结果的复杂性。<resultMap>元素的结构如下。

```
1.  <!--column 不做限制,可以为任意表的字段,而 property 须为 type 定义的 pojo 属性-->
2.  <resultMap id="唯一的标识" type="映射的 pojo 对象">
3.     <id column="表的主键字段,或者可以为查询语句中的别名字段" jdbcType="字段类型" property="映射 pojo 对象的主键属性" />
4.     <result column="表的一个字段(可以为任意表的一个字段)" jdbcType="字段类型" property="映射到 pojo 对象的一个属性(须为 type 定义的 pojo 对象中的一个属性)" />
5.     <association property="pojo 的一个对象属性" javaType="pojo 关联的 pojo 对象">
6.        <id column="关联 pojo 对象对应表的主键字段" jdbcType="字段类型" property="关联 pojo 对象的唯一标识属性"/>
7.        <result column="任意表的字段" jdbcType="字段类型" property="关联 pojo 对象的属性"/>
8.     </association>
9.     <!--集合中的 property 须为 oftype 定义的 pojo 对象的属性-->
10.    <collection property="pojo 的集合属性" ofType="集合中的 pojo 对象">
11.       <id column="集合中 pojo 对象对应的表的主键字段" jdbcType="字段类型" property="集合中 pojo 对象的唯一标识属性" />
12.       <result column="可以为任意表的字段" jdbcType="字段类型" property="集合中的 pojo 对象的属性"/>
13.    </collection>
14. </resultMap>
```

下面是一个使用<resultMap>的简单示例。

例 6-1 现有一个表 tb_stu,其结构为主键 t_id(int)、姓名 t_cname(varchar(20)),现要求针对表 tb_stu 编写映射文件,返回结果用<resultMap>元素定义一个结果映射,并运行测试类。

(1) 创建映射文件 StuMapper.xml,代码清单如下。

```
1.  <mapper namespace="com.mapper.StuMapper">
2.    <resultMap type="com.po.Stu" id="stuResultMap">
3.      <id property="id" column="t_id" />
4.        <result property="name" column="t_name" />
5.    </resultMap>
6.    <select id="selectAllStu" resultMap="stuResultMap">
7.      select * from tb_stu
8.    </select>
9.  </mapper>
```

在映射结果元素<resultMap>中完成了 Stu 类与表 tb_stu 之间的映射,同时通过<id>子元素实现了类的 oid 与表的主键的映射,通过<result>元素实现了类的属性与表的字段之间的映射。

第 6 行的<select>元素中使用了前面<resultMap>中定义的映射结果,具体实现是<select>元素中的 resultMap 的属性值与第 2 行定义的<resultMap>中的 id="stuResultMap"要一致。

(2) Stu 类的定义。

```
1.  package com.po;
2.  public class Stu {
3.    private String id;
4.    private String name;
5.    public String getId() {
6.      return id;
7.    }
8.    //setter and getter 方法
9.  }
```

这是一个 pojo 类,是与 tb_stu 表映射所必需的实体类。

(3) 同时不要忘记修改 mybatis-config.xml 文件,在<mappers>元素中注册新创建的映射文件 StuMapper.xml,代码如下。

```
1.  <mappers>
2.    <mapper resource="com/mapper/ClazzMapper.xml" />
3.    <mapper resource="com/mapper/StuMapper.xml" />
4.  </mappers>
```

(4) 创建测试类,测试 StuMapper.xml 文件中的 SQL 片段,测试结果在此省略。

例 6-1 是一个简单应用<resultMap>的示例,在实际项目中主要应用于多表关联操作中。这部分内容在第 7 章中会详细讲解。

6.2 动态 SQL

MyBatis 的强大特性之一是它的动态 SQL。如果你有使用 JDBC 或其他类似框架的经验,就能体会到根据不同的查询条件拼接复杂 SQL 语句的痛苦。例如,拼接时要确保不能忘记添加必要的空格,还要注意去掉列表最后一个列名的逗号等。利用动态 SQL 这一特性可以彻底摆脱这种痛苦。

动态 SQL 元素和 JSTL 或基于类似 XML 的文本处理器相似。在 MyBatis 之前的版本中,有很多元素需要花时间了解。MyBatis 3.0 大大精简了元素种类,现在只学习原来一半的元素便可。MyBatis 采用功能强大的基于 OGNL 的表达式淘汰其他大部分元素。

常用的动态 SQL 元素如下。
- if;
- choose(when、otherwise);
- where;
- set;
- foreach;
- bind;
- trim。

6.2.1 <if>元素

<if>元素在 MyBatis 中是最常用的元素,类似于 Java 中的 if 语句,根据实际需求给出的条件组成所需的查询语句。

在实际项目中,查询是项目必需的功能。而查询的要求是复杂、多样的,查询涉及的属性是不确定的,属性的值也是不确定的。针对这样的查询,在构建查询语句时就会变得异常复杂,但应用<if>元素可以轻松解决这个问题。下面通过一个示例说明<if>元素的使用。

例 6-2 现有 tb_teacher 表,生成数据库的脚本如下。

```
1.   DROP TABLE IF EXISTS `tb_teacher`;
2.   CREATE TABLE `tb_teacher` (
3.     `id` int(11) NOT NULL AUTO_INCREMENT,
4.     `loginname` varchar(20) NOT NULL,
5.     `password` varchar(20) NOT NULL,
6.     `tname` varchar(20) NOT NULL COMMENT '教师姓名',
7.     PRIMARY KEY (`id`)
8.   ) ENGINE=InnoDB AUTO_INCREMENT=3 DEFAULT CHARSET=utf8;
```

```
 9.   -- ----------------------------
10.   -- Records
11.   -- ----------------------------
12.   INSERT INTO `tb_teacher` VALUES ('1', 'ssh', 'admin', '史胜辉');
13.   INSERT INTO `tb_teacher` VALUES ('2', 'wcm', 'wcm', '王春明');
```

现要求对 tb_teacher 表进行查询，具体要求为

可根据 loginname 或 tname 中任意一个字段进行查询，或 loginname 和 tname 两个字段组合查询，如果不输入任何查询条件，则查询所有的记录。具体实现步骤如下。

（1）创建一个 Web 项目 chap6，添加所需的 JAR 包、日志文件，创建 mybatis-config.xml 配置文件，规划包的结构。

（2）创建与 tb_teacher 表对应的 Teacher 类。

```
1.  package com.po;
2.  public class Teacher {
3.      private Integer id;
4.      private String loginname;
5.      private String password;
6.      private String tname;
7.      public Integer getId() {
8.          return id;
9.      }
```

（3）创建映射文件 TeacherMapper.xml，编写查询语句。

```
 1.  <?xml version="1.0" encoding="UTF-8" ?>
 2.  <!DOCTYPE mapper
 3.      PUBLIC "-//mybatis.org//DTD Mapper 3.0//EN"
 4.      "http://mybatis.org/dtd/mybatis-3-mapper.dtd">
 5.  <mapper namespace="com.mapper.TeacherMapper">
 6.      <select id="selectTeacherByAny" resultType="teacher">
 7.          select * from tb_teacher where 1=1
 8.          <if test="loginname!=null">
 9.              AND loginname=#{loginname}
10.          </if>
11.          <if test="tname!=null">
12.              AND tname like concat('% ',#{tname},'% ')
13.          </if>
14.      </select>
```

第 8 行和第 11 行构建了两个查询。

如果 loginname！＝null 成立，则查询条件为 where loignname＝#{loginname}，即根据 loginname 字段进行查询。

如果 tname！＝null 成立，则查询条件为 where tname＝#{tname}，即根据 tname 字

段进行模糊查询。

如果 loginname!＝null 和 tname!＝null 同时成立,则查询条件为

```
where loginname=#{loginname} and tname=#{tname}
```

第 12 行中用到 concat()函数,此函数的作用是拼接查询字符串时防止 SQL 注入。

(4) 修改 batis-config.xml 文件,注册 TeacherMapper.xml 映射文件,在<mappers>标签下添加如下代码:

```
<mapper resource="com/mapper/TeacherMapper.xml" />
```

(5) 创建测试类,测试映射文件中的 SQL 片段。

```
1.    class TeacherTest {
2.      @Test
3.      void selectTeacherByAnyTest() {
4.        Teacher teacher=new Teacher();
5.        //teacher.setLoginname("ssh");
6.        //teacher.setLoginname("wcm");
7.        //teacher.setTname("王春明");
8.        //teacher.setTname("史");
9.        SqlSession sqlSession=MybatisUtils.getSession();
10.       List<Teacher>teachers=sqlSession.selectList("com.mapper.
          TeacherMapper.selectTeacherByAny",teacher);
11.       System.out.println(teachers.size());
12.       teachers.forEach(tea->System.out.println(tea));
13.     }
14.   }
```

在测试类中分别给对象 teacher 的属性 loginname 和 tname 赋不同的值,从控制台查看运行结果。

(6) 运行测试类,此处不再给出运行结果,读者可自行验证。

6.2.2 <choose>元素

有时不想应用到所有的条件语句,只想从中择其一项。针对这种情况,MyBatis 提供了 choose 元素,它有点像 Java 中的 switch 语句。<choose>元素下面有两个子元素:when 和 otherwise,它们的结构如下:

```
<choose>
    <when test=" 表达式">
        ...
    </when>
    <otherwise>
        ...
    </otherwise>
```

```
</choose>
```

下面通过一个示例说明<choose>元素的使用。

例 6-3　此例中用到的数据库与例 6-2 一致，但查询条件发生了变化。功能要求为根据 loginname 或 tname 其中之一字段查询，如果两者都为空，则不返回任何记录。实现步骤如下。

（1）打开 Web 项目 chap6。

（2）创建与 tb_teacher 表对应的 Teacher 类，与上例相同，此处忽略。

（3）修改映射文件 TeacherMapper.xml，添加如下代码。

```
1.   <select id="selectTeacherForChoose" parameterType="teacher" resultType="teacher">
2.       select * from tb_teacher where 1=1
3.       <choose>
4.         <when test="loginname!=null">
5.           AND loginname=#{loginname}
6.         </when>
7.         <when test="tname!=null">
8.           AND tname like concat('%',#{tname},'%')
9.         </when>
10.        <otherwise>
11.          AND 1=-1
12.        </otherwise>
13.      </choose>
14.  </select>
```

第 4 行<when>元素的 test 属性的值是一个条件判断表达式，如果表达式成立，则将元素体中的 SQL 语句添加到查询语句中，后面的条件不再进行判断。如果第 4、7 行两个<when>中的条件都不能满足，则将<otherwise>元素中的 AND 1=-1 语句添加到查询语句中，这条语句的作用是返回一个永远为 false 的逻辑值，这样就不会返回任何记录。

（4）创建测试类中的测试方法，测试映射文件中的 SQL 片段。

```
1.   @Test
2.   void selectTeacherForChooseTest() {
3.       Teacher teacher=new Teacher();
4.       teacher.setLoginname("ssh");
5.       //teacher.setTname("王春明");
6.       //teacher.setTname("史");
7.       SqlSession sqlSession=MybatisUtils.getSession();
8.       List<Teacher> teachers = sqlSession.selectList("com.mapper.
         TeacherMapper.selectTeacherForChoose",teacher);
```

```
9.        System.out.println(teachers.size());
10.       teachers.forEach(tea->System.out.println(tea));
11.    }
```

在测试类中,当给对象 teacher 的 loginname 属性赋值后,不管后面是否给 tname 属性赋值,其查询结果都是一样的。因为在映射文件中 loginname 的判断条件在先,只要 loginname 属性不为空,后面的判断条件就不会被执行,因此其他属性赋值也就没有意义了。

(5)运行测试类,此处不再给出运行结果,读者可自行验证。

6.2.3 <where>元素

在前面的映射文件中,每个 SQL 片段中都有一个 where 1=1 语句,如果没有这个语句中的"1=1",会发生什么情况?

可以修改映射文件,去掉<select>元素中的"1=1",修改后的代码如下。

```
1.  <select id="selectTeacherByAny" resultType="teacher">
2.    select * from tb_teacher where
3.    <if test="loginname!=null">
4.        loginname=#{loginname}
5.    </if>
6.    <if test="tname!=null">
7.        AND tname like concat('% ',#{tname},'% ')
8.    </if>
9.  </select>
```

在测试方法中,如果只给 loginname 属性赋值,程序可以正常运行。如果只给 tname 属性赋值,则运行程序会抛出异常,控制台显示如下信息:

select * from tb_teacher where AND tname like concat('% ',?,'% ')

很明显,这是一个非法的 SQL 语句,其中多了一个"AND"字符串。为什么发生这样的情况?原因是在此处 MyBatis 不能正确判断什么时候要加 AND、什么时候不加 AND 或 OR 这样的语句。为了解决类似这样的问题,MyBatis 给出了<where>元素,可以很好地解决类似的问题。

还是以例 6-2 为例,修改 TeacherMapper.xml 映射文件,在映射文件中添加如下的查询语句片段。

```
1.  <select id="selectTeacherForWhere" resultType="teacher">
2.    select * from tb_teacher
3.    <where>
4.        <if test="loginname!=null">
5.            AND loginname=#{loginname}
6.        </if>
```

```
7.            <if test="tname!=null">
8.                AND tname like concat('% ',#{tname},'% ')
9.            </if>
10.        </where>
11.    </select>
```

此处使用<where>元素取代了原来的 where 1=1 语句。where 元素只会在至少有一个子元素的条件返回 SQL 子句的情况下才插入 where 子句。而且,若语句的开头为 AND 或 OR,<where>元素也会将它们去除。

在测试方法中可对各种可能的情况进行测试,可以看到程序都能正常运行。这是因为<where>会为 SQL 查询语句添加 where 语句,并且能正确判断出什么时候需要 AND、什么时候不需要 AND 这样的语句。有了<where>元素,为我们正确编写查询语句提供了很大的便利。

6.2.4 <set>元素

在数据库编程中经常会遇到数据更新的问题,在程序运行前有时不能确定要更新表中哪些字段的值,这就要求程序员在编写数据更新的 SQL 片段时,能够根据用户的需求动态地组成更新表的 SQL 片段。<set>元素能很好地解决这类问题。这里还以示例的方式讲解<set>元素的使用方法。

例 6-4 利用 MyBatis 对 tb_teacher 表进行更新,要求在映射文件中用<set>元素实现动态更新。实现步骤如下。

(1) 打开 chap6 工程。

(2) 修改 TeacherMapper.xml 映射文件,添加动态更新的语句。

```
1.  <update id="teacherUpdate" parameterType="teacher">
2.      update tb_teacher
3.      <set>
4.          <if test="loginname!=null and loginname!=''">
5.            loginname=#{loginname},
6.          </if>
7.          <if test="password!=null and password!=''">
8.            password=#{password},
9.          </if>
10.           <if test="tname!=null and tname!=''">
11.            tname=#{tname}
12.          </if>
13.      </set>
14.      where id=#{id}
15.  </update>
```

这段代码是根据 id 更新 tb_teacher 表。在<set>元素中嵌入了<if>元素,根据<if>元素判断结果决定更新哪个字段的值,这样为编写 SQL 语句提供了很大的便利。

（3）在测试类中添加测试方法 updateTeacherTest()。

```
1.    @Test
2.    void updateTeacherTest() {
3.      SqlSession sqlSession=null;
4.      try {
5.        Teacher teacher=new Teacher();
6.        teacher.setId(1);
7.        teacher.setLoginname("ssh");
8.        teacher.setPassword("123456");
9.        teacher.setTname("史胜辉");
10.       sqlSession=MybatisUtils.getSession();
11.       sqlSession.update("com.mapper.TeacherMapper.teacherUpdate",
          teacher);
12.       sqlSession.commit();
13.       System.out.println(teacher);
14.     }catch(Exception e) {
15.       e.printStackTrace();
16.     }finally {
17.       sqlSession.close();
18.     }
19.   }
```

第 5~9 行构造了一个 teacher 对象，将这个对象作为参数传递给 update() 方法，在拼接 SQL 语句的过程中根据 teacher 中的属性值决定是否更新表中相关的记录。

（4）运行测试方法 updateTeacherTest()，在控制台上打印出如下信息：

DEBUG [main]-==> Preparing: update tb_teacher SET loginname=?, password=?, tname=? where id=?
DEBUG [main]-==> Parameters: ssh(String), 123456(String), 史胜辉(String), 1(Integer)
DEBUG [main]-<== Updates: 1
DEBUG [main] - Committing JDBC Connection [com.mysql.jdbc.JDBC4Connection@77fbd92c]
Teacjer [id=1, loginname=ssh, password=123456, tname=史胜辉]

从控制台信息可看到，MyBatis 将映射文件中的 <update> 元素的内容转换成了标准的 SQL 语句。

6.2.5 <foreach>元素

<foreach>元素是一个循环语句，类似于 Java 中的 for 语句，它的作用是遍历集合，它能够很好地支持数组和 List、Set 接口的集合，一般常用于 SQL 中的 in 关键字。

在数据库中，经常根据给定的集合对表中的数据进行查询操作。

例 6-5 对 tb_teacher 表,根据给定的 id 的集合查询教师的信息。具体步骤如下。
(1) 打开 chap6 工程。
(2) 修改映射文件 TeacherMapper.xml,添加根据 id 集合查询的方法。

```
1.  <select id="selectTeacherByids" parameterType="List" resultType=
    "teacher">
2.      select * from tb_teacher where id in
3.      <foreach item="id" index="index" collection="list"
4.          open="(" separator="," close=")">
5.          #{id}
6.      </foreach>
7.  </select>
```

第 3 行使用<foreach>元素对传入的集合进行遍历,其中<foreach>元素的属性的详细说明如下。

- item:配置的是循环中当前的元素。
- index:配置的是当前元素在集合的位置下标。
- collection:配置的 list 是传递过来的参数类型(首字母小写),它的值可以是 array、list(或 collection)、Map 集合的键、PO 包装类中数组或集合类型的属性名等。
- open 和 close:配置的是以什么符号将这些集合元素包装起来。
- separator:配置各个元素的分隔符。

(3) 在测试类 TeacherTest 中添加测试方法 selectTeacherByIdsTest()。

```
1.  @Test
2.  void selectTeacherByIdsTest() {
3.      List<Integer>ids=new ArrayList<Integer>();
4.      ids.add(2);
5.      ids.add(4);
6.      SqlSession sqlSession=MybatisUtils.getSession();
7.      List< Teacher > teachers = sqlSession.selectList ("com.mapper.
        TeacherMapper.selectTeacherByids",ids);
8.      teachers.forEach(tea->System.out.println(tea));
9.  }
```

第 7 行的 selectList()方法的第 2 个参数 ids 是一个 List 类型,这与映射文件中<foreach>元素的 collection 属性的类型 list 对应。

(4) 运行测试方法,在控制台上可得到如下的日志信息。

```
DEBUG [main] -==>Preparing: select * from tb_teacher where id in ( ? , ? )
DEBUG [main] -==>Parameters: 2(Integer), 4(Integer)
DEBUG [main] -<==   Total: 2
Teacjer [id=2, loginname=wcm, password=wcm, tname=王春明]
```

Teacjer [id=4, loginname=zhanghong, password=zhanghong, tname=张红]

从日志分析，MyBatis 已经对 SQL 语句进行了正确的拼接，对 ids 集合中的元素 2 和 4 进行遍历，分别查询出对应的教师信息。

习　　题

1. 什么时候需要使用<resultMap>标签？
2. 举例说明：什么时候需要使用动态 SQL 语句。
3. 动态 SQL 包括哪些标签？
4. 动态 SQL 中<if>标签与<choose>标签有哪些区别？
5. 简述动态 SQL 的执行过程。

实验 6　用动态 SQL 完成单表的修改和查询操作

1. 实验目的

通过本实验让读者了解动态 SQL 的使用场景和特点，掌握动态 SQL 常用的几个元素的使用方法，本实验主要使用了<if>等 4 个元素。

2. 实验内容

（1）创建数据库 ssm，创建 Student 表，表的结构为主键(id, Integer，要求设置为自动增长)、学号（loginname, varchar(10)）、姓名（username, varchar(20)）、年龄（age, Integer）。

（2）创建针对 Student 表的 PO 类和映射文件，映射文件中在<select>和<update>两个元素中嵌入动态 SQL，实现表的修改和查询操作。具体要求是：

- 在<update>元素中用<set>和<if>元素，根据所传递属性值是否为 null 或空字符串决定是否对 Student 表中的字段进行更新。
- 在<select>元素中使用<where>和<if>元素，根据 loginname 和 username 属性的值决定查询条件，如果都为空，则返回所有记录。

（3）创建 jUnit 测试类，分别测试映射文件中两个元素对应的 SQL 片段，完成表的修改和查询操作。

3. 实现思路及步骤

可参照实验 5 的对应部分。

第 7 章 关联映射

本章目标
1. 掌握一(多)对一的关联操作。
2. 掌握一对多的关联操作。
3. 掌握多对多的关联操作。
4. 了解 MyBatis 的缓存机制,掌握其设置方法。

第 5 章介绍的有关表的操作都是单表操作,在实际工程中一般都是多表操作,这就涉及表与表之间的关联操作。在进行关联操作之前要对表之间的关系有所了解。数据库中表与表之间主要有 3 种关系,具体情况说明如下。
- 一对一:在任意一方引入对方主键作为外键。
- 一对多:在多方添加一方的主键作为外键。
- 多对多:创建第 3 张表,另外两张表的主键作为第 3 张表的外键。

7.1 一(多)对一的关联操作

数据库中表与表之间一对一的关系或者说是实体与实体的关系都是相对的。以学生作业管理系统为例,在系统中学生与班级之间就是一对一的关系,因为一个学生只能属于一个班级,不可能属于第二个班级。当然,也可以说班级与学生是一对多的关系,因为一个班级有多个学生。这样说可能会让人产生一种混乱的感觉,搞不明白学生与班级究竟是什么关系。

在讲实体之间的关系时,一定要搞清楚你所在的角色,也就是你站在哪个角度看实体之间的关系。同样是两个实体,由于你的角色不同,可能会得出完全不同的结论。还是以学生与班级之间的关系为例,当我们从学生这个角度看,学生与班级就是一对一关系,当然有时也说是多对一关系,这在实现上是相同的。

那么,在 MyBatis 中是如何实现表与表之间的一对一关联呢?这可以通过第 6 章讲的 <resultMap> 元素中的子元素 <association> 实现。<association>元素主要有以下 4 个属性。

property：指定要映射的实体类的属性名称。
column：与另一张表关联的外键对应字段的名字。
javaType：指定要映射的实体类的属性的类型。
select：指定引入嵌套查询的 SQL 语句，这个查询的结果映射为 property 对应的属性。

下面以学生表 tb_student 与班级表 tb_clazz 为例，说明如何用 MyBaits 实现一对一的关联操作。

例 7-1 创建 homework 数据库，在数据库中创建 tb_student 表与 tb_clazz 表，并在 tb_student 表中创建一个外键与 tb_clazz 表关联，利用 MyBatis 实现表之间的一对一操作，具体要求是查询学生时能得到学生所在班级的信息。

（1）在数据库 homework 中创建 b_student 表与 tb_clazz 表，数据库脚本如下。

```sql
CREATE TABLE `tb_clazz` (
  `id` int(11) NOT NULL AUTO_INCREMENT,
  `cname` varchar(20) NOT NULL,
  PRIMARY KEY (`id`)
) ENGINE=InnoDB AUTO_INCREMENT=26 DEFAULT CHARSET=utf8;

CREATE TABLE `tb_student` (
  `id` int(11) NOT NULL AUTO_INCREMENT,
  `loginname` varchar(20) NOT NULL,
  `password` varchar(20) NOT NULL,
  `username` varchar(255) NOT NULL,
  `clazz_id` int(11) NOT NULL,
  PRIMARY KEY (`id`),
  KEY `clazz_id` (`clazz_id`),
  CONSTRAINT `tb_student_ibfk_1` FOREIGN KEY (`clazz_id`) REFERENCES
  `tb_clazz` (`id`)
) ENGINE=InnoDB AUTO_INCREMENT=529 DEFAULT CHARSET=utf8;
```

这个脚本中，tb_student 已经通过外键 clazz_id 与 tb_clazz 表的主键 id 建立了关联关系。

（2）创建与两个表对应的实体类 Clazz 与 Student。

Clazz 类：

```
1.  package com.po;
2.  import java.util.List;
3.  public class Clazz {
4.      private Integer id;
5.      private String cname;
6.      //setter and getter
7.  }
```

Student 类：

```
1.   package com.po;
2.   public class Student {
3.       private Integer id;
4.       private String loginname;
5.       private String password;
6.       private String username;
7.       private Clazz clazz;            //与关联的一方对应的实体类
8.   //setter and getter
9.   }
```

在 MyBatis 中实现 ORM 映射,要求表与类一一对应,前面 tb_student 表与 tb_clazz 表已经建立了一对一的关联,现在对应的两个实体类也要关联,为此在 Student 类中添加关联方 Clazz 类的对象作为属性,此处为 Clazz 类的对象 clazz。

(3) 分别创建两个表对应的映射文件、tb_clazz 表与 Clazz 类之间的映射文件 ClazzMapper.xml,代码如下。

```
1.   <?xml version="1.0" encoding="UTF-8" ?>
2.   <!DOCTYPE mapper
3.       PUBLIC "-//mybatis.org//DTD Mapper 3.0//EN"
4.       "http://mybatis.org/dtd/mybatis-3-mapper.dtd">
5.   <mapper namespace="com.mapper.ClazzMapper">
6.     <select id="selectClazzById" parameterType="int" resultType="clazz" >
7.       select * from tb_clazz where id=#{id}
8.     </select>
9.   </mapper>
```

第 6 行<select>元素实现了根据 id 对 tb_clazz 表的查询,其返回值是一个 Clazz 类的对象,对应的是表中的一条记录。这一段代码首先实现了 tb_clazz 表与 Clazz 类之间的一种映射,其次实现了表中记录与 Clazz 类的对象的映射。

tb_student 表与 Student 类之间的映射文件 StudentMapper.xml 代码如下。

```
1.   <?xml version="1.0" encoding="UTF-8" ?>
2.   <!DOCTYPE mapper
3.       PUBLIC "-//mybatis.org//DTD Mapper 3.0//EN"
4.       "http://mybatis.org/dtd/mybatis-3-mapper.dtd">
5.   <mapper namespace="com.mapper.StudentMapper">
6.     <select id="selectStudentById" resultMap="studentMap">
7.       select * from tb_student where id =#{id}
8.     </select>
9.   <resultMap type="student" id="studentMap">
10.      <id property="id" column="id"/>
11.      <result property="loginname" column="loginname"/>
```

```
12.        <result property="password" column="password"/>
13.        <result property="username" column="username"/>
14.        <association property="clazz" column="clazz_id"
15.          select="com.mapper.ClazzMapper.selectClazzById"
16.          javaType="clazz"/>
17.       </resultMap>
18.    </mapper>
```

第 6 行 <select> 元素实现了根据 id 对 tb_student 表的查询,其返回值是一个 resultMap,这个 resultMap 在下面有定义,其类型是一个 Student。这个 <select> 元素实际上实现了 tb_student 表与 Student 类之间的一种映射。

<resultMap> 在第 6 章讲过,当 <select> 元素查询的结果不是一个简单的类型,或不能用一个简单的 PO 类表达时,可以定义一个 <resultMap> 表达复杂的查询结果。

此处由于是关联查询,关系到另外一个表 tb_clazz,为此定义了一个 <resultMap>。其属性 type 的值为 student,说明这个映射结果为 Student 类的一个实例。在这个映射结果集中,大部分类的简单类型属性都可以直接与表的字段一一映射,唯独有 clazz 属性不能简单进行映射,它是关联方 Clazz 类的对象。这里用到 <association> 元素实现一对一的映射,此处是通过 tb_student 表的 clazz_id 与表 tb_clazz 进行了关联,将 Student 类的 clazz 属性映射为 tb_clazz 表对应的映射文件中的 selectClazzById 对应的查询结果。有了这个映射后,当查询某一个学生对象的信息时,这个学生所在的班级信息也会被添加到这个学生的对象中。

(4) 在前面的测试方法中需要指定映射文件中执行语句的 id,此时不能保证编写 id 时的正确性(只有运行时才能知道)。为此,可以使用 MyBatis 提供的另外一种编程方式,即使用 Mapper 接口编程。使用 Mapper 接口编程时要遵循以下规范。

- Mapper 接口的名称和对应的 Mapper.xml 映射文件的名称必须一致。
- Mapper.xml 文件中的 namespace 与 Mapper 接口的类路径相同(接口文件与映射文件要放在一个包中)。
- Mapper 接口中的方法名和 Mapper.xml 中定义的每个执行语句的 id 相同。
- Mapper 接口中方法的输入参数类型要和 Mapper.xml 中定义的每个 SQL 的 ParameterType 的类型相同。
- Mapper 接口方法的返回值类型要和 Mapper.xml 中定义的每个 SQL 的 resultType 的类型相同。

只要遵循了这些开发规范,MyBatis 就可以自动生成 Mapper 接口实现类的代理对象,从而简化开发。下面是与 StudentMapper.xml 对应的 StudentMapper 接口,它们要放在同一个包下。

```
1.   package com.mapper;
2.   import java.util.List;
3.   import com.po.Student;
4.   public interface StudentMapper {
5.       public Student selectStudentById(Integer id);
6.   }
```

下面是 ClazzMapper.xml 对应的 ClazzMapper 接口。

```
1.  package com.mapper;
2.  import java.util.List;
3.  import com.po.Clazz;
4.  public interface ClazzMapper {
5.      public Clazz selectClazzById(Integer id);
6.  }
```

（5）前面已经完成了表与类的映射及记录与对象的映射，接下来采用面向 Mapper 接口编程的方式分别创建两个测试类 ClazzTest 和 StudentTest。

ClazzTest 测试类代码如下。

```
1.  package com.test;
2.  import java.util.List;
3.  import org.apache.ibatis.session.SqlSession;
4.  import org.junit.jupiter.api.Test;
5.  import com.mapper.ClazzMapper;
6.  import com.po.Clazz;
7.  import com.po.Student;
8.  import com.utils.MybatisUtils;
9.  class ClazzTest {
10.     @Test
11.     void findCalzzByIdTest() {
12.         SqlSession sqlSession=MybatisUtils.getSession();
13.         ClazzMapper clazzMap=sqlSession.getMapper(ClazzMapper.class);
14.         Clazz clazz=clazzMap.selectClazzById(15);
15.         System.out.println(clazz);
16.     }
17. }
```

第 13 行采用面向 Mapper 接口编程，得到一个 ClazzMapper 接口的实例。第 14 行调用接口中的 selectClazzById() 方法，查询 id 为 15 的班级信息，并在控制台上打印 clazz 对象信息。在控制台打印的信息如下。

```
DEBUG [main] -==>Preparing: select * from tb_clazz where id=?
DEBUG [main] -==>Parameters: 15(Integer)
DEBUG [main] -<==      Total: 1
Clazz [id=15, cname=软件嵌入 171]
```

日志信息显示，根据 id 为 15 查询班级信息，显示对应的班级是"软件嵌入 171"，得到了我们期望的结果。

StudentTest 测试类代码如下。

```
1.  package com.test;
2.  import java.util.List;
3.  import org.apache.ibatis.session.SqlSession;
4.  import org.junit.jupiter.api.Test;
5.  import com.mapper.StudentMapper;
6.  import com.po.Student;
7.  import com.utils.MybatisUtils;
8.  class StudentTest {
9.    @Test
10.   void findStudentByIdTest() {
11.       SqlSession sqlSession=MybatisUtils.getSession();
12.       StudentMapper studentMap=sqlSession.getMapper(StudentMapper.
          class);
13.       Student student=studentMap.selectStudentById(275);
14.       System.out.println(student);
15.   }
16. }
```

第12行通过sqlSession.getMapper()方法,根据StudentMapper接口生成一个代理类,并得到一个StudentMapper的对象studentMap。第13行调用接口中的selectStudentById()方法,查询id为275的学生信息,并在控制台上打印student对象的信息。运行测试程序,控制台打印信息如下。

```
DEBUG [main] -==>Preparing: select * from tb_student where id=?
DEBUG [main] -==>Parameters: 275(Integer)
DEBUG [main] -<==      Total: 1
DEBUG [main] -Cache Hit Ratio [com.mapper.ClazzMapper]: 0.0
DEBUG [main] -==>Preparing: select * from tb_clazz where id=?
DEBUG [main] -==>Parameters: 15(Integer)
DEBUG [main] -<==      Total: 1
```

日志信息显示,id为"275"的学生所在的班级是"软件嵌入171",说明tb_student表与tb_clazz表的一对一的关联关系已经建立,并得到了验证。

总之,要用MyBatis操作一对一关联的两个表,首先,在某一个表建立外键与另一个表关联。其次,在有外键的表对应的实体类中有另一个类的对象作为成员变量,将两个类关联起来。最后,为两个表分别创建映射文件,在有外键的表对应的映射文件中,将<resultMap>元素中的子元素<association>与另外一个表的映射文件中有关的<select>元素关联起来,这样就可以实现一对一的关联。

7.2 一对多的关联操作

在数据库中,一对多的关联关系用得最多。例如,一个班级对应多个学生,一个学生选修多门课程等。当然,这种一对多的关系也是相对的,与一对一的关系一样,同样是两个表,由于所处的角色不同,得出的结论可能完全相反。当从班级的角色看班级与学生之间的关系就是一对多的关系。在数据库中,在一方的数据库表中不需要创建外键,只需在多方所在的表中创建外键即可。

7.2.1 一对多关联操作示例

在 MyBatis 中是通过 ORM 实现对数据的操作,在映射之前,首先要创建表和相应的实体类。所谓的一对多,就是一方表中的一条记录可能对应多方表中的多条记录,此时在相应的实体类中要有一个集合属性,这个集合属性用来存放多方的数据。下面以班级表 tb_clazz 和学生表 tb_student 为例说明一对多的关联操作。

例 7-2 在例 7-1 中已经创建了表 tb_clazz 和表 tb_student,并在两个表之间建立了关联,在此基础上利用 MyBatis 完成一对多的关联操作,通过班级 id 查询班级的所有学生。

(1) 创建数据库,创建表及表之间的关联,这一步可参照例 7-1。

(2) 创建与两个表对应的实体类,Student 类在例 7-1 中已经创建,在此不再列出。但 Clazz 类需要修改,要添加一个集合属性,用来存放学生信息。Clazz 类的代码如下。

```
1.  package com.po;
2.
3.  import java.util.List;
4.
5.  public class Clazz {
6.      private Integer id;
7.      private String cname;
8.      private List<Student> students;
9.
10.     //setter and getter 方法
11. }
```

第 8 行声明了一个 List 类型的属性变量,其 List 集合中的每个元素为 student 对象,用来存放多方的 student 对象。

(3) 分别为两个表创建映射文件,在 7.1 节中 tb_student 表与 Student 类之间的映射文件 StudentMapper.xml 已经创建,此处要添加 id 为 selectStudentByClazzId 的<select>查询。修改后的 StudentMapper.xml 文件如下。

```
1.  <?xml version="1.0" encoding="UTF-8" ?>
2.    <!DOCTYPE mapper
3.      PUBLIC "-//mybatis.org//DTD Mapper 3.0//EN"
4.      "http://mybatis.org/dtd/mybatis-3-mapper.dtd">
5.  <mapper namespace="com.mapper.StudentMapper">
6.
7.  <select id="selectStudentById" resultMap="studentMap">
8.      select * from tb_student where id =#{id}
```

```
9.      </select>
10.     <resultMap type="student" id="studentMap">
11.         <id property="id" column="id"/>
12.         <result property="loginname" column="loginname"/>
13.         <result property="password" column="password"/>
14.         <result property="username" column="username"/>
15.         <association property="clazz" column="clazz_id"
16.           select="com.mapper.ClazzMapper.selectClazzById"
17.           javaType="clazz"/>
18.     </resultMap>
19.     <select id="selectStudentByClazzId" parameterType="int" resultMap="studentMap" >
20.         select * from tb_student where clazz_id=#{id}
21.     </select>
22.     <delete id="deleteStudentById" parameterType="Integer">
23.         delete from tb_student where id=#{id}
24.     </delete>
25.
26. </mapper>
```

第 19～21 行字体加黑部分为新添加的内容，这段代码的作用是根据 clazz_id（外键，班级 id）查询 tb_student 表中的记录，查询结果是一个 resultMap，这个结果映射已经在映射文件中定义，后面的 ClazzMapper.xml 映射文件中会引用这个查询。

tb_clazz 表与 Clazz 类对应的映射文件 ClazzMapper.xml 需要修改，修改后的 ClazzMapper.xml 文件如下。

```
1.  <?xml version="1.0" encoding="UTF-8" ?>
2.  <!DOCTYPE mapper
3.      PUBLIC "-//mybatis.org //DTD Mapper 3.0//EN"
4.      "http://mybatis.org/dtd/mybatis-3-mapper.dtd">
5.  <mapper namespace="com.mapper.ClazzMapper">
6.
7.  <select id="selectClazzById" parameterType="int" resultMap="clazzResultMap" >
8.      select * from tb_clazz where id=#{id}
9.  </select>
10. <resultMap type="clazz" id="clazzResultMap">
11.     <id property="id" column="id"/>
12.     <result property="cname" column="cname"/>
13.     <collection property="students" javaType="ArrayList"
14.        column="id" ofType="student"
15.        select="com.mapper.StudentMapper.selectStudentByClazzId" fetchType="lazy">
16.        <id property="id" column="id"/>
17.        <result property="loginname" column="loginname"/>
```

```
18.            <result property="password" column="password"/>
19.            <result property="username" column="username"/>
20.         </collection>
21.     </resultMap>
22. </mapper>
```

第 7 行的<select>元素是在原有代码基础上的修改,将原来的 resultType 属性改为现在的 resultMap="clazzResultMap"。第 10 行是使用处理映射的<resultMap>元素定义 id="clazzResultMap"的一个映射结果。这个映射结果是针对 Clazz 类与 tb_clazz 表之间的映射。第 11 行<id>元素定义的是对象的唯一标识属性 id 与 tb_clazz 表的主键 id 的映射。第 12 行<result>元素定义的是 Clazz 类中的 cname 属性与 tb_clazz 表中的字段 cname 的映射。

第 13 行<collection>元素定义的是类 Clazz 中的集合属性 students 与 tb_student 表中的一个 id 为 selectStudentByClazzId 的查询之间的映射。这个查询可以在 StudentMapper.xml 中找到。第 14 行的 ofType="student"表示属性 students 中的每个元素的类型是 Student 类的实例。第 15 行的 select 属性指向了要执行的查询所在的位置,这里是指向了 StudentMapper 映射文件中 id 为 selectStudentByClazzId 的查询。<collection>的属性 fetchType 是有关查询性能的一个属性,它有如下两个值。

- lazy:表示获得当前的 PO 类对象后,延迟加载关联的数据。
- eager:表示获得当前的 PO 类对象后,立即加载关联的数据。

此处将 fetchType 的值设为 lazy 表示延迟加载关联的数据。

第 16~19 行是 Student 类的属性与查询结果集中查询结果中字段的映射。

(4) 创建 Mapper 接口类,在例 7-1 中已经创建,这里不再列出。

(5) 创建测试类 ClazzTest 测试 ClazzMapper 接口中的方法,代码如下。

```
1.  package com.test;
2.  import java.util.List;
3.  import org.apache.ibatis.session.SqlSession;
4.  import org.junit.jupiter.api.Test;
5.
6.  import com.mapper.ClazzMapper;
7.  import com.po.Clazz;
8.  import com.po.Student;
9.  import com.utils.MybatisUtils;
10. class ClazzTest {
11.     @Test
12.     void findCalzzByIdTest() {
13.         SqlSession sqlSession=MybatisUtils.getSession();
14.         ClazzMapper clazzMap=sqlSession.getMapper(ClazzMapper.class);
15.         Clazz clazz=clazzMap.selectClazzById(15);
16.         System.out.println(clazz);
```

```
17.         List<Student>students=clazz.getStudents();
18.         System.out.println(students.size());
19.         students.forEach(stu->System.out.println(stu));
20.     }
21. }
```

第17行得到id为15班级中的所有学生对象,并将结果赋值给一个students对象,第19行通过循环语句输出students中的所有学生信息。运行测试程序,控制台打印结果如下。

```
DEBUG [main] -==>Preparing: select * from tb_clazz where id=?
DEBUG [main] -==>Parameters: 15(Integer)
DEBUG [main] -<==   Total: 1
DEBUG [main] -==>Preparing: select * from tb_student where clazz_id=?
DEBUG [main] -==>Parameters: 15(Integer)
DEBUG [main] -<==   Total: 30
Clazz [id=15, cname=软件嵌入171]
30
Student [id=275, loginname=1713071001, password=nhce111, username=刘奕彤, clazz=Clazz [id=15, cname=软件嵌入171]]
Student [id=276, loginname=1713071002, password=1713071002, username=孟韵怡, clazz=Clazz [id=15, cname=软件嵌入171]]
Student [id=277, loginname=1713071003, password=1713071003, username=刘欣雨, clazz=Clazz [id=15, cname=软件嵌入171]]
```

从控制台日志信息可知,MyBatis执行了两次SQL查询语句,首先是查询tb_clazz表,其次是查询tb_student表,得到了我们期望的结果。

7.2.2 影响关联操作性能的相关配置

在多表关联操作时,我们关心的一个主要问题是性能。在加载一个表的数据时,有可能加载关联表中的数据,此时可能带来性能方面的问题。例如,在加载班级时,同时加载班级表关联的所有学生表中的学生信息,而我们此时并不需要处理学生的相关信息,只关心班级的信息,为此可能花费很长的时间,并占用大量的内存空间。在系统设计时,我们期望按需加载,只加载有用的数据,其他的可以在需要时加载。为此,MyBatis提供了相关的配置满足这样的需求。在MyBatis中影响关联操作性能的主要有两个地方:一个是在mybatis-config.xml文件的<settings>元素中设置,这是一个全局配置;另一个是在与表相关的映射文件中通过fetchType属性设置,fetchType属性出现在association和collection两个级联元素中,这是针对某一个查询的个体配置。而且映射文件中fetchType属性的设置会覆盖mybatis-config配置文件<settings>元素中的相关配置。

首先讨论mybatis-config.xml中的设置,在配置文件中可以通过下面的代码进行配置。

```
1.    <settings >
2.        <setting name="lazyLoadingEnabled" value="true"/>
3.    </settings>
```

第 2 行是设置关联操作时是否启用延迟加载,value="true"表示启用延迟加载。此时加载一个对象时,不会加载关联的信息。这里以加载 Clazz 类的实例为例,运行 ClazzTest 测试类的 findClazzById()方法,为了查看运行时变量的状态,在测试方法中设置了断点,如图 7-1 所示。

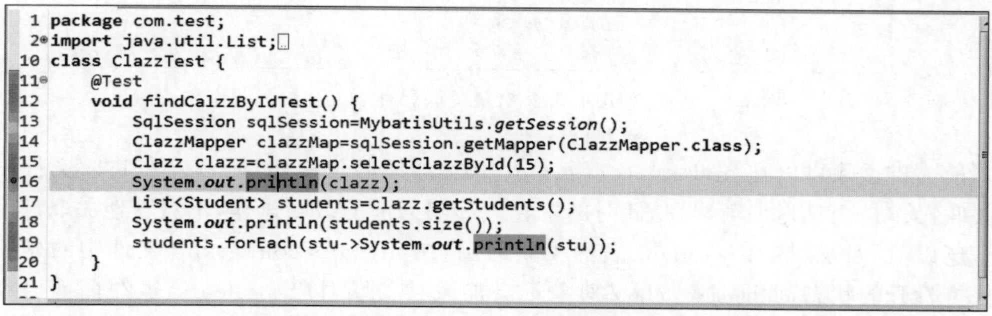

```
 1  package com.test;
 2  import java.util.List;
10  class ClazzTest {
11      @Test
12      void findCalzzByIdTest() {
13          SqlSession sqlSession=MybatisUtils.getSession();
14          ClazzMapper clazzMap=sqlSession.getMapper(ClazzMapper.class);
15          Clazz clazz=clazzMap.selectClazzById(15);
16          System.out.println(clazz);
17          List<Student> students=clazz.getStudents();
18          System.out.println(students.size());
19          students.forEach(stu->System.out.println(stu));
20      }
21  }
```

图 7-1 在 findClazzById()方法中设置断点

运行至断点处,查看当前变量的状态,如图 7-2 所示。

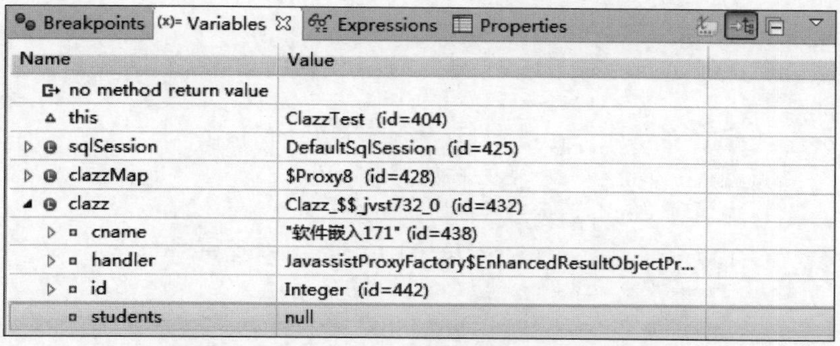

图 7-2 延迟开启时 clazz 对象关联属性 students 的状态

从当前各个变量的状态看,clazz 对象的属性 students 的值为 null,说明此时没有加载班级的关联对象 students。如果此时修改 mybaits-config.xml 中的参数,代码如下。

```
1.    <settings>
2.        <setting name="lazyLoadingEnabled" value="false"/>
3.    </settings>
```

value="false"表示此时关闭了延迟加载的开关,再次运行上面的测试用例,并查看断点状态的变量,如图 7-3 所示。

从图 7-3 中各个变量状态显示情况可以看出,clazz 的关联属性 students 已经被赋值,说明在加载 clazz 对象时,同时加载了关联对象 students。

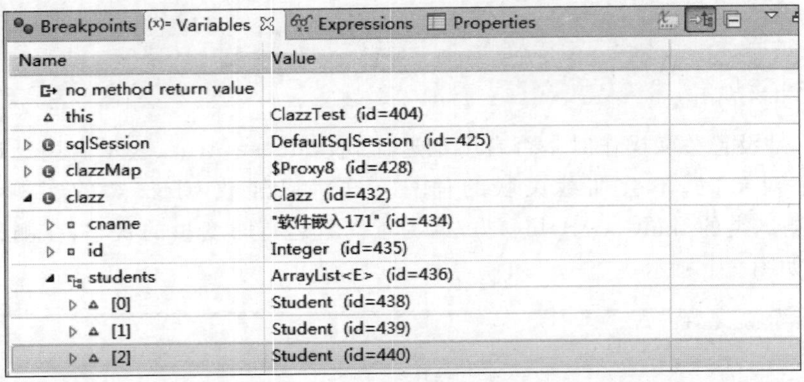

Name	Value
▷+ no method return value	
△ this	ClazzTest (id=404)
▷ ⊙ sqlSession	DefaultSqlSession (id=425)
▷ ⊙ clazzMap	$Proxy8 (id=428)
▲ ⊙ clazz	Clazz (id=432)
▷ □ cname	"软件嵌入171" (id=434)
▷ □ id	Integer (id=435)
▲ ᴿ students	ArrayList<E> (id=436)
▷ △ [0]	Student (id=438)
▷ △ [1]	Student (id=439)
▷ △ [2]	Student (id=440)

图 7-3 延迟关闭时 clazz 对象关联属性 students 的状态

前面讲过除可以在 mybatis-config.xml 中配置延迟开关外，还可以在映射文件中配置关联操作是否延迟加载，只是此时的配置只影响当前的级联操作，不对其他关联产生影响。还以 tb_clazz 与 tb_student 表的关联为例，在 mybatis-config.xml 文件中的延迟开关保持打开的状态，此时加载 clazz 对象不会加载其关联对象 students，这在前面已经验证了。现在 ClazzMapper.xml 映射文件中修改<collection>元素的 fetchType 的属性值为"eager"，修改后的 ClazzMapper 映射文件代码如下。

```
1.  <select id="selectClazzById" parameterType="int" resultMap=
    "clazzResultMap" >
2.      select * from tb_clazz where id=#{id}
3.  </select>
4.  <resultMap type="clazz" id="clazzResultMap">
5.      <id property="id" column="id"/>
6.      <result property="cname" column="cname"/>
7.      <collection property="students" javaType="ArrayList"
8.      column="id" ofType="student"
9.      select="com.mapper.StudentMapper.selectStudentByClazzId"
10.     fetchType="eager" >
11.         <id property="id" column="id"/>
12.         <result property="loginname" column="loginname"/>
13.         <result property="password" column="password"/>
14.         <result property="username" column="username"/>
15.     </collection>
16. </resultMap>
```

注意第 10 行黑体字部分的代码，此处将<collection>元素的 fetchType 属性的值设为"eager"，表示立即加载关联对象，如果设为"lazy"，则表示延迟加载。运行测试用例，查看断点处变量的状态，结果如图 7-4 所示。

由变量状态图 7-4 可以看出，此时的 students 属性已经赋值了。虽然 MyBatis 总延迟开关是打开的，但映射文件中的延迟开关是关闭的，也就是立即加载关联对象。因此可

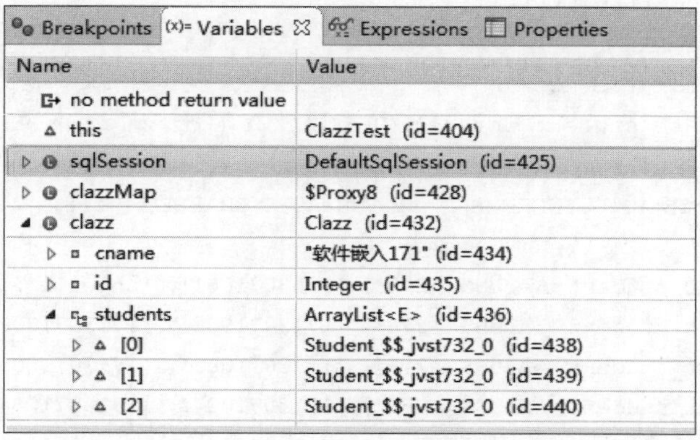

图 7-4 fetchType="eager"时 students 变量的状态

以得出结论,映射文件中的延迟开关配置将覆盖在 mybatis-config.xml 中的相关配置。

7.3 多对多的关联操作

在实际项目开发中,多对多的关联关系也是经常用到的。例如,一个学生可以选多门课程,一门课程也可以被多个学生选择,学生与课程之间就是多对多的关系。在关系数据库中不能直接创建多对多关联,但可以通过第 3 张表间接地实现多对多的关系。下面通过学生与课程之间的示例说明在 MyBatis 中如何实现多对多的关联。

例 7-3 现有 tb_student 表与课程表 tb_course,要求在两个表之间建立多对多的关联,实现查询一个学生所选的课程,同时也能查询选某门课程的学生。

(1) 创建 tb_student 表和 tb_course 表,为了建立多对多的关联,建立第 3 张表 student_course,数据库脚本如下。

```
#创建 tb_clazz 表
CREATE TABLE `tb_clazz` (
  `id` int(11) NOT NULL AUTO_INCREMENT,
  `cname` varchar(20) NOT NULL,
  PRIMARY KEY (`id`)
) ENGINE=InnoDB AUTO_INCREMENT=26 DEFAULT CHARSET=utf8;

#插入记录
INSERT INTO `tb_clazz` VALUES ('11', '软件嵌入 162');
INSERT INTO `tb_clazz` VALUES ('12', '软件嵌入 161');
#创建 tb_student 表
CREATE TABLE `tb_student` (
  `id` int(11) NOT NULL AUTO_INCREMENT,
  `loginname` varchar(20) NOT NULL,
  `password` varchar(20) NOT NULL,
  `username` varchar(255) NOT NULL,
```

```sql
  `clazz_id` int(11) NOT NULL,
  PRIMARY KEY (`id`),
  KEY `clazz_id` (`clazz_id`),
  CONSTRAINT `tb_student_ibfk_1` FOREIGN KEY (`clazz_id`) REFERENCES
`tb_clazz` (`id`)
) ENGINE=InnoDB AUTO_INCREMENT=529 DEFAULT CHARSET=utf8;
#插入记录
INSERT INTO `tb_student` VALUES ('275', '1713071001', 'nhce111', '刘奕彤', '11');
INSERT INTO `tb_student` VALUES ('276', '1713071002', '1713071002', '孟韵怡', '12');
INSERT INTO `tb_student` VALUES ('277', '1713071003', '1713071003', '刘欣雨', '11');
INSERT INTO `tb_student` VALUES ('278', '1713071004', '1713071004', '陈思', '12');
#创建 tb_course 表
CREATE TABLE `tb_course` (
  `id` int(11) NOT NULL AUTO_INCREMENT,
  `cname` varchar(40) NOT NULL,
  PRIMARY KEY (`id`)
) ENGINE=InnoDB AUTO_INCREMENT=12 DEFAULT CHARSET=utf8;
#插入记录
INSERT INTO `tb_course` VALUES('10', 'Java框架开发技术');
INSERT INTO `tb_course` VALUES('11', 'Java程序设计基础');
#创建 student_course 表
CREATE TABLE `student_course` (
  `id` int(11) NOT NULL AUTO_INCREMENT,
  `course_id` int(11) NOT NULL,
  `student_id` int(11) NOT NULL,
  PRIMARY KEY (`id`),
  KEY `course_id` (`course_id`),
  KEY `student_id` (`student_id`),
  CONSTRAINT `student_course_ibfk_1` FOREIGN KEY (`course_id`) REFERENCES `tb_course` (`id`),
  CONSTRAINT `student_course_ibfk_2` FOREIGN KEY (`student_id`) REFERENCES `tb_student` (`id`)
) ENGINE=InnoDB AUTO_INCREMENT=6 DEFAULT CHARSET=utf8;
#插入记录
INSERT INTO `student_course` VALUES('1', '10', '275');
INSERT INTO `student_course` VALUES('2', '10', '276');
INSERT INTO `student_course` VALUES('3', '11', '275');
INSERT INTO `student_course` VALUES('4', '11', '276');
INSERT INTO `student_course` VALUES('5', '11', '278');
```

student_course 表是用来关联 tb_student 表和 tb_course 表的第 3 张表,它的字段有 3 个,主键是 id,course_id 和 student_id 这两个字段是外键,分别对应 tb_course 表的主键 id 和 tb_student 表的主键 id。

(2) 创建 tb_student 表和 tb_course 表对应的两个实体类(Student 类和 Course 类),

代码如下。

Student 类：

```
1.  package com.po;
2.  import java.util.List;
3.  public class Student {
4.      private Integer id;                  //主键
5.      private String loginname;            //学号或登录账号
6.      private String password;             //密码
7.      private String username;             //学生姓名
8.      private Clazz clazz;
9.      private List<Course> courses;        //学生所选课程
10.     //setter and getter 方法
11. }
```

Course 类：

```
1.  package com.po;
2.  import java.util.List;
3.  public class Course {
4.      private Integer id;                  //主键
5.      private String cname;                //课程名称
6.      private List<Student> students;      //选择本门课程的所有学生
7.      //setter and getter 方法
8.  }
```

（3）tb_student 表和 tb_course 表对应的映射文件。

在 tb_student 表与 Student 类的映射文件 StudentMapper.xml 中，编写根据学生 id 查询学生的查询代码如下。

```
1.  <select id="selectCourseOfStudentById" parameterType="int"
2.    resultMap="studentResultMap">
3.    select * from tb_student where id=#{id}
4.  </select>
5.  <resultMap type="student" id="studentResultMap">
6.    <id property="id" column="id"/>
7.    <result property="cname" column="cname"/>
8.    <collection property="courses" javaType="ArrayList"
9.        column="id" ofType="course"
10.       select="com.mapper.CourseMapper.selectCourseByStudentId">
11.     <id property="id" column="id"/>
12.     <result property="cname" column="cname"/>
13.   </collection>
14. </resultMap>
```

第 5 行定义了一个映射结果，这里对 Student 类中的属性与 tb_student 表中的字段进行了映射。由于 Student 类的属性 courses 表示所选课程是一个 List 类型的变量，此处使用 <collection> 元素将此属性与 CourseMapper 映射文件中的 selectCourseByStudentId 查询语句进行了映射。

在 Course 类与 tb_course 表的映射文件 CourseMapper.xml 中，编写根据学生 id 查询课程的代码如下。

```xml
1. <!--根据学生 id 查询课程 -->
2. <select id="selectCourseByStudentId" parameterType="int" resultType="Course" >
3.   select * from tb_course where id in(
4.     select course_id from student_course where student_id=#{id})
5. </select>
```

第 3～5 行是一个嵌入式查询语句，首先，根据 student_id 在 student_course 表中查询所有的 course_id；其次，根据 in 语句中查询的 course_id 集合，在 tb_course 表中查询相应的课程信息。

（4）分别创建 StudentMapper 接口和 CourseMapper 接口，代码如下。

StudentMapper 接口：

```java
1.  package com.mapper;
2.  import java.util.List;
3.  import com.po.Student;
4.  public interface StudentMapper {
5.      //根据学生 id 查询学生信息,返回的是一个 Student 对象
6.      public Student selectStudentById(Integer id);
7.      //根据班级 id 查询对应班级的所有学生
8.      public List<Student> selectStudentByClazzId(Integer id);
9.      //根据学生 id 查询学生所选的课程
10.     public Student selectCourseOfStudentById(Integer id);
11. }
```

CourseMapper 接口：

```java
1.  package com.mapper;
2.  import com.po.Course;
3.  public interface CourseMapper {
4.      //根据课程 id 查询课程信息
5.      public Course selectCourseById(Integer id);
6.      //根据课程 id 查询选择此课程的所有学生信息
7.      public Course selectStudentOfCourseById(Integer id);
8.  }
```

（5）创建测试类和测试方法。

StudentTest 测试类：

```
1.   class StudentTest {
2.       //根据学生 id 查询学生所选课程
3.       @Test
4.       void findCourseOfStudentByStudentIdTest() {
5.           SqlSession sqlSession=MybatisUtils.getSession();
6.           StudentMapper studentMap = sqlSession.getMapper(StudentMapper.class);
7.           Student student=studentMap.selectCourseOfStudentById(276);
8.           List<Course>courses=student.getCourses();
9.           if(courses.size()>0)
10.              courses.forEach(cou->System.out.println(cou));
11.      }
12.  }
```

第 7 行是查询学生 id 为 276 的学生的信息，第 8 行得到这名学生所选的课程列表，第 10 行对 courses 进行遍历输出课程信息。测试方法运行结果打印在控制台，如下所示。

```
DEBUG [main] -==>Preparing: select * from tb_student where id=?
DEBUG [main] -==>Parameters: 276(Integer)
DEBUG [main] -<==    Total: 1
DEBUG [main] -==>Preparing: select * from tb_course where id in(select course_id from student_course where student_id=?)
DEBUG [main] -==>Parameters: 276(Integer)
DEBUG [main] -<==    Total: 2
Course [id=10, cname=Java 框架开发技术, students=null]
Course [id=11, cname=Java 程序设计基础, students=null]
```

从日志信息可以看出，一共执行了 3 次查询。

第一次是对 tb_student 表的查询，查询语句为

select * from tb_student where id=?

第二次是对 student_course 表的嵌入式查询，查询语句为

select course_id from student_course where student_id=?

第三次是对 tb_course 表的查询，查询语句为

select * from tb_course where id in

从日志可以看出，已经根据 id 为 276 的学生，查询到他所选课程为"Java 框架开发技术"和"Java 程序设计基础"。

CourseTest 类：

```
1.   class CourseTest {
2.       //根据课程 id 查询选择该课程的所有学生
3.       @Test
4.       void findStudentOfCourseByIdTest() {
5.           SqlSession sqlSession=MybatisUtils.getSession();
6.           CourseMapper courseMapper = sqlSession. getMapper ( CourseMapper.
             class);
7.           Course course=courseMapper.selectStudentOfCourseById(11);
8.           List<Student>students=course.getStudents();
9.           students.forEach(stu->System.out.println(stu));
10.      }
11.  }
```

第 7 行查询 id 为 11 的课程信息,第 8 行将选择此门课程的所有学生的对象赋值给 students。第 9 行对 students 进行遍历,输出学生信息。控制台输出信息如下。

```
DEBUG [main] -==>Preparing: select * from tb_course where id=?
DEBUG [main] -==>Parameters: 11(Integer)
DEBUG [main] -<==       Total: 1
DEBUG [main] -==>Preparing: select * from tb_student where id in(select student
_id from student_course where course_id=?)
DEBUG [main] -==>Parameters: 11(Integer)
DEBUG [main] -<==       Total: 3
Student [id=275, loginname=1713071001, password=nhce111, username=刘奕彤,
Student [id=276, loginname=1713071002, password=1713071002, username=孟韵怡,
Student [id=278, loginname=1713071004, password=1713071004, username=陈思,
```

日志信息显示,id 为 11 的课程一共有 3 名学生选择,控制台打印了他们的详细信息。

7.4 MyBatis 的缓存机制

在实际项目开发中,通常都对数据库的查询性能有要求,而提高数据库查询性能的最好办法是使用缓存。MyBatis 的查询缓存分为一级缓存和二级缓存。一级缓存是 SqlSession 级别的缓存,二级缓存是 mapper 级别的缓存。二级缓存是多个 SqlSession 共享的。在项目中合理配置缓存可以将数据查询性能调到最优。

7.4.1 一级缓存(SqlSession 级别)

一级缓存是 SqlSession 级别的缓存,是基于 HashMap 的本地缓存。不同的 sqlSession 之间的缓存数据区域是互相不影响的,也就是它只能作用在同一个 sqlSession 中,不同的 sqlSession 中的缓存是互相不能读取的。

一级缓存的工作原理如图 7-5 所示。用户第一次查询后,将查询结果写入一级缓存区,当用户发出第二次查询同一条记录请求时,MyBatis 先去缓存中查找是否有该数据,

如果有,直接从缓存中读取;如果没有,从数据库中查询,并将查询到的数据放入一级缓存区域,供下次查找使用。

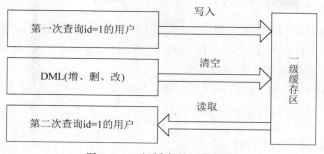

图 7-5 一级缓存的工作原理

当 sqlSession 执行 DML 操作后,即执行增、删、改操作时会清空缓存。这么做的目的是避免读取脏数据。

例 7-4 以例 7-1 中查询班级表 tb_clazz 为例,说明 MyBatis 一级缓存的工作原理。
(1) tb_clazz 表的实体类代码参照例 7-1。
(2) tb_clazz 表的映射文件代码参照例 7-1。
(3) ClazzMapper 接口参照例 7-1。
(4) 创建测试类,代码如下。

```
1.   //测试一级缓存
2.     @Test
3.     void oneLevelCacheTest() {
4.         SqlSession sqlSession=MybatisUtils.getSession();
5.         ClazzMapper clazzMap=sqlSession.getMapper(ClazzMapper.class);
6.         //第一次查询得到一个 clazz 对象
7.         Clazz clazz=clazzMap.selectClazzById(15);
8.         System.out.println(clazz);
9.         //第二次查询同一个记录
10.        Clazz clazz1=clazzMap.selectClazzById(15);
11.        System.out.println(clazz1)
12.
13.    }
```

第 7 行和第 10 行分别查询了 id=15 的记录,MyBatis 实际执行了一次查询,还是两次查询?查看控制台日志信息如下。

```
DEBUG [main] -==>Preparing: select * from tb_clazz where id=?
DEBUG [main] -==>Parameters: 15(Integer)
DEBUG [main] -<==      Total: 1
DEBUG [main] -==>Preparing: select * from tb_student where clazz_id=?
DEBUG [main] -==>Parameters: 15(Integer)
DEBUG [main] -<==      Total: 30
```

```
Clazz [id=15, cname=软件嵌入 171]
DEBUG [main] -Cache Hit Ratio [com.mapper.ClazzMapper]: 0.0
Clazz [id=15, cname=软件嵌入 171]
```

日志信息显示,第一次查询班级时执行了一次 SQL 语句,但第二次查询时没有执行 SQL 语句,而是从一级缓存中取出数据并显示。控制台显示日志信息验证了之前分析的结论。MyBatis 一级缓存默认是启动的。

如果想主动清除一级缓存,可以调用 sqlSession.clearCache()方法。当然,调用 sqlSession.close()后也可以清空一级缓存。

7.4.2 二级缓存(mapper 级别)

二级缓存是 mapper 级别的缓存,同样是基于 HashMap 进行存储,多个 SqlSession 可以共用二级缓存,其作用域是 mapper 的同一个 namespace。不同的 SqlSession 两次执行相同的 namespace 下的 SQL 语句,会执行相同的 SQL。第二次查询只会查询第一次查询时读取数据库后写到缓存的数据,不会再去数据库查询。

MyBatis 默认开启二级缓存,开启只需在配置文件中写入如下代码。

```
<settings>
    <setting name="cacheEnabled" value="true"/>
</settings>
```

二级缓存的作用如下。
- 映射文件中的所有 select 语句的结果将会被缓存。
- 映射文件中的所有 insert、update 和 delete 语句会刷新缓存。
- 缓存会使用最近最少使用(Least Recently Used,LRU)算法清除不需要的缓存。
- 缓存不会定时进行刷新(也就是说,没有刷新间隔)。
- 缓存默认会保存列表或 1024 个对象(无论查询方法返回哪种)的 1024 个引用。
- 缓存会被视为读/写缓存,这意味着获取到的对象并不是共享的,可以安全地被调用者修改,而不干扰其他调用者或线程所做的潜在修改。

这里有一点要注意,缓存只作用于 cache 标签所在的映射文件中的语句。如果混合使用 Java API 和 XML 映射文件,在共用接口中的语句将不会被默认缓存。需要使用 @CacheNamespaceRef 注解指定缓存作用域。这些属性可以通过<cache>元素的属性修改。例如:

```
<cache
  eviction="FIFO"
  flushInterval="60000"
  size="512"
  readOnly="true"/>
```

这个配置的含义为:创建了一个 FIFO 缓存,每隔 60s 刷新,最多可以存储结果对象或列表的 512 个引用,而且返回的对象被认为是只读的,因此,对它们进行修改可能会与不同线程中的调用者产生冲突。

缓存的回收策略如下。

LRU(最近最少使用)：移除最长时间不被使用的对象。

FIFO(先进先出)：按对象进入缓存的顺序移除它们。

SOFT(软引用)：基于垃圾回收器状态和软引用规则移除对象。

WEAK(弱引用)：更积极地基于垃圾回收器状态和弱引用规则移除对象。

例 7-5 以例 7-1 中查询班级表 tb_clazz 为例，说明二级缓存的使用方法。

(1) tb_clazz 表的实体类代码参照例 7-1。

(2) tb_clazz 表的映射文件 ClazzMapper.xml 修改后的代码如下。

```xml
1.  <mapper namespace="com.mapper.ClazzMapper">
2.  <cache eviction="LRU" flushInterval="60000"
3.      size="512" readOnly="true"/>
4.  <select id="selectClazzById" parameterType="int" resultMap="clazzResultMap" >
5.      select * from tb_clazz where id=#{id}
6.  </select>
7.  <resultMap type="clazz" id="clazzResultMap">
8.      <id property="id" column="id"/>
9.      <result property="cname" column="cname"/>
10.     <collection property="students" javaType="ArrayList"
11.         column="id" ofType="student"
12.         select="com.mapper.StudentMapper.selectStudentByClazzId" >
13.         <id property="id" column="id"/>
14.         <result property="loginname" column="loginname"/>
15.         <result property="password" column="password"/>
16.         <result property="username" column="username"/>
17.     </collection>
18. </resultMap>
19. </mapper>
```

请重点查看第 2、3 行黑体字部分代码，这里配置了二级缓存 cache 的一些属性参数，前面已经介绍过。

(3) ClazzMapper 接口参照例 7-1。

(4) 创建测试类和测试方法，代码如下。

```java
1.  //测试二级缓存
2.  @Test
3.  void TwoLevelCacheTest() {
4.      //第一次得到一个SqlSession
5.      SqlSession sqlSession=MybatisUtils.getSession();
6.      ClazzMapper clazzMap=sqlSession.getMapper(ClazzMapper.class);
7.      //第一次查询id=15的班级记录
```

```
8.          Clazz clazz=clazzMap.selectClazzById(15);
9.          System.out.println(clazz);
10.         //关闭sqlSession
11.         sqlSession.close();
12.         //重新得到一个SqlSession
13.         SqlSession sqlSession1=MybatisUtils.getSession();
14.         ClazzMapper clazzMap1=sqlSession1.getMapper(ClazzMapper.class);
15.         //第二次查询id=15的班级记录
16.         Clazz clazz1=clazzMap1.selectClazzById(15);
17.         System.out.println(clazz1);
18.     }
19. }
```

第 11 行语句关闭了第一次获得的 sqlSession，此时一级缓存被清空，但二级缓存是否可用呢？第 14 行重新获得一个 sqlSession，并查询 id＝15 的班级记录，与第一次查询内容相同，可以查看控制台日志信息，如果只执行了一次 SQL 语句，说明二级缓存已经启用，否则没有启用二级缓存。控制台日志信息如下。

```
DEBUG [main] - Created connection 891093184.
DEBUG [main] - Setting autocommit to false on JDBC Connection
[com.mysql.jdbc.JDBC4Connection@351d00c0]
DEBUG [main] - ==>Preparing: select * from tb_clazz where id=?
DEBUG [main] - ==>Parameters: 15(Integer)
DEBUG [main] - <==    Total: 1
DEBUG [main] - ==>Preparing: select * from tb_student where clazz_id=?
DEBUG [main] - ==>Parameters: 15(Integer)
DEBUG [main] - <==    Total: 30
Clazz [id=15, cname=软件嵌入171]
DEBUG [main] - Resetting autocommit to true on JDBC Connection
[com.mysql.jdbc.JDBC4Connection@351d00c0]
DEBUG [main] - Closing JDBC Connection
[com.mysql.jdbc.JDBC4Connection@351d00c0]
DEBUG [main] - Returned connection 891093184 to pool.
DEBUG [main] - Cache Hit Ratio [com.mapper.ClazzMapper]: 0.5
Clazz [id=15, cname=软件嵌入171]
```

日志信息显示，虽然测试类有两次查询，但实际只执行了一个 SQL 语句，第二次查询没有执行 SQL 语句，说明是从缓存中取得了所需的数据，而且日志显示 Cache 命中率为 50%。

在项目中用好一级缓存和二级缓存可以有效地提高系统的性能，结合 SQL 语句的优化，特别是关联查询语句的优化，能更好地提高查询的效率。

习 题

1. MyBatis 实现一对一有几种方式？具体怎么操作？
2. MyBatis 实现一对多有几种方式，具体怎么操作？
3. MyBatis 的一级、二级缓存有什么区别，如何开启？
4. 在 MyBatis 中如何实现多对多的关联操作？

实验 7　表的关联操作

1. 实验目的

通过本实验让读者理解表之间的关联关系，掌握一对多和多对一关联操作，具体要掌握<association>元素和<collection>元素的使用。

2. 实验内容

（1）创建数据库 ssm，创建班级表 tb_clazz、课程表 tb_course、教师表 tb_teacher 和作业表 tb_workbook。tb_clazz、tb_course、tb_teacher 这 3 张表与 tb_workbook 是一对多的关系。而且在 tb_workbook 表中有另外 3 张表的主键作为外键。数据库脚本如下。

```sql
#创建 tb_clazz 表
DROP TABLE IF EXISTS `tb_clazz`;
CREATE TABLE `tb_clazz` (
  `id` int(11) NOT NULL AUTO_INCREMENT,
  `cname` varchar(20) NOT NULL,
  PRIMARY KEY (`id`)
) ENGINE=InnoDB AUTO_INCREMENT=26 DEFAULT CHARSET=utf8;
#插入记录
INSERT INTO `tb_clazz` VALUES ('11', '软件嵌入 162');
INSERT INTO `tb_clazz` VALUES ('12', '软件嵌入 161');
INSERT INTO `tb_clazz` VALUES ('14', '软件嵌入 172');
INSERT INTO `tb_clazz` VALUES ('15', '软件嵌入 171');
#创建 tb_course 表
DROP TABLE IF EXISTS `tb_course`;
CREATE TABLE `tb_course` (
  `id` int(11) NOT NULL AUTO_INCREMENT,
  `cname` varchar(40) NOT NULL,
  PRIMARY KEY (`id`)
) ENGINE=InnoDB AUTO_INCREMENT=12 DEFAULT CHARSET=utf8;
#插入记录
INSERT INTO `tb_course` VALUES ('10', 'Java 框架开发技术');
INSERT INTO `tb_course` VALUES ('11', 'Java 程序设计基础');
```

```sql
#创建 tb_teacher 表
DROP TABLE IF EXISTS `tb_teacher`;
CREATE TABLE `tb_teacher` (
  `loginname` varchar(18) NOT NULL,
  `id` int(11) NOT NULL AUTO_INCREMENT,
  `password` varchar(18) NOT NULL,
  `username` varchar(18) DEFAULT NULL,
  `role` varchar(1) DEFAULT '2' COMMENT '1系统管理员,2为普通教师',
  PRIMARY KEY (`id`)
) ENGINE=InnoDB AUTO_INCREMENT=15 DEFAULT CHARSET=utf8;
#插入记录
INSERT INTO `tb_teacher` VALUES ('admin', '13', 'english', '史胜辉', '1');
INSERT INTO `tb_teacher` VALUES ('admin1', '14', '11', '李丽', '2');
#创建 tb_workbook 表
DROP TABLE IF EXISTS `tb_workbook`;
CREATE TABLE `tb_workbook` (
  `id` int(11) NOT NULL AUTO_INCREMENT,
  `teacher_id` int(11) NOT NULL COMMENT '与teacher表关联的外键',
  `course_id` int(11) NOT NULL COMMENT '与course表关联的外键',
  `title` varchar(30) NOT NULL COMMENT '作业名称',
  `clazz_id` int(11) NOT NULL COMMENT '与clazz表关联的外键',
  `wflag` varchar(1) NOT NULL COMMENT '表示作业是否发布,0为不发布,1为发布,2为已批改',
  `term` varchar(15) NOT NULL COMMENT '表示第几学期',
  `createdate` timestamp NOT NULL DEFAULT CURRENT_TIMESTAMP ON UPDATE CURRENT_TIMESTAMP COMMENT '创建日期',
  `fileName` varchar(50) DEFAULT NULL COMMENT '教师布置作业上传文件',
  PRIMARY KEY (`id`),
  KEY `clazz_id` (`clazz_id`),
  KEY `teacher_id` (`teacher_id`),
  KEY `course_id` (`course_id`),
  CONSTRAINT `tb_workbook_ibfk_1` FOREIGN KEY (`clazz_id`) REFERENCES `tb_clazz` (`id`),
  CONSTRAINT `tb_workbook_ibfk_2` FOREIGN KEY (`teacher_id`) REFERENCES `tb_teacher` (`id`),
  CONSTRAINT `tb_workbook_ibfk_3` FOREIGN KEY (`course_id`) REFERENCES `tb_course` (`id`)
) ENGINE=InnoDB AUTO_INCREMENT=36 DEFAULT CHARSET=utf8;
#插入记录
INSERT INTO `tb_workbook` VALUES ('3', '13', '10', 'Java框架开发技术第一次作业', '12', '2', '2018-2019-1', '2018-11-17 15:52:44', 'HelloWorld.java');
INSERT INTO `tb_workbook` VALUES ('4', '13', '10', 'Java框架开发技术第二次作业', '11', '2', '2018-2019-1', '2018-11-17 15:52:50', null);
INSERT INTO `tb_workbook` VALUES ('8', '13', '10', 'Java框架开发技术第三次作业', '11', '2', '2018-2019-1', '2018-11-17 15:52:56', null);
```

```
INSERT INTO `tb_workbook` VALUES ('11', '13', '11', 'Java程序设计第二次作业', '15',
'2', '2018-2019-1', '2018-11-17 15:53:00', null);
```

（2）创建 tb_clazz 表的 PO 类和映射文件，在映射文件中使用<collection>元素实现一对多的查询，即根据 id 查询所有关联的作业信息。其他表 tb_course 和 tb_teacher 也是如此，在相应的映射文件中实现根据课程 id 查询关联的作业信息，根据教师 id 查询关联的作业信息。

（3）创建 tb_workbook 表的 PO 类和映射文件，在映射文件中使用<association>元素实现多对一的查询，即根据作业表 id 查询作业所在班级、课程和教师的信息。

（4）创建 JUnit 测试类，每个表创建一个测试类，每个类中有一个测试方法，分别测试一对多和多对一的关联关系。

3. 实现思路及步骤

可参照实验 5 的步骤，但此处有多个映射文件，最好是先创建一方的映射文件和测试类及方法并进行测试。待一方的映射文件测试通过后，再创建多方的映射文件和相应的测试类和测试方法。首先，创建 tb_clazz、tb_course 和 tb_teacher 3 个表的 PO 类和映射文件，并进行测试。测试通过后再创建 tb_workbook 表的 PO 类和映射文件，创建测试类和测试方法并进行测试。

第 8 章　MyBatis 的注解开发

本章目标
1. 了解 MyBatis 中的常用注解。
2. 掌握基于注解实现单表的操作方法。
3. 掌握基于注解实现一对多的双向关联操作方法。
4. 掌握基于注解实现多对多的关联操作方法。
5. 掌握注解中的动态 SQL。

在 MyBatis 中对数据库的操作有两种方式：一种是本书前面介绍的基于 XML 映射文件的开发方式；另一种是基于注解的开发方式。这两种开发方式各有优劣，可根据项目开发的需要选择。

8.1　常用注解

MyBatis 中的常用注解见表 8-1。

表 8-1　MyBatis 中的常用注解

注　解	作用对象	对应的 XML 元素	描　述
@Results	方法	\<resultMap>	多个结果映射列表 属性：value，是 Result 注解的数组
@Result	方法	\<result>	在列和属性或字段之间的单独结果映射。 属性：id，column，property，javaType，jdbcType，type Handler，one，many。 one 属性用于表示一对一的关联，和\<association>相似，many 属性是对集合而言的，和\<collection>相似
@One	方法	\<association>	复杂类型的单独属性值映射。 属性：select，已映射语句（也就是映射器方法）的完全限定名，它可以加载合适类型的实例

续表

注　解	作用对象	对应的 XML 元素	描　述
@Many	方法	<collection>	复杂类型的集合属性映射。 属性：select，是映射器方法的完全限定名，它可加载合适类型的一组实例
@Options	方法	映射语句的属性	这个注解提供额外的配置选项，它们通常在映射语句上作为属性出现。 属性：useCache＝true，flushCache＝false，resultSetType＝FORWARD_ONLY，statementType＝PREPARED，fetchSize＝－1，timeout＝－1，useGeneratedKeys＝false，keyProperty＝"id"
@select @Insert @Update @Delete	方法	<select> <insert> <update> <delete>	与对应的元素功能一致。 属性：value，这是字符串数组用来组成单独的 SQL 语句
@InsertProvider @UpdateProvider @DeleteProvider @SelectProvider	方法	<insert> <update> <delete> <select> 允许创建动态 SQL	这些可选的 SQL 注解允许指定一个类名和一个方法在执行时返回运行的 SQL。基于执行的映射语句，MyBatis 会实例化这个类，然后执行由 provider 指定的方法。这个方法可以选择性地接受参数对象作为它的唯一参数，但是必须只指定该参数或者没有参数。 属性：type，method。type 属性是类的完全限定名。method 是该类中的方法名
@Param	方法	N/A	当映射器方法需多个参数时，这个注解可以被应用于映射器方法参数给每个参数一个名字，否则多参数将会以它们的顺序位置被命名。 如#{1}、#{2}等，这是默认的。 使用@Param("person")时，SQL 中的参数应该被命名为#{person}

8.2　单表的操作

下面通过一个示例说明如何用 MyBatis 的注解对表进行 CRUD 操作。

例 8-1　使用@select、@insert、@update、@delete 对表 tb_clazz 进行操作。
（1）tb_clazz 表的数据库脚本。

```
#创建数据库表 tb_clazz
CREATE TABLE `tb_clazz` (
  `id` int(11) NOT NULL AUTO_INCREMENT,
  `cname` varchar(20) NOT NULL,
  PRIMARY KEY (`id`)
) ENGINE=InnoDB AUTO_INCREMENT=26 DEFAULT CHARSET=utf8;
```

(2) 实体类 Clazz,代码省略。

(3) 创建 ClazzMapper 接口,并在接口中相应的方法前添加注解,完成表与类的映射功能。

```
1.   package com.ssm.mapper;
2.
3.   import java.util.List;
4.   import org.apache.ibatis.annotations.Delete;
5.   import org.apache.ibatis.annotations.Insert;
6.   import org.apache.ibatis.annotations.Options;
7.   import org.apache.ibatis.annotations.Param;
8.   import org.apache.ibatis.annotations.Result;
9.   import org.apache.ibatis.annotations.Results;
10.  import org.apache.ibatis.annotations.Select;
11.  import org.apache.ibatis.annotations.Update;
12.  import com.po.Clazz;
13.
14.  public interface ClazzMapper {
15.      //添加记录
16.      @Insert("insert into tb_clazz(cname) values(#{cname})")
17.      @Options(useGeneratedKeys=true,keyProperty="id")
18.      int saveClazz(Clazz clazz);
19.      //根据 id 删除记录
20.      @Delete("delete from tb_clazz where id=#{id}")
21.      int removeClazz(@Param("id") Integer id);
22.      //根据 id 更新记录
23.      @Update("update tb_clazz set cname=#{cname} where id=#{id}")
24.      int modifyClazz(Clazz clazz);
25.      //根据 id 查询记录
26.      @Select("select * from tb_clazz where id=#{id}")
27.      //tb_clazz 表中字段与 Clazz 类中的属性映射
28.      @Results({
29.          @Result(id=true,column="id",property="id"),
30.          @Result(column="cname",property="cname")
31.      })
32.      Clazz selectUserById(Integer id);
33.      //查询表中的所有记录
34.      @Select("select * from tb_clazz")
35.      List<Clazz>selectAllClazz();
36.  }
```

接口中的方法添加了相应的注解,这些注解完成的功能与第 7 章中在映射文件中完成的功能一样,只是现在将对数据库的操作命令以注解的方法写在接口中。第 28 行 @Results 注解的功能是完成表的字段与类中的属性之间的映射,如果字段名与属性名都相同,这个注解可以省略,此处去掉这个注解也是可以的,不会对查询结果有影响。

(4)在 mybatis-config.xml 中注册 CalzzMapper 接口,代码如下。

```
<mappers>
    <mapper class="com.ssm.mapper.ClazzMapper" />
</mappers>
```

注意此处注册接口与注册 XML 文件的区别。此处是以包名的格式,包与包之间以"."作为分隔符。而注册 XML 文件是以目录的格式,目录之间的分隔符是"/"。

(5)创建测试类,在类中对 ClazzMapper 接口中的每个方法都创建一个测试方法,代码如下。

```
1.   package com.ssm.test;
2.
3.   import static org.junit.Assert.assertNotNull;
4.   import static org.junit.jupiter.api.Assertions.assertEquals;
5.   import java.util.List;
6.   import org.apache.ibatis.session.SqlSession;
7.   import org.junit.jupiter.api.Test;
8.   import com.ssm.mapper.ClazzMapper;
9.   import com.ssm.po.Clazz;
10.  import com.ssm.utils.MybatisUtils;
11.
12.  class ClazzTest {
13.     //测试插入方法
14.     @Test
15.     void insertClazzTest() {
16.         SqlSession sqlSession=null;
17.         try {
18.             sqlSession=MybatisUtils.getSession();
19.             Clazz clazz=new Clazz();
20.             clazz.setCname("软件嵌入152");
21.             ClazzMapper clazzMapper=sqlSession.getMapper(ClazzMapper.class);
22.             int result=clazzMapper.saveClazz(clazz);
23.             sqlSession.commit();
24.             assertEquals(result, 1);
25.         }catch(Exception e) {
26.             e.printStackTrace();
27.             sqlSession.rollback();
28.         }finally {
29.             sqlSession.close();
30.         }
31.
```

```
32.    }
33.    //测试删除方法
34.    @Test
35.    void deleteClazzTest() {
36.        SqlSession sqlSession=null;
37.        try {
38.            sqlSession=MybatisUtils.getSession();
39.            ClazzMapper clazzMapper=sqlSession.getMapper(ClazzMapper.class);
40.            int result=clazzMapper.removeClazz(28);
41.            sqlSession.commit();
42.            assertEquals(result, 1);
43.        }catch(Exception e) {
44.            e.printStackTrace();
45.            sqlSession.rollback();
46.        }finally {
47.            sqlSession.close();
48.        }
49.
50.    }
51.    //测试修改班级方法
52.    @Test
53.    void modifyClazzTest() {
54.        SqlSession sqlSession=null;
55.        try {
56.            sqlSession=MybatisUtils.getSession();
57.            Clazz clazz=new Clazz();
58.            clazz.setId(27);
59.            clazz.setCname("软件嵌入152");
60.            ClazzMapper clazzMapper=sqlSession.getMapper(ClazzMapper.class);
61.            int result=clazzMapper.modifyClazz(clazz);
62.            sqlSession.commit();
63.            assertEquals(result, 1);
64.        }catch(Exception e) {
65.            e.printStackTrace();
66.            sqlSession.rollback();
67.        }finally {
68.            sqlSession.close();
69.        }
70.    }
71.    //测试根据班级id查询班级信息
72.    @Test
73.    void selectClazzByIdTest() {
74.        SqlSession sqlSession=null;
```

```
75.        sqlSession=MybatisUtils.getSession();
76.        ClazzMapper clazzMapper=sqlSession.getMapper(ClazzMapper.class);
77.        Clazz clazz=clazzMapper.selectUserById(15);
78.        String result=clazz.getCname();
79.        assertEquals(result,"软件嵌入 171");
80.        System.out.println(clazz);
81.    }
82.    //查询表中的所有班级
83.    @Test
84.    void selectAllClazzTest() {
85.        SqlSession sqlSession=null;
86.        sqlSession=MybatisUtils.getSession();
87.        ClazzMapper clazzMapper=sqlSession.getMapper(ClazzMapper.class);
88.        List<Clazz>clazzs=clazzMapper.selectAllClazz();
89.        assertNotNull(clazzs);
90.    }
91. }
```

类中的方法可以单独调试，这样便于发现问题和解决问题。在测试方法中使用 assertEquals()方法，判断运行的结果与预期的值是否一致。这个方法中的参数，读者可以根据 tb_clazz 表中的实际值自行调整。程序运行结果省略，读者可以自行验证。

示例中通过注解取代 xml 配置文件完成数据库表的 CRUD 操作，优点是代码简洁、方便，缺点是如果业务逻辑复杂，用注解编码不是很方便，这时还是建议用 xml 配置文件。

8.3 一对多的双向关联操作

下面通过一个示例说明如何使用 MyBatis 的注解完成一对多的多表关联操作。

例 8-2 还是以 tb_clazz 表与 tb_student 表之间的关系为例，用注解完成 tb_clazz 表对 tb_student 表的一对多查询，tb_student 表对 tb_clazz 表的一对一（或多对一）查询操作。

（1）tb_clazz 表与 tb_student 数据库脚本可参照例 7-1 中的数据库脚本。

（2）两个表的实体类 Student 类代码如下。

Student 类：

```
1. package com.ssm.po;
2.
3. public class Student {
4.     private Integer id;
5.     private String loginname;
6.     private String password;
7.     private String username;
```

```
8.        private Clazz clazz;
9.    }
```

第 8 行的 Clazz 对象的引用变量 clazz 是 tb_student 表与 tb_clazz 表建立多对一关系必需的。通过 clazz 属性可以从 Student 中得到 Clazz 的相关信息。

（3）两个表的实体类 Clazz 类代码如下。

Clazz 类：

```
1.  package com.ssm.po;
2.  import java.util.List;
3.  public class Clazz {
4.      private Integer id;
5.      private String cname;
6.      private List<Student> students;
7.  }
```

第 6 行的 students 属性是一个集合，用来存放多方 Student 对象的集合，这样可以从一方 Clazz 的对象中得到多方 Student 中的信息。

（4）创建 ClazzMapper 接口和 Student 接口，并在接口中相应的方法前添加注解，完成表与类的映射功能。

① StudentMapper 接口代码如下。

```
1.  package com.ssm.mapper;
2.  
3.  import java.util.List;
4.  import org.apache.ibatis.annotations.One;
5.  import org.apache.ibatis.annotations.Result;
6.  import org.apache.ibatis.annotations.Results;
7.  import org.apache.ibatis.annotations.Select;
8.  import org.apache.ibatis.mapping.FetchType;
9.  import com.ssm.po.Student;
10. 
11. public interface StudentMapper {
12. 
13.     //根据班级 id 查询所有学生
14.     @Select("select * from tb_student where clazz_id=#{id}")
15.     public List<Student> selectByClazzId(Integer id);
16. 
17.     //根据学生 id 查询学生及学生所在班级，一对一查询
18.     @Select("select * from tb_student where id=#{id}")
19.     @Results({
20.         @Result(id=true,column="id",property="id"),
21.         @Result(column="loginname",property="loginname"),
22.         @Result(column="password",property="password"),
```

```
23.            @Result(column="username",property="username"),
24.            @Result(column="clazz_id",property="clazz",
25.                one=@One(select="com.ssm.mapper.ClazzMapper.selectClazzById",
26.                fetchType=FetchType.EAGER))
27.        })
28.        public Student selectStudentById(Integer id);
29.    }
```

接口中有两个方法：一个方法是根据班级 id 查询所有的学生信息，返回的是一个 List；另一个方法是根据学生 id 查询学生信息，返回的是一个学生对象，这个方法中有一个一对一的关联查询。

第 25 行用 @One 注解实现了一对一的映射，即 Student 类中的 clazz 属性与 ClazzMapper 中 id 为 selectClazzById 的查询语句的映射。

@One 注解中的 fetchType＝FetchType.EAGER 表示立即加载关联的属性，此处即 clazz 对象。

② ClazzMapper 接口只给出新添加的方法，代码如下。

```
1.    //根据班级 id 查询班级及所有学生记录,一对多的关联
2.    @Select("select * from tb_clazz where id=#{id}")
3.    //tb_clazz 表中字段与 Clazz 类中的属性映射
4.    @Results({
5.        @Result(id=true,column="id",property="id"),
6.        @Result(column="cname",property="cname"),
7.        @Result(column="id",property="students",
8.            many=@Many(select="com.ssm.mapper.StudentMapper.
                selectByClazzId",
9.            fetchType=FetchType.LAZY))
10.    })
11.    Clazz selectClazzAndStudentsById(Integer id);
12. }
```

这个方法实现一对多的关联操作。第 8 行的 @Many 注解实现了 Clazz 类的 students 属性与 StudentMapper 接口中 id 为 selectByClazzId 查询的映射。

此处的 fetchType＝FetchType.LAZY 将关联属性 students 设置为延迟加载。

(5) 在 mybatis-config.xml 中注册 StudentMapper 接口，代码如下。

```
<mappers>
    <mapper class="com.ssm.mapper.ClazzMapper" />
    <mapper class="com.ssm.mapper.StudentMapper" />
</mappers>
```

(6) 创建测试类，分别为 StudentMapper 接口和 ClazzMapper 接口创建测试类，并测试接口中的所有方法。

① StudentMapper 接口的测试类 StudentTest，代码如下。

```
1.   package com.ssm.test;
2.
3.   import static org.junit.jupiter.api.Assertions.assertEquals;
4.   import org.apache.ibatis.session.SqlSession;
5.   import org.junit.jupiter.api.Test;
6.   import com.ssm.mapper.StudentMapper;
7.   import com.ssm.po.Student;
8.   import com.ssm.utils.MybatisUtils;
9.
10.  class StudentTest {
11.    //测试根据id查询某一个学生,并查询学生所在班级,这是一对一的关联查询
12.    @Test
13.    void selectStudentByIdTest() {
14.      SqlSession sqlSession=MybatisUtils.getSession();
15.      StudentMapper studentMapper = sqlSession.getMapper(StudentMapper.class);
16.      Student student=studentMapper.selectStudentById(275);
17.      String cname=student.getClazz().getCname();
18.      assertEquals(cname,"软件嵌入 171");
19.      System.out.println(student);
20.    }
21.  }
```

第 16 行查询学生 id 为 275 的学生，由于在查询语句中已经关联了 clazz 对象，所以在第 17 行通过 clazz 对象得到了这个学生所在的班级名称。通过 assertEquals() 断言方法判断查询是否成功。

② ClazzMapper 接口的测试类 ClazzTest，其他的测试方法 8.2 节已经讲过，这里是新添加的一对多的测试方法，代码如下。

```
1.   //测试根据班级 id 查询班级及所有学生信息,一对多关联
2.   @Test
3.   void selectClazzAndStudentsByIdTest() {
4.     SqlSession sqlSession=null;
5.     sqlSession=MybatisUtils.getSession();
6.     ClazzMapper clazzMapper=sqlSession.getMapper(ClazzMapper.class);
7.     Clazz clazz=clazzMapper.selectClazzAndStudentsById(15);
8.     List<Student>students=clazz.getStudents();
9.     assertNotNull(students);
10.    students.forEach(stu->System.out.println(stu));
11.  }
```

第 7 行是查询班级 id 为 15 的班级信息，所有的班级信息都封装在 clazz 对象中。第

8行通过clazz对象得到这个班级的所有学生students对应的一个List集合。第9行通过断言判断students是否为空,如果不为空,则表示查询成功。最后一行通过循环输出所有学生的信息。

8.4 多对多的关联操作

这里还是以7.3节例子中的表tb_student和表tb_course之间的关系说明基于注解的多对多的关联操作。

例8-3 学生表tb_student与课程表tb_course之间是多对多的关系,此处要求基于注解完成从学生查询所选课程,从课程查询选此课程的所有学生。

(1)数据库脚本参照例7-3中的数据库脚本。

(2)tb_course表和tb_student表对应的实体类参照例7-3中的实体类的代码。

(3)创建StudentMapper接口和CourseMapper接口,在这两个接口中添加注解,完成数据库表与类之间的映射。

创建CourseMapper接口,添加一对多的查询方法,代码如下。

```java
1.  package com.ssm.mapper;
2.
3.  import org.apache.ibatis.annotations.Many;
4.  import org.apache.ibatis.annotations.Result;
5.  import org.apache.ibatis.annotations.Results;
6.  import org.apache.ibatis.annotations.Select;
7.  import org.apache.ibatis.mapping.FetchType;
8.  import com.ssm.po.Course;
9.
10. public interface CourseMapper {
11.
12.     @Select("select * from tb_course where id=#{id}")
13.     @Results({
14.         @Result(id=true,column="id",property="id"),
15.         @Result(column="cname",property="cname"),
16.         @Result(column="id",property="students",
17.             many = @Many(select ="com.ssm.mapper.StudentMapper.
                selectStudentByCourseId",
18.             fetchType=FetchType.LAZY))
19.     })
20.     //根据课程id查询Course对象,Course对象中的students属性包含了所有学生
        //对象,这是一对多查询
21.     public Course selectCourseById(Integer id);
22.
23. }
```

第 16、17 行实现了 students 集合属性与 StudentMapper 接口中查询映射。这里用到 @Many 注解，将具体的映射指向了 id 为 selectStudentByCourseId 的查询，实现了一对多的查询。

在 StudentMapper 接口中添加根据 course_id 查询学生信息的方法，代码如下。

```
1.    @Select("select * from tb_student where id "
2.        +"in(select student_id from student_course where course_id=#{id})")
3.    //根据课程 id 查询学生信息，此处用到 in 语句，这是一个嵌入式查询
4.    public List<Student>selectStudentByCourseId(Integer id);
5.    }
```

首先运行 in 语句中的查询，根据课程 id 在 student_course 表中查询对应的所有学生 id，然后根据查询出的学生 id 在 tb_student 表中查询相应的学生信息。这里运用嵌入式查询实现了多表关联查询。

（4）在 mybatis-config.xml 中注册 CourseMapper 接口，代码如下。

```
<mappers>
    <mapper class="com.ssm.mapper.ClazzMapper"/>
    <mapper class="com.ssm.mapper.StudentMapper"/>
    <mapper class="com.ssm.mapper.CourseMapper"/>
</mappers>
```

（5）创建测试类。创建 CourseTest 测试类和测试方法，代码如下。

```
1.    package com.ssm.test;
2.    
3.    import static org.junit.Assert.assertEquals;
4.    import java.util.List;
5.    import org.apache.ibatis.session.SqlSession;
6.    import org.junit.jupiter.api.Test;
7.    import com.ssm.mapper.CourseMapper;
8.    import com.ssm.po.Course;
9.    import com.ssm.po.Student;
10.   import com.ssm.utils.MybatisUtils;
11.   
12.   class CourseTest {
13.   
14.       @Test
15.       void selectCourseByIdTest() {
16.           SqlSession sqlSession=MybatisUtils.getSession();
17.           CourseMapper courseMapper = sqlSession. getMapper (CourseMapper.
              class);
18.           Course course=courseMapper.selectCourseById(10);
19.           List<Student>students=course.getStudents();
```

```
20.        int n=students.size();
21.        assertEquals(n,2);
22.    }
23. }
```

第 18 行查询课程 id=10 的课程信息,第 19 行是得到选了这门课的所有学生的集合对象 students,通过断言 assertEquals()判断选课的人数是否为 2,这个值根据 student_course 表中的具体记录确定。

这个示例实现了查询选某门课程的学生,如果要查询某个学生选了哪些课程,读者可根据这个示例自行完成相关的程序。

8.5 注解中的动态 SQL

MyBatis 注解可以使用动态 SQL,只是要用到下面几个相关的注解:@InsertProvider、@UpdateProvider、@DeleteProvider 和@SelectProvider。当编写复杂的 SQL 语句时,在 Mapper 接口中无法完成,这时可以借助上面的 4 个注解,将 SQL 语句和 Mapper 接口分开,SQL 语句单独写在一个文件中。

上述 4 个 Provider 注解都有 type 和 method 属性。type 属性用于指定存放 SQL 语句的类,method 属性用于指定对应的方法。通过简单字符串拼接 SQL 语句有时非常困难,MyBatis 提供了一个 SQL 工具类 org.apache.ibatis.jdbc.SQL 帮助解决该问题。使用 SQL 类,简单地创建一个实例来调用方法生成 SQL 语句。下面通过一个示例说明如何使用 SQL 类生成动态 SQL 语句。

例 8-4 应用@SelectProvider 注解和 SQL 类对 tb_student 表进行动态查询。具体要求为:根据传入的不同参数生成动态查询语句。

(1)在例 8-3 基础上修改 StudentMapper 接口,添加如下代码。

```
1. //通过 type 属性指定生成动态查询 tb_student 表的动态 SQL 类和方法
2. @SelectProvider(type=StudentDynaSqlProvider.class,method="selectWithParam")
3. public List<Student> selectStudentWithParam(Map<String,Object> Param);
4. //通过 type 属性指定生成更新 tb_student 表的动态 SQL 类和方法
5. @UpdateProvider(type=StudentDynaSqlProvider.class,method="updateStudent")
6. public int updateStudent(Student student);
```

第 2 行通过@SelectProvider 注解指定生成动态查询 SQL 的类和方法,第 3 行方法的参数类型是一个 Map,这个参数中存放的是要查询的参数,它既是在外部调用方法时的参数类型,也是生成动态 SQL 类中方法的参数类型。

第 5 行通过@UpdateProvider 注解指定动态更新 SQL 的类和方法,这个方法的形参是 Student 类的对象。

(2)创建 StudentDynaSqlProvider 类,并添加动态查询方法和动态更新方法,代码如下。

```
1.   package com.ssm.mapper;
2.
3.   import java.util.Map;
4.   import org.apache.ibatis.jdbc.SQL;
5.   import com.ssm.po.Student;
6.
7.   public class StudentDynaSqlProvider {
8.       //根据 param 参数生成动态查询 SQL 语句
9.       public String selectWithParam(Map<String,Object>param) {
10.          return new SQL() {
11.              {
12.                  SELECT("*");
13.                  FROM("tb_student");
14.                  if(param.get("id")!=null) {
15.                      WHERE("id=#{id}");
16.                  }
17.                  if(param.get("loginname")!=null) {
18.                      WHERE("loginname=#{loginname}");
19.                  }
20.                  if(param.get("username")!=null) {
21.                      WHERE("username LIKE CONCAT ('%',#{username},'%')");
22.                  }
23.              }
24.          }.toString();
25.      }
26.      //根据 student 参数生成动态更新 SQL 语句
27.      public String updateStudent(Student student) {
28.          return new SQL() {
29.              {
30.                  UPDATE("tb_student");
31.                  if(student.getLoginname()!=null) {
32.                      SET("loginname=#{loginname}");
33.                  }
34.                  if(student.getPassword()!=null) {
35.                      SET("password=#{password}");
36.                  }
37.                  if(student.getUsername()!=null) {
38.                      SET("username=#{username}");
39.                  }
40.                  if(student.getClazz()!=null) {
41.                      SET("clazz=#{clazz}");
42.                  }
43.                  WHERE("id=#{id}");
```

```
44.        }
45.      }.toString();
46.    }
47.
48.  }
```

第 9 行 selectWithParam( ) 方法的形参要与 StudentMapper 接口中的 selectStudentWithParam()方法的形参一致。第 10 行调用了 SQL 类的构造方法生成动态 SQL 语句。第 14~20 行根据 param 参数中的键值是否为 null 决定生成动态 SQL 语句。

第 27 行的 updateStudent( )方法也是如此，根据形参 student 中的属性是否为 null 生成相应的动态更新 SQL 语句。

（3）在例 8-3 基础上修改 StudentTest 类，添加两个方法，代码如下。

```
1.    //测试通过注解生成动态查询的方法
2.    @Test
3.    void selectStudentWithParamTest() {
4.        SqlSession sqlSession=MybatisUtils.getSession();
5.        StudentMapper studentMapper = sqlSession.getMapper(StudentMapper.class);
6.        Map<String,Object>map=new HashMap<String,Object>();
7.        map.put("loginname","1713071010");
8.        map.put("username","王");
9.        List<Student>students=studentMapper.selectStudentWithParam(map);
10.       int n=students.size();
11.       assertEquals(n,1);
12.   }
13.   //测试通过注解生成动态更新的方法
14.   @Test
15.   void updateStudentTest() {
16.       SqlSession sqlSession=MybatisUtils.getSession();
17.       StudentMapper studentMapper = sqlSession.getMapper(StudentMapper.class);
18.       Student student=new Student();
19.       student.setId(278);
20.       student.setUsername("陈力");
21.       int result=studentMapper.updateStudent(student);
22.       sqlSession.commit();
23.       assertEquals(result,1);
24.
25.   }
```

第 6 行创建了一个 Map 对象，当调用 studentMapper 类中的 selectStudentWithParam()方法时需要这个参数，根据参数生成动态的 SQL 查询语句。

第 18 行创建了一个 student 对象,这个对象的 id 属性是必需的,其他属性是可选的,动态 SQL 是根据 id 对 tb_student 表进行更新的。

(4) 运行测试类及测试方法,在控制台可看到生成的 SQL 语句。

运行测试方法 selectStudentWithParamTest(),在控制台可看到如下的日志信息:

DEBUG [main] -==> Preparing: SELECT * FROM tb_student WHERE (loginname=? AND username=?)
DEBUG [main] -==>Parameters: 1713071010(String), 丁尧(String)
DEBUG [main] -<==   Total: 1

运行这个测试方法时,查询的参数中有 loginname 和 username 两项,所以生成的查询语句中有两个查询条件。如果在查询时,构造 stduent 对象时只有一个 loginname 属性的值,则程序运行时生成的动态查询语句就只有一个条件,如下面的日志所示。

EBUG [main] -==>   Preparing: SELECT * FROM tb_student WHERE (loginname=?)
DEBUG [main] -==>Parameters: 1713071010(String)
DEBUG [main] -<==   Total: 1

同样,运行 updateStudentTest() 方法,在控制台可看到如下的日志信息。

DEBUG [main] -==>Preparing: UPDATE tb_student SET username=? WHERE (id=?)
DEBUG [main] -==>Parameters: 陈力(String), 278(Integer)
DEBUG [main] -<== Updates: 1

在测试方法中构建 student 对象时,只给 username 属性赋了值,其他属性值为空,所以生成的动态 SQL 语句中只更新了 username 字段的值。

MyBatis 中的注解与动态 SQL 结合在一起可以非常简单、高效地完成对数据库的操作,完全可以取代基于 XML 的映射文件。当然,使用哪种方式还是因人而异,开发人员可以根据不同的项目、不同的要求选择。

习   题

1. @One 注解的作用是什么?主要用于什么地方?举例说明。
2. @Many 注解的作用是什么?主要用于什么地方?举例说明。
3. 在 MyBatis 中基于注解开发时,是否可以作用动态 SQL?如何实现?
4. 什么是 MyBatis 的接口绑定?有哪些实现方式?

实验 8 基于注解的开发

1. 实验目的

通过本次实验,让读者了解注解开发与基于 XML 映射文件开发的区别,掌握常用的注解操作数据库的方法,掌握注解中动态 SQL 操作数据库的方法。

2. 实验内容

用基于注解的开发完成实验 7 中的实验内容。

3. 实现思路及步骤

（1）可参照实验 5 的步骤，但此处不需创建映射文件，只创建与表对应的接口即可。在接口中通过注解完成一些简单数据库操作，对于复杂的动态 SQL 语句，可另外编写一个 Provider 文件。

（2）编写测试类及方法并进行测试。

第9章 SSM 框架整合

本章目标

1. 掌握基于 MyBatis 映射文件的整合 SSM 框架开发的方法。
2. 掌握基于 MyBatis 注解的整合 SSM 框架开发的方法。

前面的章节中分别讲述了 Spring MVC、Spring 和 MyBatis 3 个框架，现在要将这 3 个框架整合在一起，充分发挥每个框架的特长，更好地提高开发效率。Spring MVC 是 Spring 的一部分，因此不存在整合的问题。现在主要解决 MyBatis 与 Spring 框架的整合问题。本章分两部分讲解：一部分是基于 MyBatis 映射文件的整合；另一部分是基于 MyBatis 注解的整合。

9.1 基于 MyBatis 映射文件的整合开发

本节以学生登录系统为例，详细说明 SSM 框架的整合过程。学生登录系统要求输入的用户名和密码都正确时，跳转到 result.jsp 页面，在页面上显示"某学生登录成功"，否则显示"登录失败"。本章用到的数据库与第 8 章的相同，主要用到 tb_clazz 表和 tb_student 表。

9.1.1 创建 Web 项目

创建一个 Web 项目 chap9_1，添加 JAR 包，可以将工程 chap4 中的 JAR 包和工程 chap8 中的 JAR 包加进来，另外再添加如下的 JAR 包。

- 用于 Spring 事务管理的 JAR 包：

spring-tx-5.1.8.RELEASE.jar

- 用于建立数据库连接池的 JAR 包：

commons-dbcd2-2.1.1.jar 和 commomns-pool2-2.4.2.jar

- 用于 Spring 与 MyBatis 整合的 JAR 包：

mybatis-spring-2.0.2.jar

全部 JAR 包如下：

```
classmate-1.3.4.jar
commons-dbcp2-2.1.1.jar
commons-fileupload-1.3.2.jar
commons-io-2.5.jar
commons-logging-1.2.jar
commons-pool2-2.4.2.jar
hibernate-validator-6.0.17.Final.jar
hibernate-validator-annotation-processor-6.0.17.Final.jar
hibernate-validator-cdi-6.0.17.Final.jar
jackson-annotations-2.9.2.jar
jackson-core-2.9.2.jar
jackson-databind-2.9.2.jar
javax.el-3.0.1-b09.jar
javax.servlet.jsp.jstl-1.2.1.jar
javax.servlet.jsp.jstl-api-1.2.1.jar
jboss-logging-3.3.2.Final.jar
log4j-1.2.17.jar
log4j-api-2.3.jar
log4j-core-2.3.jar
mybatis-3.5.2.jar
mybatis-spring-2.0.2-sources.jar
mybatis-spring-2.0.2.jar
mysql-connector-java-5.1.40-bin.jar
slf4j-api-1.7.22.jar
slf4j-log4j12-1.7.22.jar
spring-aop-5.1.8.RELEASE.jar
spring-beans-5.1.8.RELEASE.jar
spring-context-5.1.8.RELEASE.jar
spring-core-5.1.8.RELEASE.jar
spring-expression-5.1.8.RELEASE.jar
spring-jdbc-5.1.8.RELEASE.jar
spring-tx-5.1.8.RELEASE.jar
spring-web-5.1.8.RELEASE.jar
spring-webmvc-5.1.8.RELEASE.jar
validation-api-2.0.1.Final.jar
```

chap9_1 工程的包结构如图 9-1 所示。

controller：包中存放 Spring MVC 的控制器。

mapper：包中存放映射文件和 Mapper 接口。

po：包中存放与表对应的实体类。

service：存放服务接口。

service.impl：存放 service 的实现类。

test：存放测试类。

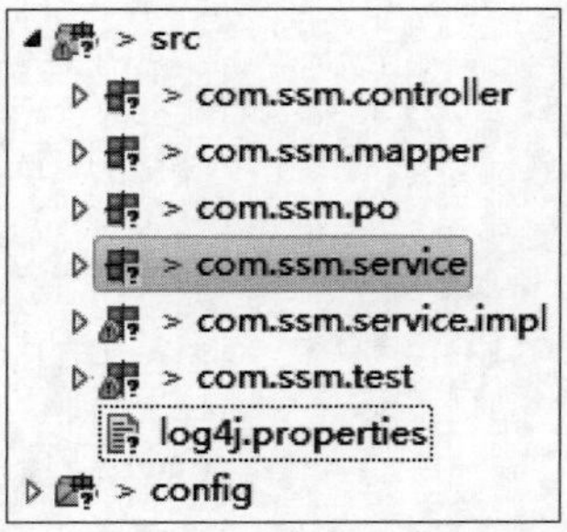

图 9-1 chap9_1 工程的包结构

config 目录：存放所有的配置文件。

9.1.2 编写配置文件

在 config 目录下分别创建 db.properties、applicationContext.xml、mybatis-config.xmlt 和 springmvc-config.xml 文件。db.properties 是配置数据库的信息，内容如下。

```
1.  jdbc.driver=com.mysql.jdbc.Driver
2.  jdbc.url=jdbc:mysql://localhost:3306/homework
3.  jdbc.username=root
4.  jdbc.password=root
5.  jdbc.maxTotal=30
6.  jdbc.maxIdle=10
7.  jdbc.initialSize=5
```

文件中除配置了数据库的 4 个基本信息外，还配置了数据库连接池的最大连接数（maxTotal）、最大空闲连接数（maxIdle）以及初始化连接数（initialSize）。

mybatis-config 是 MyBatis 的配置信息，内容如下。

```
1.  <?xml version="1.0" encoding="UTF-8" ?>
2.  <!DOCTYPE configuration
3.    PUBLIC "-//mybatis.org //DTD Config 3.0//EN"
4.    "http://mybatis.org/dtd/mybatis-3-config.dtd">
5.  <configuration>
6.
7.  <settings >
8.      <setting name="lazyLoadingEnabled" value="true"/>
9.      <setting name="aggressiveLazyLoading" value="false"/>
10. </settings>
11.
12.     <!--定义别名 -->
13.     <typeAliases>
14.         <package name="com.ssm.po" />
15.     </typeAliases>
16.
17. </configuration>
```

这里不再需要配置数据库连接信息，可以将数据库连接放在 spring 配置文件中。Spring MVC 的配置文件 springmvc-config.xml 内容如下。

```
1.  <?xml version="1.0" encoding="UTF-8"?>
2.  <beans xmlns="http://www.springframework.org/schema/beans"
3.      xmlns:mvc="http://www.springframework.org/schema/mvc"
4.      xmlns:xsi="http://www.w3.org/2001/XMLSchema-instance"
```

```
5.      xmlns:p="http://www.springframework.org/schema/p"
6.      xmlns:context="http://www.springframework.org/schema/context"
7.      xsi:schemaLocation="
8.          http://www.springframework.org/schema/beans
9.          https://www.springframework.org/schema/beans/spring-beans.xsd
10.         http://www.springframework.org/schema/context
11.         https://www.springframework.org/schema/context/spring-context.xsd
12.         http://www.springframework.org/schema/mvc
13.         https://www.springframework.org/schema/mvc/spring-mvc.xsd">
14.     <!--指定需要扫描的包 -->
15.     <context:component-scan base-package="com.ssm.controller" />
16.         <!--默认装配方案 -->
17.     <mvc:annotation-driven/>
18.     <!--静态资源处理 -->
19.     <mvc:default-servlet-handler/>
20.     <!--定义视图解析器 -->
21.     <bean id="viewResolver" class=
22.     "org.springframework.web.servlet.view.InternalResourceViewResolver">
23.         <!--设置前缀 -->
24.     <property name="prefix" value="/WEB-INF/jsp/" />
25.         <!--设置后缀 -->
26.     <property name="suffix" value=".jsp" />
27.     </bean>
28. </beans>
```

Spring 的配置文件 applicationContext.xml 的内容如下。

```
1.  <?xml version="1.0" encoding="UTF-8"?>
2.  <beans xmlns="http://www.springframework.org/schema/beans"
3.      xmlns:mvc="http://www.springframework.org/schema/mvc"
4.      xmlns:context="http://www.springframework.org/schema/context"
5.      xmlns:xsi="http://www.w3.org/2001/XMLSchema-instance"
6.      xmlns:tx="http://www.springframework.org/schema/tx"
7.      xsi:schemaLocation="
8.  http://www.springframework.org/schema/beans
9.  http://www.springframework.org/schema/beans/spring-beans.xsd
10. http://www.springframework.org/schema/mvc
11. http://www.springframework.org/schema/mvc/spring-mvc.xsd
12. http://www.springframework.org/schema/tx
13. http://www.springframework.org/schema/tx/spring-tx.xsd
14. http://www.springframework.org/schema/context
15. http://www.springframework.org/schema/context/spring-context.xsd">
```

```xml
16.     <!--读取db.properties -->
17.     <context:property-placeholder
18.         location="classpath:db.properties" />
19.     <!--配置数据源 -->
20.     <bean id="dataSource"
21.         class="org.apache.commons.dbcp2.BasicDataSource">
22.         <!--数据库驱动 -->
23.         <property name="driverClassName" value="${jdbc.driver}" />
24.         <!--连接数据库的url -->
25.         <property name="url" value="${jdbc.url}" />
26.         <!--连接数据库的用户名 -->
27.         <property name="username" value="${jdbc.username}" />
28.         <!--连接数据库的密码 -->
29.         <property name="password" value="${jdbc.password}" />
30.         <!--最大连接数 -->
31.         <property name="maxTotal" value="${jdbc.maxTotal}" />
32.         <!--最大空闲连接数 -->
33.         <property name="maxIdle" value="${jdbc.maxIdle}" />
34.         <!--初始化连接数 -->
35.         <property name="initialSize" value="${jdbc.initialSize}" />
36.     </bean>
37.     <!--事务管理器,依赖于数据源 -->
38.     <bean id="transactionManager"
39.         class="org.springframework.jdbc.datasource.DataSourceTransaction-
            Manager">
40.         <property name="dataSource" ref="dataSource" />
41.     </bean>
42.     <!--开启事务注解 -->
43.     <tx:annotation-driven
44.         transaction-manager="transactionManager" />
45.     <!--配置MyBatis工厂 -->
46.     <bean id="sqlSessionFactory"
47.         class="org.mybatis.spring.SqlSessionFactoryBean">
48.         <!--注入数据源 -->
49.         <property name="dataSource" ref="dataSource" />
50.         <!--指定核心配置文件位置 -->
51.         <property name="configLocation"
52.             value="classpath:mybatis-config.xml" />
53.     </bean>
54.     <!--实例化service -->
55.     <bean id="studentService" class="com.ssm.service.impl.
            StudentServiceImpl">
56.     <!--注入SqlSessionFactory对象实例-->
```

```
57.            <property name="sqlSessionFactory" ref="sqlSessionFactory" />
58.        </bean>
59.        <!--Mapper 代理开发(基于 MapperScannerConfigurer) -->
60.        <bean class="org.mybatis.spring.mapper.MapperScannerConfigurer">
61.            <property name="basePackage" value="com.ssm.mapper" />
62.        </bean>
63.
64.        <!--开启扫描 -->
65.        <context:component-scan
66.            base-package="com.ssm.service" />
67.
68.    </beans>
```

第 55 行定义了一个 id 为 "studentService" 的 Bean，这个 Bean 将来可以自动注入 StudentController 的属性中。在 StudentServiceImpl 中要用到 StudentMapper 接口的实例，第 60 行通过 MapperScannerConfigurer 可生成相应 Mapper 的实例。这里有一个问题是，要定义大量 service 的 Bean，其实也可以通过包扫描机制完成。9.2 节会解决这个问题。

启动 Tomcat 服务器时，第一个加载的文件是 web.xml，代码如下。

```
1.  <?xml version="1.0" encoding="UTF-8"?>
2.  <web-app xmlns:xsi="http://www.w3.org/2001/XMLSchema-instance" xmlns
    ="http://xmlns.jcp.org/xml/ns/javaee" xsi:schemaLocation="http://
    xmlns.jcp.org/xml/ns/javaee http://xmlns.jcp.org/xml/ns/javaee/web-
    app_3_1.xsd" id="WebApp_ID" version="3.1">
3.      <context-param>
4.          <param-name>contextConfigLocation</param-name>
5.          <param-value>classpath:applicationContext.xml</param-value>
6.      </context-param>
7.      <listener>
8.          <listener-class>
9.              org.springframework.web.context.ContextLoaderListener
10.         </listener-class>
11.     </listener>
12.     <filter>
13.         <filter-name>encoding</filter-name>
14.         <filter-class>
15.             org.springframework.web.filter.CharacterEncodingFilter
16.         </filter-class>
17.         <init-param>
18.             <param-name>encoding</param-name>
19.             <param-value>UTF-8</param-value>
20.         </init-param>
21.     </filter>
```

```
22.     <filter-mapping>
23.         <filter-name>encoding</filter-name>
24.         <url-pattern>/*</url-pattern>
25.     </filter-mapping>
26.     <servlet>
27.         <servlet-name>springmvc</servlet-name>
28.         <servlet-class>
29.             org.springframework.web.servlet.DispatcherServlet
30.         </servlet-class>
31.         <init-param>
32.             <param-name>contextConfigLocation</param-name>
33.             <param-value>classpath:springmvc-config.xml</param-value>
34.         </init-param>
35.         <load-on-startup>1</load-on-startup>
36.     </servlet>
37.     <servlet-mapping>
38.         <servlet-name>springmvc</servlet-name>
39.         <url-pattern>/</url-pattern>
40.     </servlet-mapping>
41. </web-app>
```

第 5 行指定要加载的 Spring 配置文件的位置,此处表示在 src 下或在 config 文件夹中。第 7～11 行定义了一个 listener,Web 服务器通过这个 listener 加载 Spring 配置文件。

第 12～21 行定义了一个过滤器(filter),作用是设置字符集为 UTF-8,主要是解决中文乱码的问题。

第 26～36 行定义了一个 servlet,作用是启用 Spring MVC 框架,第 32～33 行指定了 springmvc-config.xml 文件的位置。

完成以上配置后,当启动 Web 服务时,Spring 配置文件、MyBatis 配置文件和 Spring MVC 配置文件会自动加载。至此,所有的配置文件都已完成,下面就是开发相应的映射文件和相关的类。

9.1.3 创建映射文件与接口

学生表 tb_student 的映射文件 StudentMapper.xml 的代码如下。

```
1. <?xml version="1.0" encoding="UTF-8" ?>
2. <!DOCTYPE mapper
3.     PUBLIC "-//mybatis.org //DTD Mapper 3.0//EN"
4.     "http://mybatis.org/dtd/mybatis-3-mapper.dtd">
5. <mapper namespace="com.ssm.mapper.StudentMapper">
6.
7. <!--根据学生 id 查询学生 -->
```

```
8.    <select id="selectStudentById" resultType="student">
9.        select * from tb_student where id =#{id}
10.   </select>
11.   <!--根据学生用户名和密码查询学生 -->
12.   <select id="selectStudentByLoginnameAndPassword"
13.       parameterType="String"
14.       resultType="student">
15.       select * from tb_student where loginname =#{loginname} and password
          =#{password}
16.   </select>
17.   </mapper>
```

第 12 行的 SQL 片段是根据传递来的用户登录名 loginname 和密码 password 两个字符串进行组合查询，返回一个 Student 类的对象。其对应的接口文件 StudentMapper 代码如下。

```
1.    package com.ssm.mapper;
2.    import org.apache.ibatis.annotations.Param;
3.    import com.ssm.po.Student;
4.    public interface StudentMapper {
5.        //根据学生 ID 查询学生信息，返回的是一个 Student 对象
6.        public Student selectStudentById(Integer id);
7.        //根据学生用户名和密码查询学生信息，返回的是一个 Student 对象
8.        public Student selectStudentByLoginnameAndPassword(@Param
          ("loginname") String loginname,
9.            @Param("password")String password);
10.
11.   }
```

第 8 行的形参中用到@Param 注解，这个注解的作用是将这两个参数分别传递给映射文件中的查询语句。这里的接口我们并没有给出实现，因此在 Spring 的配置文件中要求必须有如下配置：

```
<bean class="org.mybatis.spring.mapper.MapperScannerConfigurer">
<property name="basePackage" value="com.ssm.mapper" />
</bean>
```

通过扫描包可生成所有接口对应实现类的实例，这个配置是必需的，否则无法对包中的接口进行实例化。

9.1.4 创建 Service 及其实现类

此处采用基于 Mapper 接口的编程方式。StudentService 接口的代码如下。

```
1.    package com.ssm.service;
2.
```

```
3.    import com.ssm.po.Student;
4.
5.    public interface StudentService {
6.
7.        public Student findStudentById(Integer id);
8.
9.        public Student findStudentByLoginnameAndPassword(String loginname,
          String password);
10.   }
```

其实现类 StudentServiceImpl 的代码如下。

```
1.    package com.ssm.service.impl;
2.
3.    import org.mybatis.spring.support.SqlSessionDaoSupport;
4.    import org.springframework.stereotype.Service;
5.    import org.springframework.transaction.annotation.Transactional;
6.    import com.ssm.mapper.StudentMapper;
7.    import com.ssm.po.Student;
8.    import com.ssm.service.StudentService;
9.
10.   @Service("studentService")
11.   @Transactional
12.   public class StudentServiceImpl extends SqlSessionDaoSupport implements StudentService {
13.
14.       @Override
15.       public Student findStudentById(Integer id) {
16.
17.           StudentMapper studentMapper = this.getSqlSession().getMapper(StudentMapper.class);
18.           return studentMapper.selectStudentById(id);
19.       }
20.
21.       @Override
22.       public Student findStudentByLoginnameAndPassword(String loginname,
          String password) {
23.           StudentMapper studentMapper = this.getSqlSession().getMapper(StudentMapper.class);
24.           return studentMapper.selectStudentByLoginnameAndPassword(loginname,
          password);
25.       }
26.   }
```

这个类实现了 StudentService 接口，同时继承了 SqlSessionDaoSupport 类，这个类是第三方插件 mybatis-spring 包中的类。SqlSessionDaoSupport 类中有一个 SqlSessionFactory 属性，在此并没有初始化，在 Spring 配置文件中定义 Bean 时通过注入方式初始化，其代码如下。

```xml
<!--实例化 service -->
<bean id="studentService" class="com.ssm.service.impl.StudentServiceImpl">
    <!--注入 SqlSessionFactory 对象实例-->
    <property name="sqlSessionFactory" ref="sqlSessionFactory" />
</bean>
```

第 11 行@Transaction 注解的作用是为类中的所有方法添加 Spring 事务管理。当然，此处没有数据更新操作，可以不加这个注解，加上这个注解只是为了演示注解的用法。@Transaction 注解也可以加在一个方法上，只对这个方法添加事务处理，它也有一些属性，可以对事务进行控制，在此不再细说，后面用到时再讲述。

项目进行到此，可以对 MyBatis 与 Spring 的整合进行测试。

创建一个测试类，测试 StudentServiceImpl 中的根据用户名和密码查询的方法 findStudentByLoginnameAndPassword()，代码如下。

```java
1.  package com.ssm.test;
2.
3.  import static org.junit.jupiter.api.Assertions.assertEquals;
4.  import javax.validation.constraints.NotNull;
5.  import org.junit.jupiter.api.Test;
6.  import org.springframework.context.ApplicationContext;
7.  import org.springframework.context.support.ClassPathXmlApplicationContext;
8.
9.  import com.ssm.po.Student;
10. import com.ssm.service.StudentService;
11.
12. class StudentTest {
13. //根据学生 id 查询学生信息，返回一个 Student 对象
14.     @Test
15.     void findStudentByIdtest() {
16.         ApplicationContext context=new ClassPathXmlApplicationContext
                ("applicationContext.xml");
17.         StudentService dao=(StudentService)context.getBean("studentService");
18.         Student student=dao.findStudentById(278);
19.         String loginname=student.getLoginname();
20.         assertEquals(loginname,"1713071004");
```

```
21.        }
22.        //根据学生登录名和密码查询学生信息,返回一个Student对象
23.        @Test
24.        void findStudentByLoginnameAndPasswordTest() {
25.            ApplicationContext context=new ClassPathXmlApplicationContext
                  ("applicationContext.xml");
26.            StudentService dao=(StudentService)context.getBean
                  ("studentService");
27.            Student student=dao.findStudentByLoginnameAndPassword("1713071004",
                  "1713071004");
28.            int id=student.getId();
29.            assertEquals(id,278);
30.        }
31.
32.    }
```

第16行加载的是Spring的配置文件applicationContext.xml,通过这个配置文件加载mybatis-config.xml配置文件。通过context.getBean()方法得到StudentService接口的一个实例,并测试其中的方法。

控制台打印日志信息如下。

```
DEBUG [main] -Creating a new SqlSession
DEBUG [main] - Registering transaction synchronization for SqlSession [org.
apache.ibatis.session.defaults.DefaultSqlSession@7d1cfb8b]
DEBUG [main] - JDBC Connection [461698165, URL=jdbc:mysql://localhost:3306/
homework, UserName = root @ localhost, MySQL Connector Java] will be managed
by Spring
DEBUG [main] -==> Preparing: select * from tb_student where loginname =? and
password=?
DEBUG [main] -==>Parameters: 1713071004(String), 1713071004(String)
DEBUG [main] -<==   Total: 1
```

从日志中可以看到,Spring启用了事务,查询结果显示为1,查询成功。如果将StudentServiceImpl中的@Transactional注解去掉,重新运行测试程序,控制台打印日志如下。

```
DEBUG [main] - JDBC Connection [2001321875, URL=jdbc:mysql://localhost:3306/
homework, UserName=root@localhost, MySQL Connector Java] will not be managed
by Spring
DEBUG [main] -==>Preparing: select * from tb_student where loginname =? and
password=?
DEBUG [main] -==>Parameters: 1713071004(String), 1713071004(String)
DEBUG [main] -<==   Total: 1
DEBUG [main] - Closing non transactional SqlSession [org.apache.ibatis.
session.defaults.DefaultSqlSession@39d9314d]
```

日志显示,查询结果为1,查询成功,但没有启用事务。

至此,Spring 与 MyBatis 的整合已经成功。但这里有一个问题,在 applicationContext.xml 配置文件中定义了一个 studentService,如果有多个 Service,可能就要定义多个这样的 Bean,显然这不利于管理和维护。为此,可以重新修改 StudentServiceImpl 类,将 StudentMapper 的对象注入这个实现类中,StudentServiceImpl 类不再需要继承 SqlSessionDaoSupport 父类,它们之间的耦合也就不存在了。修改后的 StudentServiceImpl 类代码如下。

```java
1.  package com.ssm.service.impl;
2.
3.  import org.springframework.beans.factory.annotation.Autowired;
4.  import org.springframework.stereotype.Service;
5.  import com.ssm.mapper.StudentMapper;
6.  import com.ssm.po.Student;
7.  import com.ssm.service.StudentService;
8.
9.  @Service("studentService")
10. public class StudentServiceImpl implements StudentService {
11.
12.     @Autowired
13.     private StudentMapper studentMapper;
14.
15.     @Override
16.     public Student findStudentByLoginnameAndPassword (String loginname,
        String password) {
17.
18.         return this.studentMapper.selectStudentByLoginnameAndPassword
            (loginname, password);
19.     }
20. }
```

第 12 行通过 @Autowired 注解将 studentMapper 对象注入进来,因为在 applicationContext.xml 配置文件中加载了 Mapper 包扫描器,因此在创建这个类的对象之前 studentMapper 对象已经被初始化。这样,在配置文件中不再需要定义 studentService 这个 Bean 了,配置文件中不会出现大量的 Bean 的定义,简化了配置文件,方便了系统的维护。

9.1.5 创建 Controller

处理用户登录请求的类是 StudentController,代码如下。

```java
1.  package com.ssm.controller;
2.
3.  import org.springframework.beans.factory.annotation.Autowired;
4.  import org.springframework.stereotype.Controller;
```

```
5.    import org.springframework.ui.Model;
6.    import org.springframework.web.bind.annotation.RequestMapping;
7.    import com.ssm.po.Student;
8.    import com.ssm.service.StudentService;
9.
10.   @Controller
11.   public class StudentController {
12.
13.       @Autowired
14.       StudentService studentService;
15.       @RequestMapping(value="/toLogin")
16.       public String toLogin() {
17.           return "login";
18.       }
19.       @RequestMapping(value="login")
20.       public String login(String loginname, String password, Model model) {
21.
22.           Student student = studentService. findStudentByLoginnameAndPassword
              (loginname, password);
23.           String msg=null;
24.           if(student!=null)
25.               msg="登录成功";
26.           else
27.               msg="登录失败";
28.           model.addAttribute("msg",msg);
29.           model.addAttribute("student",student);
30.           return "result";
31.       }
32.   }
```

第 13 行@Autowired 注解的属性 studentService 表示由 Spring 容器自动注入。这里的 StudentService 是一个接口，其实现类和实例的创建由 Spring 容器完成。

第 16 行 toLogin()方法用于转发到登录页面 WEB-INF/jsp/login.jsp。第 20 行的 login()方法用于处理登录请求。

第 22 行调用 studentService 的 findStudentByLoginnameAndPassword()方法，返回一个 Student 类的实例。如果用户名和密码正确，就会返回一个不为空的实例，否则为空，由此判断结果，决定 msg 变量中保存的内容，并将 msg 保存在 model 中，便于后面在 result.jsp 页面上显示此信息。最后，login()方法返回一个字符串"result"，表示转发到如下的 jsp 页面：WEB-INF/jsp/result.jsp。

9.1.6 创建 JSP 页面

首先创建用于登录的页面 login.jsp，页面中 form 标签内的代码如下。

```
1.  <form action="${pageContext.request.contextPath }/login"
2.  method="post" onsubmit="return check()">
3.      <br /><br />
4.  账 号：<input id="loginname" type="text" name="loginname" />
5.      <br /><br />
6.  密 码：<input id="password" type="password" name="password" />
7.      <br /><br />
8.  <center><input type="submit" value="登录" /></center>
9.  </form>
```

第4行和第6行文本框的 name 属性的值要与处理登录请求 login() 方法中的两个形参的名字一致，否则 login() 方法无法得到传递过来的参数。

其次，创建显示登录结果的页面 result.jsp，代码如下。

```
1.  <%@page language="java" contentType="text/html; charset=UTF-8"
2.      pageEncoding="UTF-8"%>
3.  <!DOCTYPE html>
4.  <html>
5.  <head>
6.  <meta charset="UTF-8">
7.  <title>登录成功</title>
8.  </head>
9.  <body>
10.     ${student.username}
11.     ${msg}
12. </body>
13. </html>
```

第10行和第11行通过 EL 表达式显示了保存在 model 对象中的两个属性的值。

9.1.7 运行程序

发布系统，启动 Tomcat 8.0，在地址栏中输入网址 http://localhost/chap9_1/toLogin，进入登录页面，输入正确的用户名和密码，显示登录成功，否则显示登录失败。

至此，已经成功对3个框架进行了整合。在上面的例子中，虽然只写了一个登录模块，但如果想添加其他的功能模块，在此基础上开发会变得非常简单、高效，这也是 SSM 框架的魅力所在。

9.2 基于 MyBatis 注解的整合开发

为了让读者掌握使用 MyBatis 注解与 Spring 整合开发的步骤，这里给出一个根据班级列表查询学生信息的示例。具体要求为

（1）在首页显示所有班级列表，列表中班级的名称有一个超链接。

（2）单击班级超链接进入显示这个班级学生的页面。

（3）在学生列表页面，每行记录的最后一项是一个删除的超链接，单击这个超链接，删除当前学生记录。

9.2.1 创建 Web 项目

创建一个 Web 项目 chap9_2，JAR 包与 chap9_1 相同。包的结构与 chap9_1 相同。

9.2.2 编写配置文件

此处配置文件与 chap_1 项目有所不同，由于项目中没有 XML 映射文件，也就无须配置 mybatis-config.xml 文件。在 applicationContext.xml 中将配置 mybaits-config.xml 文件部分去掉。

```xml
1.  <bean id="sqlSessionFactory"
2.      class="org.mybatis.spring.SqlSessionFactoryBean">
3.      <!--注入数据源 -->
4.      <property name="dataSource" ref="dataSource" />
5.      <!--指定核心配置文件位置 -->
6.      <property name="configLocation"
7.          value="classpath:mybatis-config.xml" />
8.  </bean>
```

这是原来的 Spring 配置文件 applicationContext.xml 中的部分代码，其中加黑部分可以删除。除此以外，其他的配置文件与 chap9_1 项目相同。

9.2.3 创建接口与注解

创建 tb_clazz 表的 Mapper 接口文件，代码如下。

```java
1.  package com.ssm.mapper;
2.
3.  import java.util.List;
4.
5.  import org.apache.ibatis.annotations.*;
6.  import org.apache.ibatis.mapping.FetchType;
7.  import com.ssm.po.Clazz;
8.
9.  public interface ClazzMapper {
10.
11.     //查询表中的所有记录
12.     @Select("select * from tb_clazz")
13.     List<Clazz>selectAllClazz();
14.     //根据 ID 查询班级及所有学生记录，一对多的关联
```

```
15.        @Select("select * from tb_clazz where id=#{id}")
16.        //tb_clazz 表中字段与 Clazz 类中的属性映射
17.        @Results({
18.            @Result(id=true,column="id",property="id"),
19.            @Result(column="cname",property="cname"),
20.            @Result(column="id",property="students",
21.                many=@Many(select="com.ssm.mapper.StudentMapper.select-
                   ByClazzId",
22.                fetchType=FetchType.EAGER))
23.        })
24.        Clazz selectClazzAndStudentsById(Integer id);
25.    }
```

第 12 行的注解功能是查询 tb_clazz 表中的所有记录，返回一个 List，其元素是 Clazz 类的对象。第 15 行是根据 tb_clazz 表的 id 查询班级信息，返回一个 Clazz 类的对象，对应的是一条 tb_clazz 表的记录。

第 17 行的 @Results 注解是一个映射结果集，其中有多个 result 元素，每个元素定义一个表中的字段和类中的属性映射。第 20 行映射的类的属性 students 是一 List<Student> 集合，此处用 @Many 注解完成了与 tb_student 表的 StudentMapper 中的 selectByClazzId 方法的映射。第 22 行将 fetchType 设置为 FetchType.EAGER，表示立即加载关联对象，即加载某班的所有学生信息。

创建 tb_student 表的 Mapper 接口文件，代码如下。

```
1.     package com.ssm.mapper;
2.     
3.     import java.util.List;
4.     import java.util.Map;
5.     import org.apache.ibatis.annotations.*;
6.     import org.apache.ibatis.mapping.FetchType;
7.     import com.ssm.po.Student;
8.     
9.     public interface StudentMapper {
10.    
11.        //根据班级 ID 查询所有学生
12.        @Select("select * from tb_student where clazz_id=#{id}")
13.        public List<Student> selectByClazzId(Integer id);
14.    
15.        //根据学生 ID 查询学生及学生所在班级，一对一查询
16.        @Select("select * from tb_student where id=#{id}")
17.        @Results({
18.            @Result(id=true,column="id",property="id"),
19.            @Result(column="loginname",property="loginname"),
20.            @Result(column="password",property="password"),
```

```java
21.          @Result(column="username",property="username"),
22.          @Result(column="clazz_id",property="clazz",
23.              one=@One(select="com.ssm.mapper.ClazzMapper.selectClazzById",
24.              fetchType=FetchType.EAGER))
25.      })
26.      public Student selectStudentById(Integer id);
27.
28.      @Select("select * from tb_student where id "
29.              +"in(select student_id from student_course where course_id=#{id})")
30.
31.      //通过type属性指定生成动态查询tb_student表的动态SQL类和方法
32.      @SelectProvider(type = StudentDynaSqlProvider.class, method = "selectWithParam")
33.      //动态查询
34.      public List<Student>selectStudentWithParam(Map<String,Object>Param);
35.      //通过type属性指定生成更新tb_student表的动态SQL类和方法
36.      @UpdateProvider(type=StudentDynaSqlProvider.class,method="updateStudent")
37.      //动态更新
38.      public int updateStudent(Student student);
39.          //插入学生
40.      @Insert("insert into tb_student(loignname,password,username,clazz_id) "
41.              +"values(#{loginname},#{password},#{username},#{clazz_id})")
42.      public int insertStudent(Student student);
43.          //根据ID删除学生记录
44.      @Delete("delete from tb_student where id=#{id}")
45.      public int delStudent(Integer id);
46.
47.  }
```

第12行应用@Select注解根据clazz_id查询tb_student表中的记录，返回值List<Student>，集合中存放当前班级中的学生信息。第16行是多对一的查询，也可以说是一对一的查询，查询学生及学生所在班级信息。第23行应用@One注解将Student中的clazz引用变量与ClazzMapper中的selectClazzById()方法进行映射。

第32～34行实现了动态SQL查询，其动态SQL语句写在@SelectProvider注解的type属性指定的StudentDynaSqlProvider类中，注解的method属性指定了动态SQL类中的具体方法名称。第36～38行实现了动态SQL更新的功能，其SQL语句写在与前面查询语句相同的StudentDynaSqlProvider类中，但处理的方法名为updateStudent()。

存放动态SQL的StudentDynaSqlProvider类代码如下。

```java
1.  package com.ssm.mapper;
2.
3.  import java.util.Map;
4.  import org.apache.ibatis.jdbc.SQL;
5.  import com.ssm.po.Student;
6.
7.  public class StudentDynaSqlProvider {
8.      //根据 param 参数生成动态查询 SQL 语句
9.      public String selectWithParam(Map<String,Object>param) {
10.         return new SQL() {
11.             {
12.                 SELECT("*");
13.                 FROM("tb_student");
14.                 if(param.get("id")!=null) {
15.                     WHERE("id=#{id}");
16.                 }
17.                 if(param.get("loginname")!=null) {
18.                     WHERE("loginname=#{loginname}");
19.                 }
20.                 if(param.get("username")!=null) {
21.                     WHERE("username=#{username}");
22.                 }
23.             }
24.         }.toString();
25.     }
26.     //根据 student 参数生成动态更新 SQL 语句
27.     public String updateStudent(Student student) {
28.         return new SQL() {
29.             {
30.                 UPDATE("tb_student");
31.                 if(student.getLoginname()!=null) {
32.                     SET("loginname=#{loginname}");
33.                 }
34.                 if(student.getPassword()!=null) {
35.                     SET("password=#{password}");
36.                 }
37.                 if(student.getUsername()!=null) {
38.                     SET("username=#{username}");
39.                 }
40.                 if(student.getClazz()!=null) {
41.                     SET("clazz=#{clazz}");
42.                 }
43.                 WHERE("id=#{id}");
```

```
44.            }
45.        }.toString();
46.    }
47.
48. }
```

第 9 行的方法功能是动态查询 tb_student 表,其查询条件通过 param 参数传递进来,这是一个 Map,其键和值对应的是字段的名字和要查询的值。第 12~21 行是根据 param 中的键值对动态组合成 SQL 语句,最后调用 SQL 类的 toString()方法返回一个 SQL 语句。

第 27 行定义了动态更新 tb_student 表的方法 updateStudent(), tb_student 表中是否更新某个字段的值取决于方法的形参 student 对象属性的值。第 30~41 行根据 student 中的属性值是否为空决定是否更新某个字段的值。这个方法的返回值与查询方法一样,调用 SQL 类的 toString()方法返回 SQL 语句。

9.2.4 创建 Service 及其实现类

创建 ClazzService 接口,代码如下。

```
1. package com.ssm.service;
2.
3. import java.util.List;
4. import com.ssm.po.Clazz;
5.
6. public interface ClazzService {
7.     //添加班级
8.     public int addClazz(Clazz clazz);
9.     //根据 ID 删除班级
10.    public int delClazz(int id);
11.    //查询 tb_clazz 表中的所有班级
12.    public List<Clazz>findAllClazz();
13.    //根据 clazz_id 查询 tb_clazz 表,返回的 clazz 对象中的 students 属性中存
       //放了所有学生的信息
14.    public Clazz findClazzAndStudentsById(Integer clazz_id);
15. }
```

创建 ClazzService 接口的实现类 ClazzServiceImpl,代码如下。

```
1. package com.ssm.service.impl;
2.
3. import java.util.List;
4. import org.springframework.beans.factory.annotation.Autowired;
5. import org.springframework.stereotype.Service;
```

```
6.    import org.springframework.transaction.annotation.Transactional;
7.    import com.ssm.mapper.ClazzMapper;
8.    import com.ssm.po.Clazz;
9.    import com.ssm.service.ClazzService;
10.
11.   @Service("clazzService")
12.   @Transactional
13.   public class ClazzServiceImpl implements ClazzService {
14.
15.       @Autowired
16.       private ClazzMapper clazzMapper;
17.
18.       @Override
19.       public int addClazz(Clazz clazz) {
20.           return clazzMapper.saveClazz(clazz);
21.       }
22.
23.       @Override
24.       public int delClazz(int id) {
25.           return clazzMapper.removeClazz(id);
26.       }
27.       //查询 tb_clazz 表中的所有班级
28.       @Override
29.       public List<Clazz> findAllClazz() {
30.           return clazzMapper.selectAllClazz();
31.       }
32.       //根据班级 ID 查询 Clazz 对象,对象中包含学生信息
33.       @Override
34.       public Clazz findClazzAndStudentsById(Integer clazz_id) {
35.           return clazzMapper.selectClazzAndStudentsById(clazz_id);
36.       }
37.   }
```

第 11 行的 @Service("clazzService") 注解表示这是 Service 层，同时相当于在 applicationContext.xml 中定义了一个 id 为 "clazzService" 的 Bean。第 15 行的 @Autowired 注解表示 clazzMapper 对象由 IoC 容器自动注入。

创建 StudentService 接口，代码如下。

```
1.    package com.ssm.service;
2.
3.    import java.util.List;
4.    import com.ssm.po.Student;
5.
6.    public interface StudentService {
```

```
7.      //根据ID查询学生
8.      public Student findStudentById(Integer id);
9.      //根据班级clazz_id查询学生
10.     public List<Student> findStudentsByClazzId(Integer clazz_id);
11.     //添加学生
12.     public int addStudent(Student student);
13.     //删除学生
14.     public int removeStudent(Integer id);
15. }
```

创建StudentService的实现类StudentServiceImpl,代码如下。

```
1.  package com.ssm.service.impl;
2.
3.  import java.util.List;
4.
5.  import com.ssm.mapper.StudentMapper;
6.  import com.ssm.po.Student;
7.  import com.ssm.service.StudentService;
8.
9.  @Service("studentService")
10. public class StudentServiceImpl implements StudentService {
11.
12.     @Autowired
13.     private StudentMapper studentMapper;
14.
15.     @Override
16.     public Student findStudentById(Integer id) {
17.         return studentMapper.selectStudentById(id);
18.
19.     }
20.
21.     @Override
22.     public List<Student> findStudentsByClazzId(Integer clazz_id) {
23.         return studentMapper.selectByClazzId(clazz_id);
24.
25.     }
26.
27.     @Override
28.     public int addStudent(Student student) {
29.         return studentMapper.insertstudent(student);
30.
31.     }
```

```
32.
33.        @Override
34.        public int removeStudent(Integer id) {
35.            return studentMapper.delStudent(id);
36.        }
37.    }
```

第 9 行的@Service("studentService")注解表示这是 Service 层,同时定义了一个 id 为"studentService"的 Bean。第 12 行@Autowired 注解注释的属性 studentMapper,表示由 IoC 容器自动注入。

9.2.5 创建 Controller

创建与班级处理请求相关的控制器类 ClazzController,代码如下。

```
1.  package com.ssm.controller;
2.
3.  @Controller
4.  @RequestMapping("/clazz")
5.  public class ClazzController {
6.
7.      @Autowired
8.      ClazzService clazzService;
9.      //得到所有班级,并转发 showClazz.jsp 页面显示
10.     @RequestMapping("/showClazz")
11.     public ModelAndView showAllClazz(ModelAndView mv) {
12.
13.         List<Clazz>clazzs=clazzService.findAllClazz();
14.     //将 clazzs 保存在 mv 中,在 showClazz.jsp 页面中用于显示班级信息
15.         mv.addObject("clazzs",clazzs);
16.     //设置转发的视图名,这里为 showClazz.jsp 页面
17.         mv.setViewName("showClazz");
18.
19.         return mv;
20.     }
21.     //根据班级 ID 查询 clazz 对象,对象中包含所有学生的信息 students
22.     @RequestMapping("/showStudent")
23.     public ModelAndView showStudentsByClazzId(Integer id,ModelAndView mv) {
24.         Clazz clazz=clazzService.findClazzAndStudentsById(id);
25.     //将 clazz 对象保存在 mv 对象中
26.         mv.addObject("clazz",clazz);
27.     //设置转发的页面为 showStudent.jsp
```

```
28.         mv.setViewName("showStudent");
29.         return mv;
30.     }
31. }
```

第 8 行的 clazzService 属性由 IoC 容器自动注入。ClazzController 类中定义了两个方法：一个是用于获取所有班级的 showAll()方法；另一个是根据班级 id 查询班级，返回一个 clazz 对象。这个 clazz 对象中的 students 属性存放所有学生的信息。

创建与学生处理请求相关的控制器类 StudentController，代码如下。

```
1.  package com.ssm.controller;
2.
3.  import com.ssm.service.StudentService;
4.
5.  @Controller
6.  @RequestMapping("/student")
7.  public class StudentController {
8.
9.      @Autowired
10.     private StudentService studentService;
11.     //处理删除学生记录的请求
12.     @RequestMapping("/removeStudent")
13.     public String removeStudent(Integer id, Integer clazz_id, Model model) {
14.         int n=studentService.removeStudent(id);
15.
16.         if(n>0) {
17.             model.addAttribute("id",clazz_id);
18.             return "redirect:/clazz/showStudent";
19.         }
20.         else
21.         {
22.             model.addAttribute("msg","操作失败");
23.             return "result";
24.         }
25.     }
26. }
```

第 10 行的 studentService 属性由 IoC 容器自动注入。StudentController 类中只定义了一个删除方法，读者可试着定义其他的处理请求方法。

9.2.6 创建 JSP 页面

创建 showClazz.jsp 页面，显示 tb_clazz 表中所有班级的信息，代码如下。

```
1.  <body>
2.  <c:set var="ctx" value="${pageContext.request.contextPath}"/>
3.      <!--数据展示区 -->
4.  <table width="50% " border="1" cellpadding="5" cellspacing="0" style=
    "border:#c2c6cc 1px solid; border-collapse:collapse;">
5.      <tr class="main_trbg_tit" align="center">
6.          <td>班级名</td>
7.          <td align="center">操作</td>
8.      </tr>
9.  <c:forEach items="${requestScope.clazzs}" var="clazz" varStatus="stat">
10.     <tr id="data_${stat.index}" align="center" class="main_trbg" >
11.             <td>${clazz.cname} </td>
12.             <td align="center" width="40px;">
13.     <a href="${ctx}/clazz/showStudent?id=${clazz.id}">查询</a>
14.             </td>
15.         </tr>
16. </c:forEach>
17. </table>
18. </body>
```

第 9 行应用 JSTL 的<foreach>循环标签输出所有班级信息。items 的属性是一个集合，此处的值是 clazzs，集合中存放的是 Clazz 的对象，也就是 tb_clazz 表中的所有记录。var="clazz"相当于一个循环变量，代表集合的当前元素。第 11 行的 ${clazz}是 EL 表达式，用于输出 clazz 变量中的值。

第 13 行是一个超链接，指向 clazz/showStudent 映射路径，实际提交给了 ClazzController 中的 showStudent()方法，调用这个方法时传递了班级的 id 值，根据班级 id 查询对应班级的所有学生信息。

创建 showStudent.jsp 页面，显示某个班级所有学生的信息，代码如下。

```
1.  <c:forEach items="${requestScope.clazz.students}" var="stu" varStatus=
    "stat">
2.      <tr id="data_${stat.index}" align="center" class="main_trbg" >
            <td>${stu.loginname} </td>
3.          <td>${stu.password}   </td>
4.          <td>${stu.username}   </td>
5.          <td align="center" width="40px;">
6.      <a href="${ctx}/student/removeStudent?id=${stu.id}&clazz_id=
    ${clazz.id}">删除</a>
7.          </td>
8.      </tr>
9.  </c:forEach>
```

第 1 行应用<c:forEach>循环标签输出 Clazz 对象中的 Students 集合属性的值。

Students 中存放的是 Student 对象的集合,对应 tb_student 表中的记录。第 6 行的超链接提交给 student/removeStudent 映射路径,对应的是 StudentController 中的 removeStudent()方法,执行删除 tb_student 表中记录的操作。

9.2.7 运行程序

发布服务,启动 Tomcat 服务器,在浏览器地址栏中输入网址 http://localhost/chap9_2/clazz/showClazz。浏览器中显示如图 9-2 所示的页面。

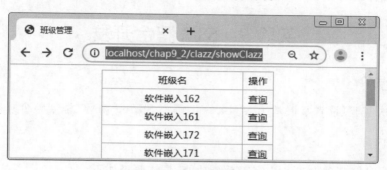

图 9-2 显示班级列表

图 9-2 中显示了 tb_clazz 表中所有班级的信息,操作列对应的查询操作是一个超链接,单击查询链接,会跳转到一个 showStudent.jsp 页面,显示要查询班级的所有学生信息。单击第 1 行的查询链接,浏览器中会显示如图 9-3 所示的画面。

图 9-3 显示学生信息

从图 9-3 中可以看到,软件嵌入 162 班所有学生的信息已经显示在页面上。单击"删除"超链接时,会执行删除当前学生的操作。至此,我们完成了基于 MyBatis 注解与 Spring 及 Spring MVC 的整合开发。

习　题

1. 在 SSM 框架整合过程中如何在 applicationContext.xml 中进行事务的配置？
2. 在 Spring MVC 中如何配置访问资源为 *.do 的格式？
3. 在 SSM 框架整合过程中，哪些内容可以从 mybatis-config.xml 文件移到 Spring 的配置文件 applicationContext.xml 中？

实验 9　SSM 整合开发

1. 实验目的

通过本次实验，读者掌握基于 MyBatis 注解与 Spring MVC 框架整合开发的方法。

2. 实验内容

在 9.2 节案例基础上进一步完善系统功能，具体要求如下。

（1）从 Web 页面添加学生信息，要求从列表框中选择班级。

（2）在 Web 页面给出两个文本框，可以根据 tb_student 表中的 loginname 字段和 username 字段中的任意字段进行组合查询，要求用动态 SQL 实现。

（3）在 Web 页面显示某一个学生的信息，可以更新 tb_student 表中 loginname、password 和 username 中的任意一个字段的值，要求用动态 SQL。

3. 实现思路及步骤

参照 9.2 节的步骤。

第 10 章 项目案例：作业管理系统

本章目标
1. 作业管理系统功能介绍。
2. 系统设计。
3. 系统环境搭建。
4. 功能模块实现。
5. 发布运行系统。

10.1 系统简介

作业管理系统是一个帮助教师管理学生作业或上机作业的系统，可支持多个教师对多门课程和多个班级布置作业。学生上交作业可在网页上直接提交，也可以文件形式提交。教师可在网页上直接批阅作业，对学生作业中的问题给出评语和评分。当作业为已批阅状态，在学生端可显示学生作业题的参考答案。本系统的另外一个功能是可以对学生提交的作业进行查重，这个功能主要是针对学生上机作业（简答题）。查重的范围可以自定义，是以班级为单位，可以是一个班级，也可以是多个班级。查重的结果是给出与某一个学生的作业的相似度超过一定程度的所有学生的姓名。如果想自己编写查重算法，只重写 ssh.homework.service.impl 包中 SimilarityImpl 类的 doSimilarity() 方法即可。

10.1.1 系统用例图

系统一共有 3 个角色：管理员、教师和学生，角色与功能之间的关系可以用下面的用例图描述，如图 10-1 描述了管理员、教师对应的功能。图 10-2 是学生角色与系统功能之间的关系。

10.1.2 系统功能框图

图 10-3 所示是教师管理功能框图，系统一级模块一共有 8 个，二级模块有 17 个。

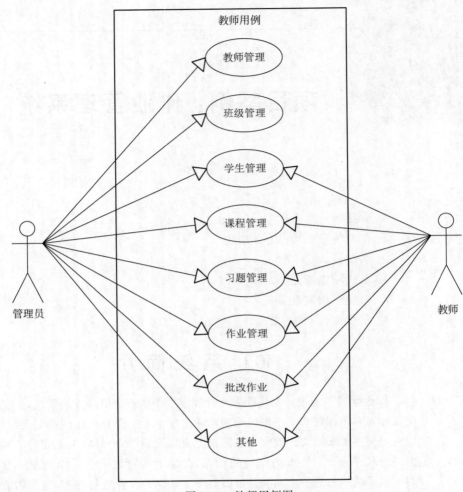

图 10-1　教师用例图

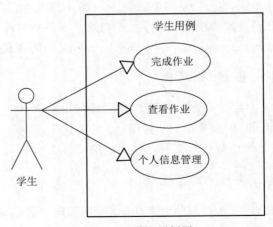

图 10-2　学生用例图

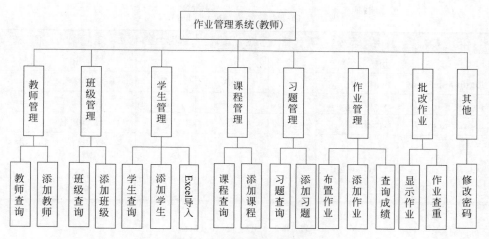

图 10-3　教师管理功能框图

学生作业管理功能框图如图 10-4 所示，一级模块有 3 个。完成作业模块部分有两个二级模块，分别为修改作业和上传文件，上传文件的功能是可选的。有时在页面上能够完成的作业不必上传文件，如果在页面上不能完成，如程序设计的代码有很多个文件，这时可将文件打包后启用上传文件功能，将学生的整个项目上传，这个功能要求必须在页面上完成作业的同时上传文件，也就是上传文件时完成作业的页面内容不能为空。

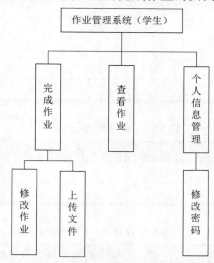

图 10-4　学生作业管理功能框图

10.2　系统设计

10.2.1　数据库设计

作业管理系统一共有 8 张表（表 10-1～表 10-8），各张表的结构如下。

表 10-1 教师表（tb_teacher）

字 段 名	类 型	长度	是否主键	是否为空	说 明
id	int	11	是	否	教师 id
loginname	varchar	18	否	否	登录名
password	varchar	18	否	否	密码
username	varchar	18	否	否	教师姓名
role	varchar	1	否	否	权限（值为 1,2,3） 1. 管理员 2. 教师 3. 学生

表 10-2 班级表（tb_clazz）

字 段 名	类 型	长度	是否主键	是否为空	说 明
id	int	11	是	否	班级 id
cname	varchar	20	否	否	班级名称

表 10-3 学生表（tb_student）

字 段 名	类 型	长度	是否主键	是否为空	说 明
id	int	11	是	否	学生 id
loginname	varchar	20	否	否	登录名
password	varchar	20	否	否	密码
username	varchar	20	否	否	学生姓名
clazz_id	int	11	否	否	班级外键

表 10-4 课程表（tb_course）

字 段 名	类 型	长度	是否主键	是否为空	说 明
id	int	11	是	否	课程 id
cname	varchar	40	否	否	课程名称

表 10-5 题库表（tb_exercise）

字 段 名	类 型	长度	是否主键	是否为空	说 明
id	int	11	是	否	习题 id
kind	varchar	1	否	否	习题类型 1. 选择题 2. 填空题 3. 问答题
content	varchar	3000	否	否	题目内容
answer	varchar	6000	否	是	习题答案
chapter	varchar	2	否	否	习题所在章节

续表

字段名	类型	长度	是否主键	是否为空	说明
course_id	int	11	否	否	课程表的外键
teacher_id	int	11	否	否	教师表的外键

表 10-6 作业表（tb_workbook）

字段名	类型	长度	是否主键	是否为空	说明
id	int	11	是	否	作业 id
clazz_id	int	11	否	否	班级 id 外键
teacher_id	int	11	否	否	教师 id 外键
course_id	int	11	否	否	课程 id 外键
title	varchar	30	否	否	作业标题
wflag	varchar	1	否	否	作业是否发布， 0. 不发布 1. 发布 2. 已批改 默认为 0
term	varchar	15	否	否	第几学期
createDate	timestamp	4	否	否	创建日期
fileName	varchar	50	否	是	教师布置作业上传文件的路径

表 10-7 发布作业表（tb_assignment）

字段名	类型	长度	是否主键	是否为空	说明
id	int	11	是	否	发布作业 id
workbook_id	int	11	否	否	作业 id 外键
exercise_id	int	11	否	否	题库 id 外键
grade	int	11	否	是	作业题的分值

表 10-8 学生作业表（student_workbook）

字段名	类型	长度	是否主键	是否为空	说明
id	int	11	是	否	作业 id
workbook_id	int	11	否	否	作业 id 外键
student_id	int	11	否	否	学生 id 外键
exercise_id	int	11	否	否	题库 id 外键
studentAnswer	varchar	6000	否	是	学生作业答案

续表

字 段 名	类 型	长度	是否主键	是否为空	说　明
grade	float		否	是	每个习题的分值
score	float		否	是	学生得分
notes			否	否	教师批注
rate	int	11	否	是	最高的查重率
studentRate	varchar	200	否	是	存放多个学生的查重率，中间用","分开
instructions	varchar	200	否	是	与 studentRate 查重率对应的学生的 id，中间用","分开
fileName	varchar	100	否	是	学生上传作业文件的路径

10.2.2　实体类的设计

现在设计与 8 个表对应的实体类及类之间的关联关系，此处用 UML 类图表示，如图 10-5 所示。

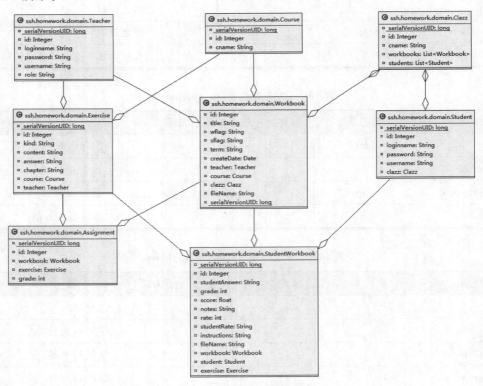

图 10-5　实体类之间的关系

图 10-5 中关联最多的两个类是 Workbook 和 StudentWorkbook。作业 Workbook 类是与 tb_workbook 表对应的实体类,它分别与教师 Teacher、课程 Course 和班级 Clazz 3 个类关联,是多对一的关联,这正好是 tb_workbook 与 tb_teacher、tb_course、tb_clazz 3 表之间关系在类的映射上的体现。

StudentWorkbook 类与题库 Exercise、作业 Workbook、学生 Student 3 个类之间的关系也是多对一的关联,这正好是 student_workbook 与 tb_exercise、tb_workbook、tb_student 3 个表之间的关系在类的映射上的具体体现。

同样,Assignment 类与 Exercise、Workbook 两个类之间是多对一的关系,与它们对应表之间的关系是一致的。

具体实体类的设计如下。

与班级表 tb_clazz 对应的实体类 Clazz 代码如下。

```java
1.   package ssh.homework.domain;
2.
3.    import java.io.Serializable;
4.   import java.util.List;
5.   //与 tb_clazz 表对应的实体类
6.   public class Clazz implements Serializable {
7.
8.       private static final long serialVersionUID=1L;
9.       private Integer id;              //班级 id
10.      private String cname;            //班级名称
11.      //与班级关联的作业列表,这是一对多的映射
12.      private List<Workbook> workbooks;
13.      //与班级关联的学生列表,这是一对多的映射
14.      private List<Student> students;
15.      //setter and getter 方法
16.  }
```

与学生表 tb_student 对应的实体类 Student 代码如下。

```java
1.   package ssh.homework.domain;
2.
3.   import java.io.Serializable;
4.
5.   public class Student implements Serializable {
6.
7.       private static final long serialVersionUID=1L;
8.       private Integer id;                //学生 id
9.       private String loginname;          //学生登录用户名
10.      private String password;           //密码
11.      private String username;           //学生姓名
12.      private Clazz clazz;               //学生所在班级,这是多对一的关联所需属性
```

```
13.    //setter and getter 方法
14. }
```

与教师表 tb_teacher 对应的实体类 Teacher 代码如下。

```
1.  package ssh.homework.domain;
2.
3.  import java.io.Serializable;
4.
5.  public class Teacher implements Serializable{
6.
7.      private static final long serialVersionUID=1L;
8.      private Integer id;              //教师 id
9.      private String loginname;        //教师登录名
10.     private String password;         //密码
11.     private String username;         //教师姓名
12.     private String role;             //教师角色
13.     //setter and getter 方法
14. }
```

与题库表 tb_exercise 对应的实体类 Exercise 代码如下。

```
1.  package ssh.homework.domain;
2.
3.  import java.io.Serializable;
4.
5.  public class Exercise implements Serializable {
6.
7.      private static final long serialVersionUID=1L;
8.      private Integer id;              //习题 id
9.      private String kind;             //题型
10.     private String content;          //习题内容
11.     private String answer;           //习题答案
12.     private String chapter;          //章节
13.     private Course course;           //习题所在的课程,这是多对一的关联
14.     private Teacher teacher;         //出此习题的教师,这是多对一的关联
15. //setter and getter 方法
16. }
```

与课程表 tb_course 对应的实体类 Course 代码如下。

```
1.  package ssh.homework.domain;
2.
3.  import java.io.Serializable;
4.
```

```
5.    public class Course implements Serializable{
6.
7.        private static final long serialVersionUID=1L;
8.        private Integer id;           //课程 id
9.        private String cname;         //课程名称
10.   //setter and getter 方法
11.   }
```

与作业表 tb_workbook 对应的实体类 Workbook 代码如下。

```
1.    package ssh.homework.domain;
2.
3.    import java.io.Serializable;
4.    import java.util.Date;
5.    public class Workbook implements Serializable{
6.
7.        private static final long serialVersionUID=1L;
8.        private Integer id;              //作业 id
9.        private String title;            //作业标题
10.       private String wflag;            //是否发布作业
11.       private String sflag;            //作业是否已批改
12.       private String term;             //第几学期的第几次作业
13.       private Date createDate;         //创建日期
14.       private Teacher teacher;         //布置作业的教师,是多对一的关联
15.       private Course course;           //作业所在课程,是多对一的关联
16.       private Clazz clazz;             //与作业关联的班级,是多对一的关联
17.       private String fileName;         //教师发布给学生文件的路径
18.   //setter and getter 方法
19.   }
```

与发布作业表 tb_assignment 对应的实体类 Assignment 代码如下。

```
1.    package ssh.homework.domain;
2.
3.    import java.io.Serializable;
4.
5.    public class Assignment implements Serializable {
6.
7.        private static final long serialVersionUID=1L;
8.        private Integer id;                    //发布作业 id
9.        private Workbook workbook;             //与发布作业 id 对应的作业对象
10.       private Exercise exercise;             //与发布作业 id 对应的题库对象
11.       private int grade;                     //当前作业题的分值
```

```
12.         //setter and getter 方法
13.    }
```

与学生作业表 student_workbook 对应的实体类 StudentWorkbook 代码如下。

```
1.   package ssh.homework.domain;
2.
3.   import java.io.Serializable;
4.
5.   public class StudentWorkbook implements Serializable {
6.
7.       private static final long serialVersionUID=1L;
8.       private Integer id;                     //学生作业的 id
9.       private String studentAnswer;           //学生当前习题作业的答案
10.      private int grade;                      //每个题的分值
11.      private float score;                    //学生当前习题得分
12.      private String notes;                   //教师批注
13.      private int rate;                       //设置的查重率
14.      //如果学生与其他同学有超标情况,此变量存放超标的查重率,中间用","分开
15.      private String studentRate;
16.      //与 studentRate 对应的学生的 ID,中间用","分开
17.      private String instructions;
18.      private String fileName;                //学生上传的作业文件路径
19.      private Workbook workbook;              //当前习题对应的作业对象
20.      private Student student;                //当前学生
21.      private Exercise exercise;              //当前作业中的习题
22.         //setter and getter 方法
23.   }
```

以上就是与 8 个表对应的 8 个实体类,后面的代码中将会经常用到这些实体类,通过实体类封装数据,并在不同的层之间进行数据的传递。

10.2.3 系统结构设计

本系统采用多层体系结构设计,主要分为以下 4 个层次。

- 持久层:该层由多个实体类组成,10.2.2 节介绍的就是这部分内容。
- 数据访问层(DAO 层):该层由多个 DAO 接口和 MyBatis 映射文件组成,本系统采用了注解方式,不再需要映射文件。接口类的名称统一以 Dao 结尾。
- 业务逻辑层(Service 层):该层由多个 Service 接口和实现类组成。在本系统中,业务逻辑层的接口统一用 Service 结尾,其实现类名称统一在接口名后加 Impl。该层主要用于实现系统的业务逻辑。
- Web 表现层:该层主要包括 Spring MVC 中的 Controller 类和 JSP 页面。Controller 类主要负责拦截用户请求,并调用业务逻辑层中相应组件的业务逻辑

方法处理用户请求,然后将相应的结果返回给JSP页面。

为了让读者进一步了解系统的设计,掌握这种多层体系结构的设计方法,在功能模块实现部分我们给出了具体功能模块的类图设计。

10.3 系统环境的搭建

10.3.1 所需JAR包

由于本系统使用了SSM框架开发,因此需要准备三大框架的JAR包,除此之外,项目中还涉及数据库连接、数据库连接池、Excel文件操作、文件上传与下载和JSTL标签库等,所以还有其他一些JAR包。具体内容如下。

```
ant-1.9.6.jar
ant-launcher-1.9.6.jar
asm-5.2.jar
aspectjrt.jar
aspectjtools.jar
aspectjweaver.jar
c3p0-0.9.5.2.jar
cglib-3.2.5.jar
commons-codec-1.5.jar
commons-fileupload-1.3.1.jar
commons-fileupload-1.3.3.jar
commons-io-2.2.jar
commons-io-2.6.jar
commons-logging-1.2.jar
dom4j-1.6.1.jar
hibernate-c3p0-5.2.10.Final.jar
javassist-3.22.0-CR2.jar
javax.servlet.jsp.jstl-1.2.1.jar
javax.servlet.jsp.jstl-api-1.2.1.jar
log4j-1.2.17.jar
log4j-api-2.3.jar
log4j-core-2.3.jar
mchange-commons-java-0.2.11.jar
mybatis-3.4.5.jar
mybatis-spring-1.3.1.jar
mysql-connector-java-5.1.44-bin.jar
ognl-3.1.15.jar
org.aspectj.matcher.jar
poi-3.9.jar
poi-ooxml-3.9.jar
poi-ooxml-schemas-3.9.jar
```

```
slf4j-api-1.7.25.jar
slf4j-log4j12-1.7.25.jar
spring-aop-5.0.1.RELEASE.jar
spring-aspects-5.0.1.RELEASE.jar
spring-beans-5.0.1.RELEASE.jar
spring-context-5.0.1.RELEASE.jar
spring-context-indexer-5.0.1.RELEASE.jar
spring-context-support-5.0.1.RELEASE.jar
spring-core-5.0.1.RELEASE.jar
spring-expression-5.0.1.RELEASE.jar
spring-instrument-5.0.1.RELEASE.jar
spring-jcl-5.0.1.RELEASE.jar
spring-jdbc-5.0.1.RELEASE.jar
spring-jms-5.0.1.RELEASE.jar
spring-messaging-5.0.1.RELEASE.jar
spring-orm-5.0.1.RELEASE.jar
spring-oxm-5.0.1.RELEASE.jar
spring-test-5.0.1.RELEASE.jar
spring-tx-5.0.1.RELEASE.jar
spring-web-5.0.1.RELEASE.jar
spring-webflux-5.0.1.RELEASE.jar
spring-webmvc-5.0.1.RELEASE.jar
spring-websocket-5.0.1.RELEASE.jar
stax-api-1.0.1.jar
xml-apis-1.0.b2.jar
xmlbeans-2.3.0.jar
```

10.3.2 创建数据库

在 MySQL 中创建数据库 homework，在数据库管理程序 navicat 8.0 中，或者在 MySQL 的命令行状态运行数据库脚本，生成完整的数据库和表。数据库脚本如下所示。

```
------------------------------
--student_workbook 表的结构
------------------------------
DROP TABLE IF EXISTS `student_workbook`;
CREATE TABLE `student_workbook` (
  `id` int(11) NOT NULL AUTO_INCREMENT,
  `workbook_id` int(11) NOT NULL,
  `student_id` int(11) NOT NULL,
  `exercise_id` int(11) NOT NULL,
  `studentAnswer` varchar(6000) DEFAULT NULL COMMENT '学生作业答案',
  `grade` int(11) DEFAULT NULL COMMENT '每个题的分值',
  `score` float DEFAULT NULL COMMENT '本题的得分',
```

```
  `notes` varchar(200) DEFAULT NULL COMMENT '教师批注',
  `rate` int(11) DEFAULT NULL COMMENT '最高的查重率',
  `studentRate` varchar(200) DEFAULT NULL COMMENT '存放多个学生的查重率,中间用","分开',
  `instructions` varchar(200) DEFAULT NULL COMMENT '与rate对应的学生的ID,中间用","分开',
  `fileName` varchar(100) DEFAULT NULL COMMENT '学生上传作业文件',
  PRIMARY KEY (`id`),
  KEY `exercise_id` (`exercise_id`),
  KEY `student_id` (`student_id`),
  KEY `workbook_id` (`workbook_id`),
  CONSTRAINT `student_workbook_ibfk_3` FOREIGN KEY (`exercise_id`) REFERENCES `tb_exercise` (`id`),
  CONSTRAINT `student_workbook_ibfk_6` FOREIGN KEY (`student_id`) REFERENCES `tb_student` (`id`),
  CONSTRAINT `student_workbook_ibfk_7` FOREIGN KEY (`workbook_id`) REFERENCES `tb_workbook` (`id`) ON DELETE CASCADE
) ENGINE=InnoDB AUTO_INCREMENT=4356 DEFAULT CHARSET=utf8;

-------------------------------
--tb_assignment 表的结构
-------------------------------
DROP TABLE IF EXISTS `tb_assignment`;
CREATE TABLE `tb_assignment` (
  `id` int(11) NOT NULL AUTO_INCREMENT,
  `workbook_id` int(11) NOT NULL,
  `exercise_id` int(11) NOT NULL,
  `grade` int(11) DEFAULT NULL,
  PRIMARY KEY (`id`),
  KEY `exercise_id` (`exercise_id`),
  KEY `workbook_id` (`workbook_id`),
  CONSTRAINT `tb_assignment_ibfk_1` FOREIGN KEY (`exercise_id`) REFERENCES `tb_exercise` (`id`),
  CONSTRAINT `tb_assignment_ibfk_3` FOREIGN KEY (`workbook_id`) REFERENCES `tb_workbook` (`id`) ON DELETE CASCADE
) ENGINE=InnoDB AUTO_INCREMENT=200 DEFAULT CHARSET=utf8;

-------------------------------
--tb_clazz 表的结构
-------------------------------
DROP TABLE IF EXISTS `tb_clazz`;
CREATE TABLE `tb_clazz` (
  `id` int(11) NOT NULL AUTO_INCREMENT,
  `cname` varchar(20) NOT NULL,
```

```sql
  PRIMARY KEY (`id`)
) ENGINE=InnoDB AUTO_INCREMENT=24 DEFAULT CHARSET=utf8;
```

-- **tb_course** 表的结构

```sql
DROP TABLE IF EXISTS `tb_course`;
CREATE TABLE `tb_course` (
  `id` int(11) NOT NULL AUTO_INCREMENT,
  `cname` varchar(40) NOT NULL,
  PRIMARY KEY (`id`)
) ENGINE=InnoDB AUTO_INCREMENT=12 DEFAULT CHARSET=utf8;
```

-- **tb_exercise** 表的结构

```sql
DROP TABLE IF EXISTS `tb_exercise`;
CREATE TABLE `tb_exercise` (
  `id` int(11) NOT NULL AUTO_INCREMENT,
  `kind` varchar(1) NOT NULL COMMENT '题的类型：1为选择题,2为填空题,3为问答题',
  `content` varchar(3000) NOT NULL COMMENT '题目内容',
  `answer` varchar(6000) DEFAULT NULL COMMENT '题目答案',
  `chapter` varchar(2) NOT NULL COMMENT '章节',
  `course_id` int(11) NOT NULL COMMENT '课程表关联外键',
  `tea_id` int(11) NOT NULL COMMENT 'tb_teacher 表的外键',
  PRIMARY KEY (`id`),
  KEY `tea_id` (`tea_id`),
  KEY `tb_exercise_couse` (`course_id`),
  CONSTRAINT `tb_exercise_couse` FOREIGN KEY (`course_id`) REFERENCES `tb_course` (`id`),
  CONSTRAINT `tb_exercise_ibfk_1` FOREIGN KEY (`tea_id`) REFERENCES `tb_teacher` (`id`)
) ENGINE=InnoDB AUTO_INCREMENT=132 DEFAULT CHARSET=utf8;
```

-- **tb_student** 表的结构

```sql
DROP TABLE IF EXISTS `tb_student`;
CREATE TABLE `tb_student` (
  `id` int(11) NOT NULL AUTO_INCREMENT,
  `loginname` varchar(20) NOT NULL,
  `password` varchar(20) NOT NULL,
  `username` varchar(255) NOT NULL,
  `clazz_id` int(11) NOT NULL,
```

```sql
  PRIMARY KEY (`id`),
  KEY `clazz_id` (`clazz_id`),
  CONSTRAINT `tb_student_ibfk_1` FOREIGN KEY (`clazz_id`) REFERENCES `tb_clazz` (`id`)
) ENGINE=InnoDB AUTO_INCREMENT=529 DEFAULT CHARSET=utf8;
```

-- **tb_teacher** 表的结构

```sql
DROP TABLE IF EXISTS `tb_teacher`;
CREATE TABLE `tb_teacher` (
  `loginname` varchar(18) NOT NULL,
  `id` int(11) NOT NULL AUTO_INCREMENT,
  `password` varchar(18) NOT NULL,
  `username` varchar(18) DEFAULT NULL,
  `role` varchar(1) DEFAULT '2' COMMENT '1为系统管理员,2为普通教师',
  PRIMARY KEY (`id`)
) ENGINE=InnoDB AUTO_INCREMENT=15 DEFAULT CHARSET=utf8;
```

-- **tb_workbook** 表的结构

```sql
DROP TABLE IF EXISTS `tb_workbook`;
CREATE TABLE `tb_workbook` (
  `id` int(11) NOT NULL AUTO_INCREMENT,
  `teacher_id` int(11) NOT NULL COMMENT '与teacher表关联的外键',
  `course_id` int(11) NOT NULL COMMENT '与course表关联的外键',
  `title` varchar(30) NOT NULL COMMENT '作业名称',
  `clazz_id` int(11) NOT NULL COMMENT '与clazz表关联的外键',
  `wflag` varchar(1) NOT NULL DEFAULT '0' COMMENT '表示作业是否发布,0为不发布,1为发布,2为已批改',
  `term` varchar(15) NOT NULL COMMENT '表示第几学期',
  `createdate` timestamp NOT NULL DEFAULT CURRENT_TIMESTAMP ON UPDATE CURRENT_TIMESTAMP COMMENT '创建日期',
  `fileName` varchar(50) DEFAULT NULL COMMENT '教师布置作业上传文件',
  PRIMARY KEY (`id`),
  KEY `clazz_id` (`clazz_id`),
  KEY `teacher_id` (`teacher_id`),
  KEY `course_id` (`course_id`),
  CONSTRAINT `tb_workbook_ibfk_1` FOREIGN KEY (`clazz_id`) REFERENCES `tb_clazz` (`id`),
  CONSTRAINT `tb_workbook_ibfk_2` FOREIGN KEY (`teacher_id`) REFERENCES `tb_teacher` (`id`),
  CONSTRAINT `tb_workbook_ibfk_3` FOREIGN KEY (`course_id`) REFERENCES `tb_
```

```
course` (`id`)
) ENGINE=InnoDB AUTO_INCREMENT=39 DEFAULT CHARSET=utf8;
```

10.3.3 创建 Web 项目

（1）创建名为 homework 的 Web 项目，添加所需的 JAR 包。

（2）规划包和目录结构，项目包的结构和项目目录的结构分别如图 10-6 和图 10-7 所示。

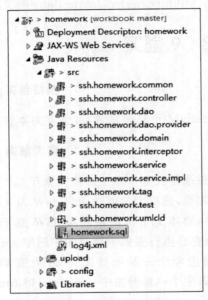

图 10-6　项目包的结构　　　　图 10-7　项目目录的结构

（3）创建配置文件。

所有的配置文件都存放在 config 文件夹中，具体内容如下。

① 创建存放数据库信息的属性文件 db.properties，代码如下。

```
1.  dataSource.driverClass=com.mysql.jdbc.Driver
2.  dataSource.jdbcUrl=jdbc:mysql://127.0.0.1:3306/homework?useUnicode=
    true&characterEncoding=utf-8&useSSL=false
3.  dataSource.user=root
4.  dataSource.password=root
5.  dataSource.maxPoolSize=20
6.  dataSource.maxIdleTime=1000
7.  dataSource.minPoolSize=6
8.  dataSource.initialPoolSize=5
```

② Spring 配置文件 applicationContext.xml，代码如下。

```
1.  <?xml version="1.0" encoding="UTF-8"?>
2.  <beans xmlns="http://www.springframework.org/schema/beans"
```

```
3.      xmlns:mybatis="http://mybatis.org/schema/mybatis-spring"
4.      xmlns:xsi="http://www.w3.org/2001/XMLSchema-instance"
5.      xmlns:p="http://www.springframework.org/schema/p"
6.      xmlns:context="http://www.springframework.org/schema/context"
7.      xmlns:aop="http://www.springframework.org/schema/aop"
8.      xmlns:mvc="http://www.springframework.org/schema/mvc"
9.      xmlns:tx="http://www.springframework.org/schema/tx"
10.     xsi:schemaLocation="http://www.springframework.org/schema/beans
11.                         http://www.springframework.org/schema/beans/spring
                            -beans.xsd
12.                         http://www.springframework.org/schema/context
13.                         http://www.springframework.org/schema/context/
                            spring-context.xsd
14.                         http://www.springframework.org/schema/mvc
15.                         http://www.springframework.org/schema/mvc/spring-
                            mvc.xsd
16.                         http://www.springframework.org/schema/aop
17.                         http://www.springframework.org/schema/aop/spring-
                            aop.xsd
18.                         http://www.springframework.org/schema/tx
19.                         http://www.springframework.org/schema/tx/spring-
                            tx.xsd
20.                         http://mybatis.org/schema/mybatis-spring
21.                         http://mybatis.org/schema/mybatis-spring.xsd">
22.
23.     <!--mybatis:scan 会扫描 ssh.homework.dao 包里的所有接口当作 Spring 的
        bean 配置，之后可以进行依赖注入 -->
24.     <mybatis:scan base-package="ssh.homework.dao" />
25.
26.     <!--扫描 ssh.homework 包下面的 Java 文件，如果有 Spring 的相关注解的类，则
        把这些类注册为 Spring 的 bean -->
27.     <context:component-scan
28.         base-package="ssh.homework" />
29.
30.     <!--使用 PropertyOverrideConfigurer 后处理器加载数据源参数 -->
31.     <context:property-override
32.         location="classpath:db.properties" />
33.
34.     <!--配置 c3p0 数据源 -->
35.     <bean id="dataSource"
36.         class="com.mchange.v2.c3p0.ComboPooledDataSource" />
37.
38.     <!--配置 SqlSessionFactory, org.mybatis.spring.SqlSessionFactoryBean 是
        MyBatis 社区开发用于整合 Spring 的 bean -->
39.     <bean id="sqlSessionFactory"
40.         class="org.mybatis.spring.SqlSessionFactoryBean"
```

```
41.        p:dataSource-ref="dataSource" />
42.
43.    <!--JDBC 事务管理器 -->
44.    <bean id="transactionManager"
45.        class="org.springframework.jdbc.datasource.DataSourceTransaction-
        Manager"
46.        p:dataSource-ref="dataSource" />
47.    <!--通知 -->
48.    <tx:advice id="txAdvice"
49.        transaction-manager="transactionManager">
50.        <tx:attributes>
51.            <!--传播行为 -->
52.            <tx:method name="save*" propagation="REQUIRED" />
53.            <tx:method name="add*" propagation="REQUIRED" />
54.            <tx:method name="modify*" propagation="REQUIRED" />
55.            <tx:method name="remove*" propagation="REQUIRED" />
56.            <tx:method name="del*" propagation="REQUIRED" />
57.            <tx:method name="find*" propagation="SUPPORTS"
58.                read-only="true" />
59.            <tx:method name="select*" propagation="SUPPORTS"
60.                read-only="true" />
61.            <tx:method name="get*" propagation="SUPPORTS"
62.                read-only="true" />
63.        </tx:attributes>
64.    </tx:advice>
65.    <!--切面 -->
66.    <aop:config>
67.        <aop:advisor advice-ref="txAdvice"
68.            pointcut="execution(* ssh.homework.service.*.*(..))" />
69.    </aop:config>
70.    <!--启用支持 annotation 注解方式事务管理 -->
71.    <tx:annotation-driven
72.        transaction-manager="transactionManager" />
73.
74. </beans>
```

第 44 行定义了一个事务管理器。第 48～64 行启用了 Spring 中的 AOP，<tx:advice>标签定义了一个通知，通知中编写了切入点和具体执行事务的细节，如以 add、save、modify、del、remove 开头的方法进行正常事务管理，将以 find、select、get 开头的方法设为只读事务。在 Spring 配置文件中对事务做详细的配置，在程序中就无须配置任何事务，只要严格按照前面指定的方法名称对方法进行定义，执行数据库操作时会启用相应的事务进行管理，这样可以节省大量的代码，也便于对事务的管理。

③ SpringMVC 配置文件 springmvc-config.xml，代码如下。

```xml
1.  <?xml version="1.0" encoding="UTF-8"?>
2.  <beans xmlns="http://www.springframework.org/schema/beans"
3.    xmlns:xsi="http://www.w3.org/2001/XMLSchema-instance"
4.    xmlns:mvc="http://www.springframework.org/schema/mvc"
5.    xmlns:p="http://www.springframework.org/schema/p"
6.    xmlns:context="http://www.springframework.org/schema/context"
7.    xsi:schemaLocation="
8.      http://www.springframework.org/schema/beans
9.      http://www.springframework.org/schema/beans/spring-beans.xsd
10.     http://www.springframework.org/schema/mvc
11.     http://www.springframework.org/schema/mvc/spring-mvc.xsd
12.     http://www.springframework.org/schema/context
13.     http://www.springframework.org/schema/context/spring-context.xsd">
14.   <!--自动扫描该包,Spring MVC会将包下用@controller注解的类注册为Spring的controller -->
15.   <context:component-scan base-package="ssh.homework.controller"/>
16.   <!--设置默认配置方案 -->
17.   <mvc:annotation-driven/>
18.   <!--使用默认的Servlet响应静态文件 -->
19.   <mvc:default-servlet-handler/>
20.
21.   <!--定义Spring MVC的拦截器 -->
22.   <mvc:interceptors>
23.       <mvc:interceptor>
24.           <mvc:mapping path="/*"/>
25.           <!--自定义判断用户权限的拦截类 -->
26.           <bean class="ssh.homework.interceptor.AuthorizedInterceptor"/>
27.       </mvc:interceptor>
28.   </mvc:interceptors>
29.   <!--视图解析器 -->
30.       <bean id="viewResolver"
31.           class="org.springframework.web.servlet.view.InternalResourceViewResolver"
32.           p:prefix="/WEB-INF/jsp/" p:suffix=".jsp"/>
33.       <!--文件上传下载 -->
34.       <bean id="multipartResolver"
35.           class=" org.springframework.web.multipart.commons.CommonsMultipartResolver">
36.       <!--上传文件大小上限,单位为字节(1000MB) -->
37.       <property name="maxUploadSize">
38.           <value>10485760</value>
39.       </property>
40.       <!--请求的编码格式,必须和JSP的pageEncoding属性一致,以便正确读取表单的内容,默认为ISO-8859-1 -->
```

```
41.         <property name="defaultEncoding">
42.             <value>UTF-8</value>
43.         </property>
44.     <!--开启 resolveLazily,该参数表示延迟解析 -->
45.         <property name="resolveLazily" value="true" />
46.     </bean>
47.     <!--Spring MVC 提供的简单异常处理器-->
48.     <bean class="org.springframework.web.servlet.handler.SimpleMapping-
    ExceptionResolver"
49.         p:defaultErrorView="error"
50.             p:exceptionAttribute="ex">
51.     </bean>
52.   </beans>
```

第 22~28 行定义了一个拦截器,用于权限检查。如果用户访问某个资源时,首先通过拦截器判断用户是否已经登录,如果没有登录,则转发到登录页面;如果已经登录,则转发到要访问的资源,这样对资源可起到保护作用。

第 34~46 行定义了一个 id 为 multipartResolver 的 Bean,其作用是完成文件的上传,其中的<property>元素用来设置上传文件中的一些限制,如上传文件的大小、文件名的字符集等。

第 48~51 行定义了一个异常处理器,作用是当程序发生异常时,转发到 error.jsp 页面。

④ Tomcat 服务器启动时,第一个加载的文件是 web.xml,这里的大部分代码与第 9 章中的 web.xml 相同,此处给出额外添加的代码,具体代码如下。

```
1.  <!--JSP 的配置 -->
2.  <jsp-config>
3.      <jsp-property-group>
4.          <!--配置拦截所有的 JSP 页面 -->
5.          <url-pattern>*.jsp</url-pattern>
6.          <!--可以使用 EL 表达式 -->
7.          <el-ignored>false</el-ignored>
8.          <!--不能在页面使用 Java 脚本 -->
9.          <scripting-invalid>true</scripting-invalid>
10.         <!--给所有的 JSP 页面导入要依赖的库,tablib.jsp 就是一个全局的标签库文
    件 -->
11.         <include-prelude>/WEB-INF/jsp/taglib.jsp</include-prelude>
12.     </jsp-property-group>
13. </jsp-config>
14.
15. <error-page>
```

```
16.            <error-code>404</error-code>
17.            <location>/404.html</location>
18.        </error-page>
19.        <welcome-file-list>
20.            <welcome-file>index.jsp</welcome-file>
21.        </welcome-file-list>
```

第 2～13 行使用 <jsp-config> 标签对 JSP 页面做了全局配置，其中第 11 行 <include-prelude> 标签加载了一个 taglib.jsp 页面，这是一个全局的标签库文件，这个 JSP 文件会在加载所有的 JSP 页面之前加载。

taglib.jsp 页面代码如下。

```
1.  <%@page language="java" contentType="text/html; charset=UTF-8"
2.       pageEncoding="UTF-8"%>
3.  <%@taglib prefix="c" uri="http://java.sun.com/jsp/jstl/core" %>
4.  <!--设置一个项目路径的变量 -->
5.  <c:set var="ctx" value="${pageContext.request.contextPath}"></c:set>
6.  <!--配置分页标签 -->
7.  <%@taglib prefix="fkjava" uri="/pager-tags" %>
```

第 5 行设置了一个 ctx 变量，这个变量的值为项目的路径，将来可以在所有的 JSP 页面中通过 ${ctx} 表达式加以引用。第 7 行导入了一个自定义的标签库，主要是分页标签。至此，所有的配置文件已经编写完成。

10.4 功能模块实现

在本系统的数据库操作部分，在 MyBatis 数据操作接口类中通过注解编写 SQL 语句，对于一些复杂的数据操作，采用动态 SQL 语句，这部分语句写在与接口类对应的 provider 类中。

10.4.1 教师管理模块

教师管理模块的功能包括：教师登录和对教师表 tb_teacher 的增、删、改、查功能。此部分功能一共涉及 4 个主要的类和接口，它们是 TeacherController、TeacherService、TeacherServiceImpl 和 TeacherDao。它们之间的关系通过图 10-8 所示类图加以描述。

从图 10-8 可以看出，TeacherService 的对象 teacherService 是 TeacherController 类的一部分，TeacherController 类中的所有业务逻辑都通过调用 teacherService 的方法实现，teacherService 是 TeacherController 类的重要组成部分。teacherService 对象通过 IoC 容器自动注入控制器中。

TeacherDao 接口实例化也是通过 IoC 容器完成的，并自动注入到 TeacherServiceImpl 类中，TeacherDao 的对象是 TeacherServiceImpl 类的成员变量，当然也是类的重要组成部分。TeacherServiceImpl 类中的所有数据库操作都是通过 teacherDao 对象完成的。

Java Web 框架开发技术(Spring+Spring MVC+MyBatis)

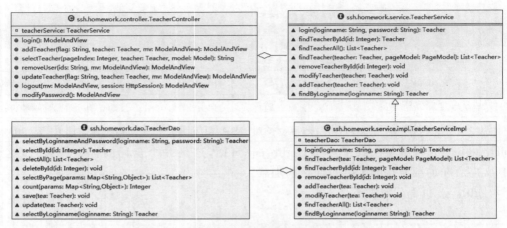

图 10-8 教师管理模块类图

这个模块涉及 4 个类的详细情况如下。

1. 数据访问层

TeacherDao 接口的主要功能是操作数据库,其代码如下。

```
1.   package ssh.homework.dao;
2.
3.   import ssh.homework.dao.provider.TeacherDynaSqlProvider;
4.   import ssh.homework.domain.Teacher;
5.   import static ssh.homework.common.HomeworkConstants.TEACHERTABLE;
6.   /**
7.    * @Description:TeacherDao 接口
8.    */
9.   public interface TeacherDao{
10.    /**
11.     * @Description: 根据登录名和密码查询员工 */
12.    @Select(" select * from " + TEACHERTABLE +" where loginname = #
           {loginname} and password = #{password}")
13.    Teacher selectByLoginnameAndPassword(
14.        @Param("loginname") String loginname,
15.        @Param("password") String password);
16.    /** 根据 id 查询用户 */
17.    @Select("select * from "+TEACHERTABLE+" where id = #{id}")
18.    Teacher selectById(Integer id);
19.
20.    /** 查询所有用户  */
```

```java
21.     @Select("select * from "+TEACHERTABLE+" ")
22.     List<Teacher> selectAll();
23.
24.     /** 根据id删除用户 */
25.     @Delete(" delete from "+TEACHERTABLE+" where id =#{id} ")
26.     void deleteById(@Param("id") Integer id);
27.     /** 动态查询 */
28.     @SelectProvider(type = TeacherDynaSqlProvider.class, method = "selectWhitParam")
29.     List<Teacher> selectByPage(Map<String, Object> params);
30.
31.     //根据参数查询用户总数
32.     @SelectProvider(type=TeacherDynaSqlProvider.class,method="count")
33.     Integer count(Map<String, Object> params);
34.
35.     //动态插入用户
36.     @SelectProvider(type = TeacherDynaSqlProvider.class, method = "insertTea")
37.     void save(Teacher tea);
38.     //动态更新用户
39.     @SelectProvider(type = TeacherDynaSqlProvider.class, method = "updateTeacher")
40.     void update(Teacher tea);
41.     //根据登录用户名查询用户
42.      @Select("select * from "+TEACHERTABLE+" where loginname = #{loginname}")
43.     Teacher selectByLoginname(String loginname);
44. }
```

TeacherDao 接口中的一些方法涉及动态 SQL 查询,相关动态查询语句存放在另一个类 TeacherDynaSqlProvider 中,其代码如下。

```java
1.  package ssh.homework.dao.provider;
2.
3.  import java.util.Map;
4.  import org.apache.ibatis.jdbc.SQL;
5.
6.  import ssh.homework.domain.Teacher;
7.  import static ssh.homework.common.HomeworkConstants.TEACHERTABLE;
8.
9.  /** 教师动态SQL语句提供类 */
10. public class TeacherDynaSqlProvider {
11.     /** 动态分页查询 */
12.     public String selectWhitParam(Map<String, Object> params) {
```

```
13.        String sql=new SQL() {
14.            {
15.                SELECT("*");
16.                FROM(TEACHERTABLE);
17.                if (params.get("teacher")!=null) {
18.                    Teacher tea=(Teacher) params.get("teacher");
19.                    if (tea.getUsername()!=null && !tea.getUsername()
                          .equals("")) {
20.                      WHERE(" username LIKE CONCAT ('% ',#{teacher
                            .username},'% ') ");
21.                    }
22.                    if (tea.getLoginname()!=null && !tea.getLoginname()
                          .equals("")) {
23.                        WHERE(" loginname LIKE CONCAT ('% ',#{teacher
                              .loginname},'% ') ");
24.                    }
25.
26.                }
27.            }
28.        }.toString();
29.
30.        if (params.get("pageModel") !=null) {
31.            sql +=" limit #{pageModel.firstLimitParam} , #{pageModel
                  .pageSize} ";
32.        }
33.
34.        return sql;
35.    }
36.
37.    /** 动态查询总数量 */
38.    public String count(Map<String, Object>params) {
39.        return new SQL() {
40.            {
41.                SELECT("count(*)");
42.                FROM(TEACHERTABLE);
43.                if (params.get("teacher") !=null) {
44.                    Teacher tea=(Teacher) params.get("teacher");
45.                    if (tea.getLoginname() !=null && !tea.getLoginname()
                          .equals("")) {
46.                        WHERE(" loginname LIKE CONCAT ('% ',#{tea.loginname},
                              '% ') ");
47.                    }
```

```java
48.                if (tea.getUsername() != null && !tea.getUsername()
                        .equals("")) {
49.                    WHERE(" username LIKE CONCAT ('% ',#{tea
                        .username},'% ') ");
50.                }
51.
52.            }
53.        }
54.    }.toString();
55. }
56.
57. /** 动态插入 */
58. public String insertTea(Teacher tea) {
59.     return new SQL() {
60.         {
61.             INSERT_INTO(TEACHERTABLE);
62.             if (tea.getLoginname() !=null && !tea.getLoginname()
                    .equals("")) {
63.                 VALUES("loginname", "#{loginname}");
64.             }
65.             if (tea.getUsername() !=null && !tea.getUsername()
                    .equals("")) {
66.                 VALUES("username", "#{username}");
67.             }
68.             if (tea.getPassword() !=null && !tea.getPassword()
                    .equals("")) {
69.                 VALUES("password", "#{password}");
70.             }
71.              if (tea.getRole() !=null && !tea.getRole().equals("")) {
72.                 VALUES("role", "#{role}");
73.             }
74.         }
75.     }.toString();
76. }
77.
78. /** 动态更新 */
79. public String updateTeacher(Teacher tea) {
80.
81.     return new SQL() {
82.         {
83.             UPDATE(TEACHERTABLE);
84.             if (tea.getLoginname() !=null) {
85.                 SET(" loginname =#{loginname} ");
```

```
86.                    }
87.                    if (tea.getUsername() !=null) {
88.                        SET("username =#{username} ");
89.                    }
90.                    if (tea.getPassword() !=null) {
91.                        SET(" password =#{password} ");
92.                    }
93.                    if (tea.getRole() !=null) {
94.                        SET(" role =#{role} ");
95.                    }
96.
97.                    WHERE(" id =#{id} ");
98.                }
99.            }.toString();
100.       }
101.   }
```

TeacherDynaSqlProvider 类中提供了动态 SQL 语句，实现对 tb_teacher 表的动态更新和查询。其中第 12 行的 selectWhitParam() 是动态查询方法，根据传递的 params 参数，判断这个 Map 中的键值是否为空，决定是否将查询条件添加到 SQL 语句中，组成动态查询语句。第 38 行的 count() 方法是得到查询记录的总数量，用于分页显示时计算每页的行数和一共有多少要显示的页。coutn() 方法的形参应该与 selectWhitParam() 方法的形参完全一致，这样才能保证正确显示查询结果。

2. 业务逻辑层

教师管理模块的第二层是 Service 层，对应的接口是 TeacherService，代码如下。

```
1.  package ssh.homework.service;
2.
3.  import java.util.List;
4.  import ssh.homework.domain.Teacher;
5.  import ssh.homework.tag.PageModel;
6.  /** 教师管理服务层接口 */
7.  public interface TeacherService {
8.      /** 教师登录 */
9.      Teacher login(String loginname,String password);
10.
11.     /** 根据 id 查询教师
12.      * @param id
13.      * @return 教师对象
14.      * */
15.     Teacher findTeacherById(Integer id);
16.     List<Teacher>findTeacherAll();
```

```
17.
18.      /**
19.       * 教师动态查询
20.       * @return Teacher 对象的 List 集合
21.       **/
22.      List<Teacher> findTeacher(Teacher teacher,PageModel pageModel);
23.
24.      /**
25.       * 根据 id 删除教师
26.       * @param id
27.       **/
28.      void removeTeacherById(Integer id);
29.
30.      /**
31.       * 修改教师
32.       **/
33.      void modifyTeacher(Teacher teacher);
34.
35.      /**
36.       * 添加教师
37.       *
38.       **/
39.      void addTeacher(Teacher teacher);
40.      /**
41.       * 查询用户
42.       * loginname 登录用户名
43.       **/
44.      Teacher findByLoginname(String loginname);
45.  }
```

接口 TeacherService 的实现类是 TeacherServiceImpl，代码如下。

```
1.   package ssh.homework.service.impl;
2.
3.   import ssh.homework.dao.TeacherDao;
4.   import ssh.homework.domain.Teacher;
5.   import ssh.homework.service.TeacherService;
6.   import ssh.homework.tag.PageModel;
7.
8.   /** 教师管理模块服务层接口 TeacherService 实现类   */
9.   @Transactional(propagation = Propagation.REQUIRED, isolation = Isolation.DEFAULT)
10.  @Service("teacherService")
11.  public class TeacherServiceImpl implements TeacherService{
12.
```

```java
13.     /**
14.      * 自动注入持久层Dao对象
15.      **/
16.     @Autowired
17.     private TeacherDao teacherDao;
18.
19.     /**
20.      * TeacherService接口login()方法实现
21.      **/
22.     @Override
23.     public Teacher login(String loginname, String password) {
24.   //    System.out.println("TeacherServiceImpl login -->>");
25.         return teacherDao.selectByLoginnameAndPassword(loginname, password);
26.     }
27.
28.     /** 根据tea对象中的属性值动态查询 */
29.     @Override
30.     public List<Teacher> findTeacher(Teacher tea, PageModel pageModel) {
31.         /** 当前需要分页的总数据条数 */
32.         Map<String,Object> params=new HashMap<>();
33.         params.put("teacher", tea);
34.         int recordCount=teacherDao.count(params);
35.         pageModel.setRecordCount(recordCount);
36.         if(recordCount >0) {
37.             /** 开始分页查询数据：查询第几页的数据 */
38.             params.put("pageModel", pageModel);
39.         }
40.         List<Teacher> teas=teacherDao.selectByPage(params);
41.
42.         return teas;
43.     }
44.
45.     /**根据id查询教师 */
46.     @Override
47.     public Teacher findTeacherById(Integer id) {
48.         return teacherDao.selectById(id);
49.     }
50.
51.     /**
52.      * TeacherService接口removeUserById方法实现
53.      * @see { TeacherService }
54.      **/
55.     @Override
```

```
56.    public void removeTeacherById(Integer id) {
57.        teacherDao.deleteById(id);
58.
59.    }
60.    /**
61.     * TeacherService 接口 addTeacher 方法实现
62.     * @see { TeachermService }
63.     **/
64.    @Override
65.    public void addTeacher(Teacher tea) {
66.        if(findByLoginname(tea.getLoginname())==null)
67.            teacherDao.save(tea);
68.
69.    }
70.    /**
71.     * TeacherService 接口 modifyTeacher 方法实现
72.     * @see { TeachermService }
73.     **/
74.    @Override
75.    public void modifyTeacher(Teacher tea) {
76.        teacherDao.update(tea);
77.    }
78.    /**
79.     * TeacherService 接口 findTeacherAll 方法实现
80.     * @see { TeachermService }
81.     **/
82.    @Override
83.    public List<Teacher> findTeacherAll() {
84.        return teacherDao.selectAll();
85.    }
86.    /**
87.     * TeacherService 接口 findByLoginname 方法实现
88.     * @see { TeachermService }
89.     **/
90.    @Override
91.    public Teacher findByLoginname(String loginname) {
92.        return teacherDao.selectByLoginname(loginname);
93.    }
94. }
```

在 TeacherServiceImpl 类中，第 16 行 @Autowired 注解通过 IoC 自动注入了 TeacherDao 接口的实例。

3. Web 表示层

教师管理模块的第三层是 Web 表现层，这一层主要由 Controller 和 JSP 页面组成。为了让读者对系统工作流程有一个更清晰的了解，下面给出系统登录页面和登录流程图。

图 10-9 所示为系统登录页面，其中有一个下拉列表框，这个列表框可以选择登录的角色：教师或学生。单击"登录"按钮后提交给 TeacherController 类的 login() 方法处理请求，在 login() 方法中判断登录角色，如果是学生，则转发给 StudentController 类的 login() 方法继续处理登录请求，否则在当前的方法中处理教师登录请求。系统登录流程图如图 10-10 所示。

图 10-9　系统登录页面

系统登录流程图显示，如果是教师登录成功后转发到 main.jsp 页面，则显示教师作业管理功能，否则转发到 studentMain.jsp 页面显示学生作业管理功能。

这个模块主要涉及控制器 TeacherController 类和控制器 FormController 类。

（1）控制器 TeacherController 类代码如下。

```
1.   package ssh.homework.controller;
2.
3.   import ssh.homework.common.HomeworkConstants;
4.   import ssh.homework.domain.Teacher;
5.   import ssh.homework.service.TeacherService;
6.   import ssh.homework.tag.PageModel;
7.   /** 有关教师请求的处理类*/
8.   @Controller
9.   public class TeacherController {
10.      //由 IoC 容器自动注入 teacherService 对象
11.      @Autowired
12.      @Qualifier("teacherService")
13.      private TeacherService teacherService;
```

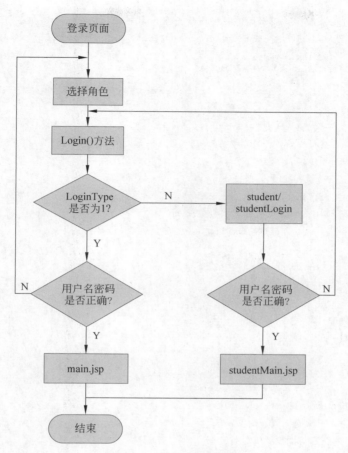

图 10-10 系统登录流程图

```
14.      /** 处理教师登录请求 */
15.      @RequestMapping(value="/login")
16.      public ModelAndView login(
17.          @RequestParam("loginname") String loginname,
18.          @RequestParam("password") String password,
19.          HttpSession session,
20.          @RequestParam("loginType") String loginType,
21.          ModelAndView mv) {
22.      /** 判断是学生,还是教师,如果loginType=2,则转发到学生登录页面,否则略过
         这段代码执行下面的语句 */
23.          if(loginType.equals("2")) {
24.              mv.addObject("loginname",loginname);
25.              mv.addObject("password",password);
26.              mv.setViewName("forward:/student/studentLogin");
27.              return mv;
28.          }
```

```java
29.        //调用业务逻辑组件判断用户是否可以登录
30.        Teacher teacher=teacherService.login(loginname, password);
31.        if(teacher !=null){
32.            //将用户保存到 HttpSession 中
33.            session.setAttribute(HomeworkConstants.TEACHER_SESSION,
                   teacher);
34.            //客户端跳转到 main 页面
35.            mv.setViewName("redirect:/main");
36.        }else{
37.            //设置登录失败提示信息
38.            mv.addObject("message","登录名或密码错误!请重新输入");
39.            //服务器内部跳转到登录页面
40.            mv.setViewName("forward:/loginForm");
41.        }
42.        return mv;
43.
44.    }
45.    //添加教师方法
46.    @RequestMapping(value="/teacher/addTeacher")
47.    public ModelAndView addTeacher(
48.        String flag,
49.        @ModelAttribute Teacher teacher,
50.        ModelAndView mv){
51.        if(flag.equals("1")){
52.            //设置跳转到添加页面
53.            mv.setViewName("teacher/showAddTeacher");
54.        }else{
55.            //执行添加操作
56.            teacherService.addTeacher(teacher);
57.            //设置客户端跳转到查询请求
58.            mv.setViewName("redirect:/teacher/selectTeacher");
59.        }
60.        //返回
61.        return mv;
62.    }
63.    //查询教师方法,查询条件根据 teacher 对象中属性的值动态生成 SQL 语句
64.    @RequestMapping(value="/teacher/selectTeacher")
65.
66.    public String selectTeacher(Integer pageIndex,
67.        @ModelAttribute Teacher teacher,
68.        Model model){
69.        PageModel pageModel=new PageModel();
70.        if(pageIndex !=null){
```

```java
71.            pageModel.setPageIndex(pageIndex);
72.        }
73.        /** 查询用户信息   */
74.        List<Teacher> teachers = teacherService.findTeacher(teacher,
            pageModel);
75.        model.addAttribute("teachers", teachers);
76.        model.addAttribute("pageModel", pageModel);
77.        return "teacher/teacher";
78.
79.    }
80.    //根据id删除教师
81.    @RequestMapping(value="/teacher/removeTeacher")
82.    public ModelAndView removeUser(String ids,
83.        ModelAndView mv){
84.        //分解id字符串
85.        String[] idArray=ids.split(",");
86.
87.    for(String id : idArray){
88.        try {
89.        //根据id删除员工
90.        teacherService.removeTeacherById(Integer.parseInt(id));
91.        }catch(Exception ep) {
92.            mv.addObject("error","教师不能删除!");
93.            mv.setViewName("error/error");
94.            return mv;
95.        }
96.    }
97.    //设置客户端跳转到查询请求
98.        mv.setViewName("redirect:/teacher/selectTeacher");
99.    //返回ModelAndView
100.        return mv;
101.    }
102.    //根据teacher对象中属性的值生成动态更新SQL语句
103.    @RequestMapping(value="/teacher/updateTeacher")
104.    public ModelAndView updateTeacher(
105.        String flag,
106.        @ModelAttribute Teacher teacher,
107.        ModelAndView mv){
108.        if(flag.equals("1")){
109.        //根据id查询用户
110.        Teacher target = teacherService.findTeacherById(teacher.
            getId());
111.        //设置Model数据
```

```
112.            mv.addObject("teacher", target);
113.            //返回修改员工页面
114.            mv.setViewName("teacher/showUpdateTeacher");
115.        }else{
116.            //执行修改操作
117.            teacherService.modifyTeacher(teacher);
118.            //设置客户端跳转到查询请求
119.            mv.setViewName("redirect:/teacher/selectTeacher");
120.        }
121.        //返回
122.        return mv;
123.    }
124.    //注销方法
125.    @RequestMapping(value="/logout")
126.    public ModelAndView logout(
127.            ModelAndView mv,
128.            HttpSession session) {
129.        //注销 session
130.        session.invalidate();
131.        //跳转到登录页面
132.        mv.setViewName("redirect:/loginForm");
133.        return mv;
134.    }
135.    //修改教师密码
136.    @RequestMapping(value="/teacher/modifyPassword")
137.    public ModelAndView modifyPassword(
138.            String flag,
139.            String oldPassword,
140.            String password,
141.            HttpSession session,
142.            ModelAndView mv) {
143.        if(flag.equals("1")) {
144.
145.            mv.setViewName("/teacher/modifyPassword");
146.
147.        }
148.        else
149.        {
150.            //执行更新
151.            //从 session 中得到登录学生的信息
152.            Teacher teacher=(Teacher)session.getAttribute(TEACHER_SESSION);
153.            //重新查询 Teacher,得到 Teacher 关联对象的信息
154.            teacher=teacherService.findTeacherById(teacher.getId());
```

```
155.        if(oldPassword.equals(teacher.getPassword())) {
156.            teacher.setPassword(password);
157.            teacherService.modifyTeacher(teacher);
158.            mv.addObject("message","修改成功!");
159.            mv.setViewName("success");
160.        }
161.        else
162.        {
163.            mv.addObject("error","原密码不对!");
164.            mv.setViewName("error/error");
165.
166.        }
167.    }
168.
169.    return mv;
170.    }
171.
172. }
```

TeacherController 类中一共有 7 个和教师操作有关的方法,具体内容如下。
- loign():教师登录方法。
- addTeacher():添加教师方法。
- updateTeacher():修改教师方法。
- modifyPassword():修改教师密码方法。
- removeUser():删除教师方法。
- selectTeacher():根据查询参数动态查询教师方法。
- logout():注销方法。

(2) 控制器 FormController 类代码如下。

```
1.  /**
2.   * 动态页面跳转控制器
3.   **/
4.  @Controller
5.  public class FormController{
6.
7.      @RequestMapping(value="/{formName}")
8.      public String loginForm(@PathVariable String formName){
9.          //动态跳转页面
10.         return formName;
11.     }
12. }
```

这个控制器的作用是响应 main.jsp 页面中的超链接,根据超链接中的请求转发到相应的页面。注意,第 7 行的 value="/{formName}"中的{}不能省略,这代表占位符,可响应多个请求,根据传递过来的字符串转发到相应的页面。如发送的请求为"left",则转发页面为 left.jsp。

(3)系统主页 main.jsp。

学生作业管理系统教师模块的主页是一个框架页面,整个页面由 4 个框架组成,具体情况如下。

main.jsp:主页面。

top.jsp:顶部页面,显示登录用户的信息。

left.jsp:左侧显示菜单页面。

light:右侧显示功能操作页面。

用户只有登录成功后,才会显示该页面。如果没有登录,则会转发到登录页面。系统管理员登录成功后显示的页面如图 10-11 所示。

图 10-11　学生作业管理系统主界面

main.jsp 页面代码如下。

```
1.    <body>
2.    <frameset rows="80,*" cols="*" frameborder="no" border="0" framespacing="0">
3.      <frame src="${ctx}/top" name="title" scrolling="no" noresize="noresize"
4.        <frameset cols="220,*" frameborder="no" border="0" framespacing="0">
5.          <frame src="${ctx}/left" name="tree" scrolling="no" marginheight=
              "0"
6.            marginwidth="0">
7.          <frame src="${ctx}/right" name="main" scrolling="yes" frameborder="0"
8.            marginwidth="0" marginheight="0" noresize="noresize">
9.        </frameset>
10.   </frameset>
11.   <body>
```

main.jsp 页面代码由 3 个 <frame> 组成,每个 <frame> 都有一个 src 属性,代表用于显示的页面,3 个 src 属性分别指向 3 个页面,${ctx}\top.jsp、${ctx}left.jsp 和 ${ctx}right.jsp。${ctx} 是在 taglib.jsp 中定义的一个变量,其代码如下。

```
<c:set var="ctx" value="${pageContext.request.contextPath}"></c:set>
```

这段代码的含义是用 EL 表达式定义一个变量 ctx,它代表的是部署应用程序名。这样,不管如何部署,所用路径都是正确的。

这 3 个页面中最主要的是 left.jsp 页面,该页面显示了系统中所有的功能菜单。下面分别介绍 top.jsp 和 left.jsp 页面代码。

top.jsp 页面代码如下。

```
1.   <body>
2.   <table width="100% " border="0" cellpadding="0" cellspacing="0">
3.     <tr><td height="50" class="toplink" align="right">
4.     < img src="${ctx}/images/top_exit.gif"><a href="${ctx}/logout;" id
       ="exit">注销退出</a>   </td>
5.         </tr>   <tr>
6.   <td class="topnavlh" align="left"><img src="${ctx}/images/StatBar_
       admin.gif">当前用户:【${sessionScope.teacher_session.username}】</td>
7.   <td class="topnavlh" align="right"><img src="${ctx}/images/StatBar_
       time.gif"><span id="nowTime"></span>
8.   </td></tr>
9.   </table>
10.  </body>
```

第 4 行有一个"注销退出"超链接,超链接指向 logout 映射请求处理。第 6 行应用 EL 表达式 ${sessionScope.teacher_session.username} 显示当前教师的姓名。其中 teacher_session.username 是在登录成功时保存在 session 中的变量。

left.jsp 页面代码如下。

```
1.   <body>
2.        <c:if test="${teacher_session.role==1}">
3.     <tr><td class="left_nav_top"><div class="font1">用户管理
4.     </td></tr>
5.        <tr valign="top">
6.        <td class="left_nav_bgshw" height="50">
7.   <p class="left_nav_link"><img src="${ctx}/images/left_nav_arrow.gif">
8.   <a href="teacher/selectTeacher" target="main">用户查询</a></img></p>
9.   <p class="left_nav_link"><img src="${ctx}/images/left_nav_arrow.gif">
10.  <a href="teacher/addTeacher?flag=1" target="main">添加用户</a></img></p>
```

```
11.        </td></tr>
12.     <tr><td height="2"></td></tr>
13.     <tr><td id="navbg1" class="left_nav_closed" ><div class="font1">班级管理</td></tr>
14.     <tr valign="top" id="submenu1" style="display: none">
15.     <td class="left_nav_bgshw" height="50">
16.     <p class="left_nav_link"><img src="${ctx}/images/left_nav_arrow.gif">
17.     <a href="clazz/selectClazz" target="main">班级查询</a></img></p>
18.     <p class="left_nav_link"><img src="${ctx}/images/left_nav_arrow.gif">
19.     <a href="clazz/addClazz?flag=1" target="main">添加班级</a></img></p>
20.        </td></tr>
21.     </c:if>
22.     <tr><td height="2"></td></tr>
23.     <tr><td id="navbg2" class="left_nav_closed" ><div class="font1">学生管理</td></tr>
24.         <tr valign="top" id="submenu2" style="display: none">
25.         <td class="left_nav_bgshw" height="50">
26.     <p class="left_nav_link"><img src="${ctx}/images/left_nav_arrow.gif">
27.     <a href="student/selectStudent" target="main">学生查询</a></img></p>
28.     <p class="left_nav_link"><img src="${ctx}/images/left_nav_arrow.gif">
29.     <a href="student/addStudent?flag=1" target="main">添加学生</a></img>
        </p>
30.     <p class="left_nav_link"><img src="${ctx}/images/left_nav_arrow.gif">
31.     <a href="student/leadStudentExcel?flag=1" target="main">EXCEL 导入学生</a></img></p>
32.         </td></tr>
33.         <tr><td height="2"></td></tr>
34.     <tr><td id="navbg3" class="left_nav_closed" ><div class="font1">课程管理</td></tr>
35.     <tr valign="top" id="submenu3" style="display: none">
36.         <td class="left_nav_bgshw" height="50">
37.     <p class="left_nav_link"><img src="${ctx}/images/left_nav_arrow.gif">
38.     <a href="course/selectCourse" target="main">课程查询</a></img></p>
39.     <p class="left_nav_link"><img src="${ctx}/images/left_nav_arrow.gif">
40.     <a href="course/addCourse?flag=1" target="main">添加课程</a></img></p>
41.         </td></tr>
42.         <tr><td height="2"></td></tr>
43.     <tr><td id="navbg4" class="left_nav_closed" ><div class="font1">习题管理</div></td></tr>
44.     <tr valign="top" id="submenu4" style="display: none">
45.       <td class="left_nav_bgshw tdbtmline" height="50">
```

```
46.     <p class="left_nav_link"><img src="${ctx}/images/left_nav_arrow.gif">
47.     <a href="${ctx}/exercise/selectExercise" target="main">习题查询</a>
        </img></p>
48.     <p class="left_nav_link"><img src="${ctx}/images/left_nav_arrow.gif">
49.     <a href="${ctx }/exercise/addExercise?flag=1" target="main">添加习题
        </a></img></p>
50.         </td></tr>
51.         <tr><td height="2"></td></tr>
52.     <tr><td id="navbg5" class="left_nav_closed" onclick="showsubmenu(5)">
        <div class="font1">作业管理</div></td> </tr>
53.     <tr valign="top" id="submenu5" style="display: none">
54.         <td class="left_nav_bgshw tdbtmline" height="50">
55.     <p class="left_nav_link"><img src="${ctx}/images/left_nav_arrow.gif">
56.     <a href="${ctx}/workbook/selectWorkbook" target="main">布置作业</a>
        </img></p>
57.     <p class="left_nav_link"><img src="${ctx}/images/left_nav_arrow.gif">
58.     <a href="${ctx }/workbook/addWorkbook?flag=1" target="main">添加作业</a>
        </img></p>
59.     <p class="left_nav_link"><img src="${ctx}/images/left_nav_arrow.gif">
60.     <a href="${ctx}/workbook/showScoreSelect" target="main">查询成绩</a>
        </img></p>
61.         </td>    </tr>
62.         <tr><td height="2"></td></tr>
63.     <tr><td id="navbg6" class="left_nav_closed"><div class="font1">批改
        作业</div></td></tr>
64.         <tr valign="top" id="submenu6" style="display: none">
65.         <td class="left_nav_bgshw tdbtmline" height="50">
66.     <p class="left_nav_link"><img src="${ctx}/images/left_nav_arrow.gif">
67.     <a href="${ctx}/correctWorkbook/selectCorrectWorkbook?isSimilarity=
        1" target="main">显示作业</a></img></p>
68.     <p class="left_nav_link"><img src="${ctx}/images/left_nav_arrow.gif">
69.     <a href="${ctx}/correctWorkbook/selectCorrectWorkbook?isSimilarity=
        2" target="main">查重</a></img></p>
70.         </td></tr>
71.         <tr><td height="2"></td></tr>
72.     <tr><td id="navbg7" class="left_nav_closed" ><div class="font1">其他
        </div></td></tr>
73.         <tr valign="top" id="submenu7" style="display: none">
74.          <td class="left_nav_bgshw" height="50">
75.     <p class="left_nav_link"><img src="${ctx}/images/left_nav_arrow.gif">
76.     <a href="${ctx}/teacher/modifyPassword?flag=1" target="main">修改密
        码</a></img></p>
```

```
77.            </td></tr>
78.            <tr><td height="2"></td></tr>
79.    </table>
80.    </body>
```

第2~21行是只有系统管理员才有的功能菜单,因此,将这部分内容放在<c:if>标签体中,只有满足判断条件的用户,才显示这部分菜单。此处的判断条件是保存在session中的变量role的值是否为1,如果为1,表示系统管理员,显示<c:if>标签体中的菜单,否则不显示这部分功能菜单。

这里所有菜单都是通过超链接向服务器发出处理请求,其中有的超链接中带有一个flag参数,这个参数的作用是简化处理请求方法的数量,将两个请求处理方法简化为一个方法。例如,单击"添加教师"的请求时,如果flag=1,表示转发到添加教师的页面showAddTeacher.jsp。如果flag=2,表示执行添加教师的操作,这样可以将两个请求处理方法变为一个方法。

(4) 登录页面loginForm.jsp。

登录页面中有两个文本框,文本框的name值要与TeacherCotroller类中login()方法的两个形参一致,即loginname和password。

loginForm.jsp页面代码如下。

```
1.  <script type="text/javascript">
2.  $(function(){
3.      /** 按了回车键 */
4.  $(document).keydown(function(event){
5.      if(event.keyCode ==13){
6.          $("#login-submit-btn").trigger("click");
7.      }
8.  })
9.  /** 给登录按钮绑定单击事件 */
10. $("#login-submit-btn").on("click",function(){
11.     /** 校验登录参数 ctrl+K */
12.     var loginname =$("#loginname").val();
13.     var password =$("#password").val();
14.     var msg ="";
15.     if ($.trim(loginname.val()) ==""){
16.         msg ="用户名不能为空!";
17.         loginname.focus();
18.     }
19.      else if($.trim(password.val()) ==""){
20.         msg ="密码不能为空!";
21.         password.focus();
22.     }
```

```
23.        if(msg!=""){
24.            $.ligerDialog.error(msg);
25.            return false;
26.        }
27.        else return true;
28.        /** 提交表单 */
29.        $("#loginForm").submit();
30.    })
31. })
32. </script>
33. <form action="login" method="post" id="loginForm">
34.
35. <input type="text" placeholder="账号"
36.        id="loginname" name="loginname" value="${loginname}">
37. <input class="m-wrap placeholder-no-fix" type="password"
                    placeholder="密码" id="password" name="password"
38.                          value="${password}">
39. <select name="loginType" d="loginType"
40.                >
41.        <option value="2">学生</option>
42.        <option value="1">教师</option>
43. </select>
44. <!--单击登录 -->
45. <button type="submit" id="login-submit-btn" class="btn green"
46.         style="margin-left: 20px">
47. 登录<i class="m-icon-swapright m-icon-white"></i>
48. </button>
49. </form>
```

第 1～32 行应用 JavaScript 函数完成对表单文本框控件的数据校验功能。第 33 行的 action 属性的值为 login，表示将请求提交给 TeacherController 类中的 login() 方法。两个文本框 name 的值是 loginname 和 password，这两个控件的值提交给 login() 方法的两个形参 loginname 和 password。

第 39～43 行的下拉列表框中的值表示登录者的角色是教师，还是学生，列表框中选中的值会提交给 login() 方法的 loginType 变量，登录方法中会根据 loginType 的值判断当前登录的角色是学生，还是教师。如果 loginType 的值为 2，则转发到 studentLogin 处理学生登录。

在 login.jsp 页面输入用户名和密码后，提交给 TeacherController 类的 login() 方法，如果登录成功，则转发到 main.jsp 页面，main.jsp 是系统网站的主页，前面已经介绍过。

(5) 添加教师页面 addTeacher.jsp。

从左边菜单栏中单击"添加用户"超链接可进入添加教师页面，此超链接带有一个参数 flag，在处理请求的 addTeacher() 方法中根据 flag 的值进行相应处理。如果 flag 的值是 1，则表示转发到 addTeacher.jsp 页面；如果 flag 的值为 2，则执行插入数据库记录操作。

addTeacher.jsp 页面代码如下。

```
1.   <form action="${ctx}/teacher/addTeacher" id="teacherForm" method="post">
2.       <!--隐藏表单,flag 表示添加标记 -->
3.       <input type="hidden" name="flag" value="2">
4.   <table width="100% " border="0" cellpadding="0" cellspacing="10" class="main_tab">
5.       <tr><td class="font3 fftd">
6.   <table>
7.   <tr><td class="font3 fftd">登录名：<input name="loginname" id="loginname" size="20" /></td>
8.       <td class="font3 fftd">密 码：<input name="password" id="password" size="20" /></td>
9.    </tr>
10.  <tr><td class="font3 fftd">姓 名：<input type="text" name="username" id="username" size="20"/></td>
11.      <td class="font3 fftd">角 色：<input type="text" name="role" id="role" size="20"/></td>
12.  </tr>
13.  </table>
14.     </td></tr>
15.     <tr><td class="main_tdbor"></td></tr>
16.     <tr><td align="left" class="fftd">
17.  <input type="submit" value="添加"><input type="reset" value="取 消 ">
18.     </td></tr>
19.     </table>
20.  </form>
```

此页面的数据校验是由 JavaScript 函数实现的，由于代码较长，此处略去。添加教师处理请求的映射是 teacher/addTeacher，对应的方法是 TeacherController 类中的 addTeacher()方法。该方法有两个形参接收页面传递的参数：一个是 String 类型的 flag；另一个是 Teacher 类型的 teacher 变量。页面上的 4 个文本框对应 teacher 变量中的 4 个属性变量，要求文本框的 name 属性的名字与 Teacher 类中的属性名字一致，这样，在提交请求时文本框的值自动添充进 teacher 变量中。flag 的值此处为 2，表示在 addTeacher()方法中执行添加记录的操作。

（6）显示教师信息页面 teacher.jsp。

在左侧菜单栏中单击"教师查询"超链接，映射的处理请求是 TeacherController 类中的 teacher/selectTeacher，对应的处理方法是 selectTeacher()。在此方法中调用了 teacherService.findTeacher()方法，将查询的结果封装在一个 List 中，通过 model 对象传递到 JSP 页面，在 select.jsp 页面通过<c:forEach>循环标签输出保存在 requestCope 中的 teachers 集合中的教师信息,显示页面如图 10-12 所示。

图 10-12　显示教师信息

进入 teacher.jsp 页面默认显示全部教师,最上面的两个文本框用来输入教师姓名和登录名,可以根据用户名和教师姓名进行查询。教师信息列表框最后一列的图标是一个超链接,单击这个超链接可以显示修改教师的页面,对教师信息进行修改。

teacher.jsp 页面代码有两部分:前半部分是 JavaScript 函数,主要执行删除教师的操作;后半部分是<body>的内容。JavaScript 函数代码如下。

```
1.    <script type="text/javascript">
2.        $(function(){
3.            /** 获取上一次选中的部门数据 */
4.            var boxs=$("input[type='checkbox'][id^='box_']");
5.            /** 给数据行绑定鼠标覆盖以及鼠标移开事件 */
6.            $("tr[id^='data_']").hover(function(){
7.                $(this).css("backgroundColor","#eeccff");
8.            },function(){
9.                $(this).css("backgroundColor","#ffffff");
10.           })
11.           /** 删除员工绑定单击事件 */
12.           $("#delete").click(function(){
13.               /** 获取到用户选中的复选框 */
14.               var checkedBoxs=boxs.filter(":checked");
15.               if(checkedBoxs.length <1){
16.                   $.ligerDialog.error("请选择一个需要删除的用户!");
17.               }else{
18.                   /** 得到用户选中的所有需要删除的 ids */
19.                   var ids=checkedBoxs.map(function(){
20.                       return this.value;
21.                   })
```

```
22.            $.ligerDialog.confirm("确认要删除吗?","删除用户",function(r){
23.                if(r){
24.                    //alert("删除: "+ids.get());
25.                    //发送请求
26.                    window.location="${ctx}/teacher/removeTeacher?ids
                        ="+ids.get();
27.                }
28.            });
29.        }
30.    })
31.    })
32.    </script>
```

这个 JavaScript 函数的功能是：将页面复选框选中教师的 id 组成字符串，然后将要删除的 id 以字符串的方式发送到服务器端的 teacher/removeTeacher 映射处理请求，对应这个处理请求的是 TeacherController 类的 deleteTeacher() 方法，根据 id 对 tb_teacher 表执行删除操作。

teacher.jsp 页面的<body>部分代码如下。

```
1.   <body>
2.     <!--导航 -->
3.     <table width="100%" border="0" cellpadding="0" cellspacing="0">
4.       <tr><td height="10"></td></tr>
5.       <tr>
6.         <td width="15" height="32"><img src="${ctx}/images/main_
            locleft.gif" width="15" height="32"></td>
7.         <td class="main_locbg font2"><img src="${ctx}/images/pointer.
            gif">当前位置：用户管理 >用户查询</td>
8.         <td width="15" height="32"><img src="${ctx}/images/main_
            locright.gif" width="15" height="32"></td>
9.       </tr>
10.    </table>
11.
12.    <table width="100%" height="90%" border="0" cellpadding="5"
         cellspacing="0" class="main_tabbor">
13.      <!--查询区 -->
14.      <tr valign="top">
15.        <td height="30">
16.          <table width="100%" border="0" cellpadding="0"
               cellspacing="10" class="main_tab">
17.            <tr>
18.              <td class="fftd">
```

```
19.    <form name="empform" method="post" id="empform"
20.      action="${ctx}/teacher/selectTeacher">
21.          <table width="100% " border="0" cellpadding="0" cellspacing="0">
22.              <tr>
23.                  <td class="font3">
24.                      姓名：<input type="text" name="username">
25.                      登录名：<input type="text" name="loginname">
26.                      <input type="submit" value="搜索"/>
27.                      <input id="delete" type="button" value="删除"/>
28.                  </td>
29.              </tr>
30.          </table>
31.      </form>
32.    </td>
33.  </tr>
34. </table>
35. </td>
36. </tr>
37.
38. <!--数据展示区 -->
39.  <tr valign="top">
40.    <td height="20">
41.      <table width="100% " border="1" cellpadding="5" cellspacing="0" style="border:#c2c6cc 1px solid; border-collapse:collapse;">
42.        <tr class="main_trbg_tit" align="center">
43.          <td><input type="checkbox" name="checkAll" id="checkAll"></td>
44.          <td>登录名</td>
45.          <td>密码</td>
46.          <td>用户名</td>
47.          <td>角色</td>
48.
49.          <td align="center">操作</td>
50.        </tr>
51.        <c:forEach items="${requestScope.teachers}" var="teacher" varStatus="stat">
52.          <tr id="data_${stat.index}" align="center" class="main_trbg" onMouseOver="move(this);" onMouseOut="out(this);">
53.            <td><input type="checkbox" id="box_${stat.index}" value="${teacher.id}"></td>
54.            <td>${teacher.loginname}</td>
55.            <td>${teacher.password}</td>
```

```
56.                    <td>${teacher.username}</td>
57.                    <td>${teacher.role}</td>
58.
59.                     <td align="center" width="40px;"><a href="${ctx}/
    teacher/updateTeacher?flag=1&id=${teacher.id}">
60.                     <img title="修改" src="${ctx}/images/update.gif"/></a>
61.                    </td>
62.                </tr>
63.            </c:forEach>
64.        </table>
65.        </td>
66.    </tr>
67.    <!--分页标签-->
68.        <tr valign="top"><td align="center" class="font3">
69.        <fkjava:pager
70.            pageIndex="${requestScope.pageModel.pageIndex}"
71.            pageSize="${requestScope.pageModel.pageSize}"
72.            recordCount="${requestScope.pageModel.recordCount}"
73.            style="digg"
74.            submitUrl="${ctx}/teacher/selectTeacher?pageIndex={0}"/>
75.        </td></tr>
76.    </table>
77.    <div style="height:10px;"></div>
78. </body>
```

第 20 行<form>标签的 action 属性指向 teacher/selectTeacher 请求处理映射的地址，其映射的是 TeacherController 类的 selectTeacher()方法，这个方法有两个参数接收页面传递变量的值：一个参数是 pageIndex，这是一个 Integer 类型，这个参数的值是通过分页标签<fkjava:pager>中的 pageIndex 属性的值传递的，表示当前显示的是第几页；另一个参数是 teacher，这是 Teacher 类的对象，teacher 对象中属性的值是由第 24 行和第 25 行两个文本框中的值决定的，而 teacher 对象中属性的值决定了动态查询 SQL 语句的查询条件。单击最后一列的修改图标超链接，进入修改页面，显示当前 id 对应的教师信息。

(7) 更新教师页面 updateTeacher.jsp。

图 10-12 显示了所有教师的信息，如果想修改某个教师的信息，可以在要修改教师信息所在行最后一列单击修改图标，进入修改当前教师信息的页面，如图 10-13 所示。

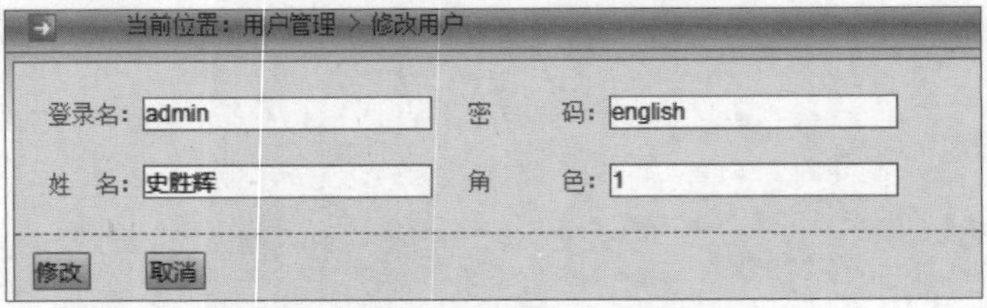

图 10-13　教师信息修改页面

updateTeacher.jsp 页面代码如下。

```jsp
1.     <form action="${ctx}/teacher/updateTeacher" id="userForm" method="post">
2.         <!--隐藏表单,flag表示添加标记-->
3.         <input type="hidden" name="flag" value="2">
4.         <input type="hidden" name="id" value="${teacher.id}">
5.     <table width="100%" border="0" cellpadding="0" cellspacing="10" class="main_tab">
6.         <tr><td class="font3 fftd">
7.     <table>
8.         <tr>
9.             <td class="font3 fftd">登录名:
10.            <input name="loginname" id="loginname" size="20"
               value="${teacher.loginname}"/></td>
11.                <td class="font3 fftd">密  码:<input name="password" id="password" size="20"
12.            value="${teacher.password}"/></td>
13.        </tr>
14.        <tr>
15.            <td class="font3 fftd">姓 名:
    <input type="text" name="username" id="username" size="20"
16.            value="${teacher.username}"/></td>
17.        <td class="font3 fftd">角 色:
    <input type="text" name="role" id="role" size="20"
18.            value="${teacher.role}"/></td>
19.            </tr>
20.    </table>
21.        </td></tr>
22.        <tr><td class="main_tdbor"></td></tr>
23.        <tr><td align="left" class="fftd">
24.    <input type="submit" value="修改">
25.        <input type="reset" value="取消"></td></tr>
26.    </table>
27.        </form>
```

此页面是更新教师信息页面,要求在更新前显示当前 id 教师的详细信息,并在此基础上进行修改。从 updateTecher() 方法转发到 updateTeacher.jsp 页面时,已经将教师信息封装在 teacher 对象中,并将 teacher 对象保存在 model 对象,为了在页面的文本框显示教师信息,在 updateTeacher.jsp 页面应用 EL 表达式 ${} 显示 teacher 对象的各属性值,如 ${teacher.username} 表示教师姓名。教师信息修改完成后再提交给 teacher/updateTeacher 请求处理,此处由于将隐藏变量 flag 设置为 2,表示在 updateTeacher() 方法中执行更新操作。

10.4.2 班级管理模块

班级管理模块主要涉及 tb_clazz 表的增、删、改、查功能,本模块涉及的类除实体类 Clazz 外,主要包括 ClazzDao 接口、ClazzService 接口、ClazzServiceImpl 类和控制器 CalzzController 类。这几个类之间的关系可通过下面的类图描述。班级管理模块的 UML 类图如图 10-14 所示。

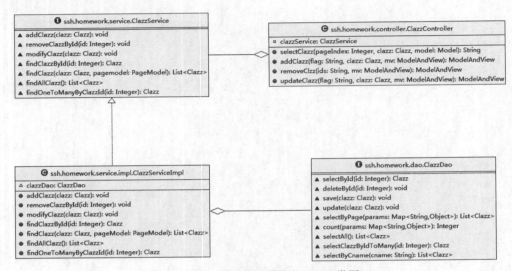

图 10-14 班级管理模块的 UML 类图

从图 10-14 可以看出,ClazzService 是 ClazzController 的组成部分,所有的业务逻辑都由 ClazzService 完成。而 ClazzService 类的对象 clazzService 通过 IoC 容器自动注入控制器中。

同样,ClazzDao 是 ClazzServiceImpl 的组成部分,ClazzDao 对象 clazzDao 也是通过 IoC 容器注入 ClazzServiceImpl 类中。在和班级相关的业务逻辑中,与数据库相关的操作全部由 clazzDao 完成。这里的 ClazzDao 也是一个接口,其主要功能是执行数据库相关操作,其实例化也是由 Spring 的 IoC 容器完成。

在 ClazzController 中是面向 ClazzService 接口编程,在 ClazzServiceImpl 中是面向 ClazzDao 接口编程。采用面向接口编程使得系统的灵活性大大增强,这也是应用 Spring 的 IoC 容器必须满足的条件。

1. 数据访问层

数据访问层主要的类是 ClazzDao 接口,其中应用 MyBatis 的注解完成一些简单的数据库操作,涉及复杂的数据库操作,应用动态 SQL 语句完成,这部分代码写在相应的 provider 类中。

ClazzDao 接口代码如下。

```java
1.   package ssh.homework.dao;
2.
3.   import static ssh.homework.common.HomeworkConstants.CLAZZTABLE;
4.   import ssh.homework.dao.provider.ClazzDynaSqlProvider;
5.   import ssh.homework.domain.Clazz;
6.
7.   public interface ClazzDao {
8.       //根据id查询班级
9.       @Select("select * from " +CLAZZTABLE +" where id=#{id}")
10.      Clazz selectById(Integer id);
11.
12.      //根据id删除班级
13.      @Delete(" delete from " +CLAZZTABLE +" where id=#{id} ")
14.      void deleteById(Integer id);
15.
16.      //插入班级
17.      @Insert("insert into " +CLAZZTABLE +" (cname) values(" +"#{cname})")
18.      @Options(useGeneratedKeys=true, keyProperty="id")
19.      void save(Clazz clazz);
20.
21.      //更新班级
22.      @Update("update " +CLAZZTABLE +" SET CNAME=#{cname} WHERE id=#{id}")
23.      void update(Clazz clazz);
24.
25.      //动态查询
26.      @SelectProvider(type =ClazzDynaSqlProvider.class, method =
         "selectWhitParam")
27.      List<Clazz>selectByPage(Map<String, Object>params);
28.
29.      //根据参数查询班级总数
30.      @SelectProvider(type=ClazzDynaSqlProvider.class, method="count")
31.      Integer count(Map<String, Object>params);
32.
33.      //根据班级id查询所有的workbook
34.      @Select("select * from " +CLAZZTABLE +" ")
35.      List<Clazz>selectAll();
36.
37.      //根据班级id查询班级的同时查询相应的一对多的关联对象workbook
38.      @Select("select * from " +CLAZZTABLE +" where id=#{id}")
39.      @Results({@Result(id=true, column="id", property="id"), @Result
         (column="cname", property="cname"),
40.          @Result(column="id", property="workbooks", many=@Many(select=
     "ssh.homework.dao.WorkbookDao.selectWorkbooksByClazzId", fetchType=
     FetchType.LAZY)
```

```
41.
42.         ),
43.         @Result(column="id", property="students", many=@Many(select=
    "ssh.homework.dao.StudentDao.selectStudentsByClazzId", fetchType=
    FetchType.LAZY))
44.
45.     })
46.     Clazz selectClazzByIdToMany(Integer id);
47.
48.     //根据班级名称 cname 查询班级
49.     @Select("select * from " +CLAZZTABLE +" where cname =#{cname}")
50.     List<Clazz>selectByCname(@Param("cname") String cname);
51.     }
```

ClazzDao 接口中的一些方法涉及动态 SQL 查询,相关动态查询语句存放在另一个类 ClazzDynaSqlProvider 中,其代码如下。

```
1.  package ssh.homework.dao.provider;
2.
3.  import static ssh.homework.common.HomeworkConstants.CLAZZTABLE;
4.  import ssh.homework.domain.Clazz;
5.
6.  public class ClazzDynaSqlProvider {
7.      //分页动态查询
8.      public String selectWhitParam(Map<String, Object>params) {
9.        String sql=new SQL() {
10.       {
11.           SELECT("*");
12.           FROM(CLAZZTABLE);
13.           if (params.get("clazz") !=null) {
14.               Clazz clazz=(Clazz) params.get("clazz");
15.               if (clazz.getCname() !=null && !clazz
                  .getCname().equals("")) {
16.                  WHERE(" cname LIKE CONCAT ('% ',#{clazz.cname},'% ') ");
17.               }
18.
19.           }
20.       }
21.       }.toString();
22.
23.       if (params.get("pageModel") !=null) {
24.           sql +=" limit #{pageModel.firstLimitParam} , #{pageModel
              .pageSize} ";
```

```
25.        }
26.
27.        return sql;
28.    }
29.
30.    // 动态查询总数量
31.    public String count(Map<String, Object>params) {
32.      return new SQL() {
33.        {
34.            SELECT("count(*)");
35.            FROM(CLAZZTABLE);
36.            if (params.get("clazz") !=null) {
37.                Clazz tea =(Clazz) params.get("clazz");
38.                if (tea.getCname() !=null && !tea.getCname().equals("")) {
39.                    WHERE(" cname LIKE CONCAT('% ',#{clazz.cname},'%') ");
40.                }
41.
42.            }
43.         }
44.      }.toString();
45.    }
46.
47.  }
```

ClazzDynaSqlProvider 类中提供了动态 SQL 语句，实现对 tb_clazz 表的动态更新和查询。其中第 8 行的 selectWhitParam() 是动态查询方法，根据传递的 params 参数，判断这个 Map 中的键值是否为空，决定是否将查询条件添加到 SQL 语句中，组成动态查询语句。第 31 行的 count() 方法是得到查询记录的总数量，用于分页显示时计算每页的行数和一共有多少要显示的页。Coutn() 方法的形参应该与 selectWhitParam() 方法的形参完全一致，这样才能保证正确显示查询结果。

2. 业务逻辑层

班级管理模块的第二层是 Service 层，对应的类是 ClazzService，代码如下。

```
1.  package ssh.homework.service;
2.
3.  import ssh.homework.domain.Clazz;
4.  import ssh.homework.tag.PageModel;
5.
6.  public interface ClazzService {
7.    //添加班级
8.    void addClazz(Clazz clazz);
9.    //根据 id 删除班级
```

```
10.     void removeClazzById(Integer id);
11.     //修改班级
12.     void modifyClazz(Clazz clazz);
13.     //根据id查询班级,返回的是一个Clazz对象
14.     Clazz findClazzById(Integer id);
15.     //根据Clazz对象中属性的值进行动态查询,返回值是一个List
16.     List<Clazz>findClazz(Clazz clazz,PageModel pagemodel);
17.     //查询班级表中的所有记录,返回值是一个List
18.     List<Clazz>findAllClazz();
19.     //这是一对多查询,查询班级,同时查询班级中所有学生的信息
20.     Clazz findOneToManyByClazzId(Integer id);
21. }
```

ClazzService 接口的实现类是 ClazzServiceImpl,其代码如下。

```
1.  package ssh.homework.service.impl;
2.
3.  import ssh.homework.dao.ClazzDao;
4.  import ssh.homework.domain.Clazz;
5.  import ssh.homework.service.ClazzService;
6.  import ssh.homework.tag.PageModel;
7.  //在IoC容器中注册,创建一个id="clazzService"的Bean
8.  @Service("clazzService")
9.  //处理班级相关业务逻辑的服务类
10. public class ClazzServiceImpl implements ClazzService {
11.     @Autowired
12.     ClazzDao clazzDao;
13.     //添加班级
14.     @Override
15.     public void addClazz(Clazz clazz) {
16.         if (clazzDao.selectByCname(clazz.getCname()).size() <=0)
                                                    //班级名称不能重复
17.             clazzDao.save(clazz);
18.     }
19.     //根据id删除班级
20.     @Override
21.     public void removeClazzById(Integer id) {
22.         clazzDao.deleteById(id);
23.     }
24.     //修改班级
25.     @Override
26.     public void modifyClazz(Clazz clazz) {
27.         clazzDao.update(clazz);
```

```
28.          }
29.      }
30.      //根据id查询班级,返回一个Clazz对象
31.      @Override
32.      public Clazz findClazzById(Integer id) {
33.          return clazzDao.selectById(id);
34.      }
35.      //根据Clazz对象中属性的值进行动态查询
36.      @Override
37.      public List<Clazz>findClazz(Clazz clazz, PageModel pageModel) {
38.          /** 当前需要分页的总数据条数 */
39.          Map<String, Object>params=new HashMap<>();
40.          params.put("clazz", clazz);
41.          int recordCount=clazzDao.count(params);
42.          pageModel.setRecordCount(recordCount);
43.          if (recordCount >0) {
44.              /** 开始分页查询数据:查询第几页的数据 */
45.              params.put("pageModel", pageModel);
46.          }
47.          List<Clazz>clazzs=clazzDao.selectByPage(params);
48.          return clazzs;
49.      }
50.      //返回班级的所有对象,即tb_clazz表中的所有记录
51.      @Override
52.      public List<Clazz>findAllClazz() {
53.          return clazzDao.selectAll();
54.      }
55.      // 根据班级查询关联的对象,即查询班级及班级所有学生的信息
56.      @Override
57.      public Clazz findOneToManyByClazzId(Integer id) {
58.          return clazzDao.selectClazzByIdToMany(id);
59.      }
60. }
```

ClazzService类的主要功能是对tb_clazz表的操作,类的属性clazzDao由IoC自动注入,大部分功能通过调用clazzDao对象的相关方法完成。由于在applicationContext.xml文件中已经配置了事务,此处无须再配置,以find开头的方法是只读事务,其他方法都会启用事务管理。

3. Web表示层

ClazzController类是班级管理模块的控制器,所有的操作都需经过它指挥完成,它是这个模块的核心类。这个类中主要有以下4个方法。

- addClazz():添加班级方法,返回类型为ModelAndView。

- removeClazz()：删除班级方法，返回类型为 ModelAndView。
- updateClazz()：更新班级方法，返回类型为 ModelAndView。
- selectClazz()：查询班级方法，返回类型为 String。

ClazzController 类的代码如下。

```
1.   package ssh.homework.controller;
2.
3.   import ssh.homework.domain.Clazz;
4.   import ssh.homework.service.ClazzService;
5.   import ssh.homework.tag.PageModel;
6.   //处理班级表 tb_clazz 的所有请求控制器
7.   @Controller
8.   public class ClazzController {
9.       //班级表 tb_clazz 的业务逻辑处理类 ClazzService 的实例，由 IoC 容器注入
10.      @Autowired
11.      @Qualifier("clazzService")
12.      private ClazzService clazzService;
13.      //查询班级请求处理方法
14.      @RequestMapping(value="/clazz/selectClazz")
15.      public String selectClazz(Integer pageIndex,
16.          @ModelAttribute Clazz clazz,
17.          Model model){
18.          //控制分页的对象
19.          PageModel pageModel=new PageModel();
20.          if(pageIndex !=null){
21.              pageModel.setPageIndex(pageIndex);
22.          }
23.          /** 查询班级信息,clazzs存放满足查询条件的班级对象   */
24.          List<Clazz>clazzs=clazzService.findClazz(clazz, pageModel);
25.          /** clazzs 存放在 model 中用,用于在 JSP 页面中显示   */
26.          model.addAttribute("clazzs", clazzs);
27.          model.addAttribute("pageModel", pageModel);
28.          //转发到 clazz 目录下的 clazz.jsp 页面
29.          return "clazz/clazz";
30.
31.      }
32.      /** 添加班级请求处理,flag=1 表示显示 addClazz.jsp 页面
33.       * flag=2 表示对 tb_clazz 表执行添加操作*/
34.      @RequestMapping(value="/clazz/addClazz")
35.      public ModelAndView addClazz(
36.          String flag,
37.          @ModelAttribute Clazz clazz,
38.          ModelAndView mv){
```

```java
39.         if(flag.equals("1")){
40.             //设置跳转到添加页面
41.             mv.setViewName("clazz/addClazz");
42.         }else{
43.             //执行添加操作
44.             clazzService.addClazz(clazz);
45.             //设置客户端跳转到查询班级请求
46.             mv.setViewName("redirect:/clazz/selectClazz");
47.         }
48.         //返回
49.         return mv;
50.     }
51.     /** 删除班级请求处理方法,ids 参数是一个字符串数组,存放的是要删除的班级id*/
52.     @RequestMapping(value="/clazz/removeClazz")
53.     public ModelAndView removeClzz(String ids,ModelAndView mv){
54.         //分解 id 字符串
55.         String[] idArray=ids.split(",");
56.
57.         for(String id : idArray){
58.             //根据 id 删除员工
59.             try {
60.                 /** 根据 id 删除班级表 tb_clazz 中的记录*/
61.                 clazzService.removeClazzById(Integer.parseInt(id));
62.                 }catch(Exception ec) {
63.                     //如果有异常,则将错误信息保存在 error 变量中
64.                     mv.addObject("error","班级不可删除");
65.                     //设置转发到 error/error.jsp 页面
66.                     mv.setViewName("error/error");
67.                     return mv;
68.                 }
69.         }
70.         //设置客户端跳转到查询请求
71.         mv.setViewName("redirect:/clazz/selectClazz");
72.         //返回 ModelAndView
73.         return mv;
74.     }
75.     /** 更新班级请求处理方法,clazz 参数存放要更新的班级信息*/
76.     @RequestMapping(value="/clazz/updateClazz")
77.     public ModelAndView updateClazz(
78.             String flag,
79.             @ModelAttribute Clazz clazz,
80.             ModelAndView mv){
```

```
81.    //根据 flag 的值是 1 还是 2,决定是返回更新页面,还是执行更新操作
82.    if(flag.equals("1")){
83.        //根据 id 查询用户
84.        Clazz target =clazzService.findClazzById(clazz.getId());
85.        //设置 Model 数据
86.        mv.addObject("clazz", target);
87.        //返回修改员工页面
88.        mv.setViewName("clazz/showUpdateClazz");
89.    }else{
90.        //执行修改操作
91.        clazzService.modifyClazz(clazz);
92.        //设置客户端跳转到查询请求
93.        mv.setViewName("redirect:/clazz/selectClazz");
94.    }
95.    // 返回
96.    return mv;
97.    }
98.
99. }
```

班级管理模块的 JSP 页面部分与教师管理模块相似,读者可参照教师管理模块的 JSP 页面分析,此处不再列出。

10.4.3 学生管理模块

学生管理模块主要涉及 tb_student 表的增、删、改、查功能,本模块涉及的类除了实体类 Student 外,主要包括以下的类和接口。

- ClazzDao 接口、ClazzService 接口和实现类 ClazzServiceImpl。
  由于学生表 tb_student 与班级表 tb_clazz 是多对一的关联,这组类的作用是实现对班级的关联操作。
- StudentDao 接口、StudentService 接口和实现类 StudentServiceImpl。
  这组类实现与 tb_student 相关的所有操作。
- ImportExcelUtilService 接口和实现类 importExcelUtilServiceImpl。
  这组类的功能实现通过 Excel 成批导入学生信息。
- StudentController 类。
  这是学生管理模块的核心控制类,所有操作都由这个类分配完成。

这几个类之间的关系可通过下面的类图加以描述。学生管理模块的 UML 类图如图 10-15 所示。

从图 10-15 可以看出,StudentController 控制器与 ClazzService、StudentService 和 ImportExcelUtilService 3 个组件相关联,这 3 个组件是 StudentController 的一部分,它们的实例都是由 IoC 容器自动注入控制器中。

第 10 章 项目案例：作业管理系统

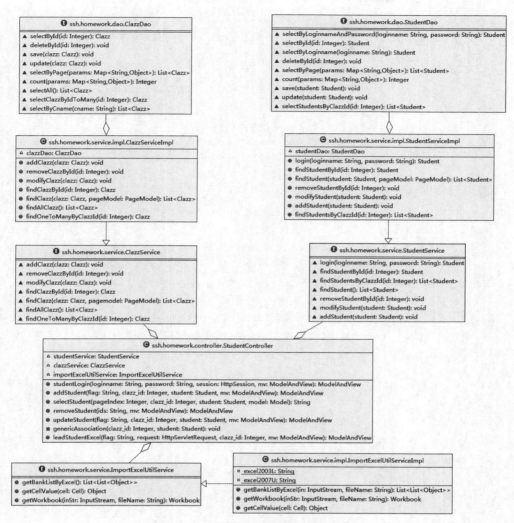

图 10-15　学生管理模块的 UML 类图

1. 数据访问层

数据访问层主要的组件是 StudentDao 接口，其中应用 MyBatis 的注解完成一些简单的数据库操作，涉及复杂的数据库操作，应用动态 SQL 语句完成，这部分代码写在了相应的 provider 类中。

StudentDao 接口代码如下。

```
1.    package ssh.homework.dao;
2.
3.    import static ssh.homework.common.HomeworkConstants.STUDENTTABLE;
4.    import ssh.homework.dao.provider.StudentDynaSqlProvider;
5.    import ssh.homework.domain.Student;
```

```java
6.
7.      /** 数据库表 tb_student 的操作类*/
8.      public interface StudentDao {
9.          //根据登录名和密码查询学生
10.             @Select("select * from "+STUDENTTABLE+" where loginname = #{loginname} and password = #{password}")
11.             Student selectByLoginnameAndPassword(
12.                 @Param("loginname") String loginname,
13.                 @Param("password") String password);
14.
15.         //根据 id 查询学生
16.         @Select("select * from "+STUDENTTABLE+" where id = #{id}")
17.         @Results({
18.             @Result(id=true,column="id",property="id"),
19.             @Result(column="loginname",property="loginname"),
20.             @Result(column="password",property="password"),
21.             @Result(column="username",property="username"),
22.             @Result(column="clazz_id", property="clazz",
23.             one=@One(select="ssh.homework.dao.ClazzDao.selectById",
24.               fetchType=FetchType.EAGER))
25.             })
26.         Student selectById(Integer id);
27.
28.             //根据登录名查询学生
29.          @Select("select * from "+STUDENTTABLE+" where loginname = #{loginname}")
30.         Student selectByLoginname(String loginname);
31.         //根据 id 删除学生
32.         @Delete("delete from "+STUDENTTABLE+" where id = #{id}")
33.         void deleteById(@Param("id") Integer id);
34.
35.         // 动态查询
36.          @SelectProvider(type=StudentDynaSqlProvider.class,method="selectWhitParam")
37.         @Results({
38.             @Result(id=true,column="id",property="id"),
39.             @Result(column="loginname",property="loginname"),
40.             @Result(column="password",property="password"),
41.             @Result(column="username",property="username"),
42.             @Result(column="clazz_id", property="clazz",
43.             one=@One(select="ssh.homework.dao.ClazzDao.selectById",
44.               fetchType=FetchType.EAGER))
45.             })
```

```
46.        List<Student>selectByPage(Map<String,Object>params);
47.
48.        //根据参数查询学生总数
49.        @SelectProvider(type=StudentDynaSqlProvider.class,method="count")
50.        Integer count(Map<String,Object>params);
51.
52.        //动态插入学生
53.          @SelectProvider(type=StudentDynaSqlProvider.class,method=
             "insert")
54.        void save(Student student);
55.        //动态更新学生
56.           @SelectProvider(type=StudentDynaSqlProvider.class,method=
             "update")
57.        void update(Student student);
58.        //根据班级id查询学生
59.        @Select("select * from "+STUDENTTABLE+" where clazz_id=#{id}")
60.        @Results({
61.           @Result(id=true,column="id",property="id"),
62.           @Result(column="loginname",property="loginname"),
63.           @Result(column="password",property="password"),
64.           @Result(column="username",property="username"),
65.           @Result(column="clazz_id", property="clazz",
66.              one=@One(select="ssh.homework.dao.ClazzDao.selectById",
67.           fetchType=FetchType.EAGER))
68.           })
69.        List<Student>selectStudentsByClazzId(Integer id);
70.    }
```

第 17 行 @Results 注解实现 Student 类中的属性与 tb_student 表中字段的映射。其中，tb_student 表与 tb_clazz 表之间存在多对一的关系，也可以说是一对一的关系，体现在 Student 类中的是 clazz 属性，此处通过 @One 注解实现了 clazz 属性与班级表的数据操作类 ClazzDao 中的 selectById() 方法之间的关联。有了这个关联，当调用第 26 行的查询方法 selectById() 时，返回的 Student 对象中的 clazz 属性会包含学生所在班级的信息，如此实现一对一的查询。

第 36 行的 @SelectProvider 注解取代了 @Select 注解，这个注解的作用是指定一个 provider 类及方法存放动态查询语句，这个方法返回一个 SQL 查询字符串。

同样，第 53、56 行也是通过 @SelectProvider 注解指定了相应的存放动态 SQL 语句的类和方法。

存放动态 SQL 语句的 StudentDynaSqlProvider 类与 ClazzDynaSqlProvider 类中的代码类似，此处不再列出。

2. 业务逻辑层

业务逻辑层主要有两个接口：StudentService 和 ImportExcelUtilService。

（1）StudentService 接口的代码。

```
1.  package ssh.homework.service;
2.
3.  import ssh.homework.domain.Student;
4.  import ssh.homework.tag.PageModel;
5.
6.  public interface StudentService {
7.
8.    /**
9.     * 学生登录
10.    * @param loginname
11.    * @param password
12.    * @return User 对象
13.    */
14.   Student login(String loginname,String password);
15.
16.   /**
17.    * 根据 id 查询学生
18.    * @param id
19.    * @return 学生对象
20.    */
21.   Student findStudentById(Integer id);
22.
23.   /**
24.    * 根据班级 id 查询学生
25.    * @param id
26.    * @return 学生对象
27.    */
28.   List<Student> findStudentsByClazzId(Integer id);
29.   /**
30.    * 获得所有学生
31.    * @return User 对象的 List 集合
32.    */
33.   List<Student> findStudent(Student student,PageModel pageModel);
34.
35.   /**
36.    * 根据 id 删除学生
37.    * @param id
38.    */
```

```
39.     void removeStudentById(Integer id);
40.
41.     /**
42.      * 修改学生
43.      * @param Student 学生对象
44.      * */
45.     void modifyStudent(Student student);
46.
47.     /**
48.      * 添加学生
49.      * @param Student 学生对象
50.      * */
51.     void addStudent(Student student);
52.
53. }
```

（2）StudentServiceImpl 类代码。

StudentService 接口的实现类 StudentServiceImpl 代码如下。

```
1.  package ssh.homework.service.impl;
2.
3.  import ssh.homework.dao.StudentDao;
4.  import ssh.homework.domain.Student;
5.  import ssh.homework.service.StudentService;
6.  import ssh.homework.tag.PageModel;
7.  //StudentService 接口实现类,功能是操作 tb_student 表
8.  @Service("studentService")
9.  public class StudentServiceImpl implements StudentService {
10.     @Autowired
11.     StudentDao studentDao;
12.     @Override
13.     //处理学生登录
14.     public Student login(String loginname, String password) {
15.         Student student=studentDao.selectByLoginnameAndPassword
                (loginname,password);
16.         return student;
17.     }
18.     @Override
19.     //根据学生 id 查询学生对象
20.     public Student findStudentById(Integer id) {
21.         //TODO Auto-generated method stub
22.         return studentDao.selectById(id);
```

```
23.    }
24.    @Override
25.    //根据student对象查询tb_student表
26.    public List < Student > findStudent ( Student student, PageModel
         pageModel) {
27.    /** 当前需要分页的总数据条数 */
28.    Map<String,Object>params=new HashMap<>();
29.    params.put("student", student);
30.    int recordCount=studentDao.count(params);
31.    pageModel.setRecordCount(recordCount);
32.    if(recordCount >0){
33.        /** 开始分页查询数据：查询第几页的数据 */
34.        params.put("pageModel", pageModel);
35.    }
36.    List<Student>students=studentDao.selectByPage(params);
37.
38.        return students;
39.    }
40.    @Override
41.    //根据学生id删除学生
42.    public void removeStudentById(Integer id) {
43.        studentDao.deleteById(id);
44.
45.    }
46.    @Override
47.    //根据student对象修改tb_student表请求
48.    public void modifyStudent(Student student) {
49.        studentDao.update(student);
50.
51.    }
52.    @Override
53.    //添加学生到tb_student表
54.    public void addStudent(Student student) {
55.
56.        if(studentDao.selectByLoginname(student.getLoginname())==null)
57.        studentDao.save(student);
58.
59.    }
60.    //根据班级id查询学生
61.    @Override
62.    public List<Student>findStudentsByClazzId(Integer id) {
63.
```

```
64.        return studentDao.selectStudentsByClazzId(id);
65.    }
66. }
```

(3) ImportExcelUtilService 接口。

ImportExcelUtilService 接口代码如下。

```
1.  package ssh.homework.service;
2.  
3.  //处理 Excel 文件的接口
4.  public interface ImportExcelUtilService {
5.      public List<List<Object>> getBankListByExcel(InputStream in, String fileName) throws Exception;
6.      public Object getCellValue(Cell cell);
7.      public Workbook getWorkbook(InputStream inStr, String fileName) throws Exception;
8.  }
```

(4) ImportExcelUtilServiceImpl 类。

ImportExcelUtilServiceImpl 类代码如下。

```
1.  package ssh.homework.service.impl;
2.  
3.  import java.io.IOException;
4.  import java.io.InputStream;
5.  import java.text.DecimalFormat;
6.  import java.text.SimpleDateFormat;
7.  import java.util.ArrayList;
8.  import java.util.List;
9.  
10. import org.apache.poi.hssf.usermodel.HSSFWorkbook;
11. import org.apache.poi.ss.usermodel.Cell;
12. import org.apache.poi.ss.usermodel.Row;
13. import org.apache.poi.ss.usermodel.Sheet;
14. import org.apache.poi.ss.usermodel.Workbook;
15. import org.apache.poi.xssf.usermodel.XSSFWorkbook;
16. import org.springframework.stereotype.Service;
17. 
18. import ssh.homework.service.ImportExcelUtilService;
19. //ImportExcelUtilService 的实现类用来处理 Excel 文件
20. @Service("importExcelUtilService")
21. public class ImportExcelUtilServiceImpl implements ImportExcelUtilService {
```

```java
22.     private final static String excel2003L = ".xls";   //2003-版本的 excel
23.     private final static String excel2007U = ".xlsx"; //2007+版本的 excel
24.     /**
25.      * 描述：获取IO流中的数据,组装成 List<List<Object>>对象
26.      * @param in,fileName
27.      * @return
28.      * @throws IOException
29.      */
30.     @Override
31.     public List<List<Object>>getBankListByExcel(InputStream in,String fileName) throws Exception{
32.         List<List<Object>>list=null;
33.
34.         //创建 Excel 工作簿
35.         Workbook work=this.getWorkbook(in,fileName);
36.         if(null==work){
37.             throw new Exception("创建 Excel 工作簿为空!");
38.         }
39.         Sheet sheet=null;
40.         Row row=null;
41.         Cell cell=null;
42.
43.         list=new ArrayList<List<Object>>();
44.         //遍历 Excel 中所有的 sheet
45.         for (int i=0; i<work.getNumberOfSheets(); i++) {
46.             sheet=work.getSheetAt(i);
47.             if(sheet==null){continue;}
48.
49.             //遍历当前 sheet 中的所有行
50.             for (int j=sheet.getFirstRowNum(); j<sheet.getLastRowNum()+1; j++) {
51.                 row=sheet.getRow(j);
52.                 if(row==null||row.getFirstCellNum()==j)
53.                 {continue;}
54.
55.                 //遍历所有列
56.                 List<Object>li=new ArrayList<Object>();
57.                 for (int y=row.getFirstCellNum(); y<row.getLastCellNum(); y++) {
58.                     cell =row.getCell(y);
59.                     li.add(this.getCellValue(cell));
60.                 }
61.                 list.add(li);
```

```
62.            }
63.        }
64.        in.close();
65.        return list;
66.    }
67.
68.    /**
69.     * 描述：根据文件后缀，自适应上传文件的版本
70.     * @param inStr,fileName
71.     * @return Workbook
72.     * @throws Exception
73.     */
74.    @Override
75.    public Workbook getWorkbook(InputStream inStr, String fileName) throws Exception{
76.        Workbook wb=null;
77.        String fileType=fileName.substring(fileName.lastIndexOf("."));
78.        if(excel2003L.equals(fileType)){
79.            wb=new HSSFWorkbook(inStr);            //2003-
80.        }else if(excel2007U.equals(fileType)){
81.            wb=new XSSFWorkbook(inStr);            //2007+
82.        }else{
83.            throw new Exception("解析的文件格式有误!");
84.        }
85.        return wb;
86.    }
87.
88.    /**
89.     * 描述：对表格中的数值进行格式化
90.     * @param cell
91.     * @return Object
92.     */
93.    @Override
94.    public Object getCellValue(Cell cell){
95.        Object value=null;
96.        DecimalFormat df=new DecimalFormat("0");    //格式化 number String 字符
97.        SimpleDateFormat sdf=new SimpleDateFormat("yyy-MM-dd");
                                                    //日期格式化
98.        DecimalFormat df2=new DecimalFormat("0.00");    //格式化数字
99.
100.       switch (cell.getCellType()){
```

```
101.         case Cell.CELL_TYPE_STRING:
102.             value=cell.getRichStringCellValue().getString();
103.             break;
104.         case Cell.CELL_TYPE_NUMERIC:
105.             if("General".equals(cell.getCellStyle()
                 .getDataFormatString())){
106.                 value=df.format(cell.getNumericCellValue());
107.             }else if("m/d/yy".equals(cell.getCellStyle().getDataFormat-
                 String())){
108.                 value=sdf.format(cell.getDateCellValue());
109.             }else{
110.                 value=df2.format(cell.getNumericCellValue());
111.             }
112.             break;
113.         case Cell.CELL_TYPE_BOOLEAN:
114.             value=cell.getBooleanCellValue();
115.             break;
116.         case Cell.CELL_TYPE_BLANK:
117.             value = "";
118.             break;
119.         default:
120.             break;
121.         }
122.         return value;
123.     }
124.
125. }
```

这个类的功能是将 Excel 文档转为一个 List，其元素为 Object，为后面数据库操作做准备。

3. Web 表示层

StudentController 类是学生管理模块的控制器，所有操作都需经过它控制完成，它是这个模块的核心类。这个类中主要有以下 6 个方法。

- addStudent()：添加学生方法，返回类型为 ModelAndView。
- removeStudent()：删除学生方法，返回类型为 ModelAndView。
- updateStudent()：更新学生方法，返回类型为 ModelAndView。
- selectStudent()：查询学生方法，返回类型为 String。
- leadStudentExcel()：通过 Excel 文档导入学生名单，返回类型为 ModelAndView。
- StudentLogin()：学生登录方法，返回类型为 ModelAndView。

（1）StudentController 类。

StudentController 类代码如下。

```java
1.  package ssh.homework.controller;
2.  
3.  import ssh.homework.common.HomeworkConstants;
4.  import ssh.homework.domain.Clazz;
5.  import ssh.homework.domain.Student;
6.  import ssh.homework.service.ClazzService;
7.  import ssh.homework.service.ImportExcelUtilService;
8.  import ssh.homework.service.StudentService;
9.  import ssh.homework.tag.PageModel;
10. 
11. @Controller
12. public class StudentController {
13.     @Autowired
14.     @Qualifier("studentService")
15.     StudentService studentService;
16.     @Autowired
17.     @Qualifier("clazzService")
18.     ClazzService clazzService;
19. 
20.     @Autowired
21.     @Qualifier("importExcelUtilService")
22.     ImportExcelUtilService importExcelUtilService;
23.     //学生登录
24.     @RequestMapping(value="/student/studentLogin")
25.     public ModelAndView studentLogin (@RequestParam ("loginname")
        String loginname,
26.         @RequestParam("password") String password,
27.         HttpSession session,
28.         ModelAndView mv) {
29.     Student student=studentService.login(loginname, password);
30.     if(student!=null) {
31.         session.setAttribute(HomeworkConstants.STUDENT_SESSION,student);
32.         mv.setViewName("redirect:/studentMain");
33. 
34.     }
35.     else {
36.             //设置登录失败提示信息
37.             mv.addObject("message", "登录名或密码错误!请重新输入");
38.             //服务器内部跳转到登录页面
39.             mv.setViewName("forward:/loginForm");}
40.     return mv;
41.     }
```

```java
42.
43.     /**
44.      * 处理添加学生请求
45.      * @param String flag 标记,1表示跳转到添加页面,2表示执行添加操作
46.      * @param String clazz_id 班级 id
47.      * @param Student student 接收添加参数
48.      * @param ModelAndView mv
49.      **/
50.     @RequestMapping(value="/student/addStudent")
51.      public ModelAndView addStudent(
52.          String flag,
53.          Integer clazz_id,
54.          @ModelAttribute Student student,
55.          ModelAndView mv){
56.         if(flag.equals("1")){
57.             //查询班级信息
58.             List<Clazz>clazzs=clazzService.findAllClazz();
59.
60.             //设置 Model 数据
61.             mv.addObject("clazzs", clazzs);
62.
63.             //返回添加员工页面
64.             mv.setViewName("student/showAddStudent");
65.         }else{
66.             //判断是否有关联对象传递,如果有,则创建关联对象
67.             this.genericAssociation(clazz_id, student);
68.             //添加操作
69.             studentService.addStudent(student);
70.
71.             //设置客户端跳转到查询请求
72.             mv.setViewName("redirect:/student/selectStudent");
73.         }
74.         //返回
75.         return mv;
76.
77.     }
78.
79.     /**
80.      * 处理查询请求
81.      * @param pageIndex 请求的是第几页
82.      * @param String clazz_id 班级 id
83.
84.      * @param student 模糊查询参数
```

```java
85.        * @param Model model
86.        **/
87.       @RequestMapping(value="/student/selectStudent")
88.       public String selectStudent(Integer pageIndex,
89.               Integer clazz_id,
90.               @ModelAttribute Student student,
91.               Model model){
92.           //模糊查询时判断是否有关联对象传递,如果有,则创建并封装关联对象
93.           this.genericAssociation(clazz_id, student);
94.           //创建分页对象
95.           PageModel pageModel=new PageModel();
96.           pageModel.setPageSize(50);
97.           //如果参数 pageIndex 不为 null,则设置 pageIndex,即显示第几页
98.           if(pageIndex !=null){
99.               pageModel.setPageIndex(pageIndex);
100.          }
101.          //查询班级信息,用于模糊查询
102.          List<Clazz> clazzs=clazzService.findAllClazz();
103.
104.          //查询员工信息
105.          List<Student> students=studentService.findStudent(student,
                  pageModel);
106.          //设置 Model 数据
107.          model.addAttribute("students", students);
108.          model.addAttribute("clazzs", clazzs);
109.
110.          model.addAttribute("pageModel", pageModel);
111.          //返回员工页面
112.          return "student/student";
113.
114.      }
115.      /**
116.       * 处理删除学生请求
117.       * @param String ids 需要删除的 id 字符串
118.       * @param ModelAndView mv
119.       **/
120.      @RequestMapping(value="/student/removeStudent")
121.      public ModelAndView removeStudent(String ids,ModelAndView mv) {
122.          String[] idArray=ids.split(",");
123.          for(String id : idArray){
124.              try {
125.                  //根据 id 删除学生
126.                  studentService.removeStudentById(Integer.parseInt(id));
```

```java
127.            }catch(Exception ep) {
128.                mv.addObject("error","学生不能删除!");
129.                mv.setViewName("error/error");
130.                return mv;
131.            }
132.        }
133.        //设置客户端跳转到查询请求
134.        mv.setViewName("redirect:/student/selectStudent");
135.
136.        return mv;
137.
138.    }
139.    /**
140.     * 处理修改学生请求
141.     * @param String flag 标记,1表示跳转到修改页面,2表示执行修改操作
142.     * @param String clazz_id 班级 id 号
143.     * @param Student student 要修改学生的对象
144.     * @param ModelAndView mv
145.     **/
146.    @RequestMapping(value="/student/updateStudent")
147.    public ModelAndView updateStudent(
148.        String flag,
149.        Integer clazz_id,
150.        @ModelAttribute Student student,
151.        ModelAndView mv) {
152.        if(flag.equals("1")) {
153.        //根据 id 查找要修改的学生
154.        Student target=studentService.findStudentById(student.getId());
155.        //查询所有班级
156.        List<Clazz>clazzs=clazzService.findAllClazz();
157.
158.        mv.addObject("student",target);
159.        mv.addObject("clazzs",clazzs);
160.        mv.setViewName("student/showUpdateStudent");
161.
162.        }
163.        else
164.        { //封装关联对象
165.            this.genericAssociation(clazz_id, student);
166.            //执行更新
167.            studentService.modifyStudent(student);
168.
169.            mv.setViewName("redirect:/student/selectStudent");
```

```
170.        }
171.
172.        return mv;
173.    }
174.    /**
175.     * 由于班级在 Student 中是对象关联映射,
176.     * 所以不能直接接收参数,需要创建 Clazz 对象
177.     * */
178.    private void genericAssociation(Integer clazz_id,Student student){
179.        if(clazz_id !=null){
180.            Clazz clazz=new Clazz();
181.            clazz.setId(clazz_id);
182.            student.setClazz(clazz);
183.        }
184.
185.    }
186.    //通过 Excel 导入学生信息
187.    @RequestMapping(value="/student/leadStudentExcel")
188.    public ModelAndView leadStudentExcel(String flag,
189.        HttpServletRequest request,
190.        Integer clazz_id,
191.        ModelAndView mv) throws Exception
192.    {
193.
194.        if(flag.equals("1")) {
195.            //查询班级信息,用于选择班级
196.            List<Clazz>clazzs=clazzService.findAllClazz();
197.            mv.addObject("clazzs",clazzs);
198.            mv.setViewName("student/leadStudentExcel");
199.        }
200.        else
201.        {
202.            MultipartHttpServletRequest  multipartRequest  =
                (MultipartHttpServletRequest) request;
203.
204.            InputStream in=null;
205.            List<List<Object>>listob=null;
206.            MultipartFile file=multipartRequest.getFile("upfile");
207.
208.            if(file.isEmpty()){
209.                throw new Exception("文件不存在!");
210.            }
211.            in=file.getInputStream();
```

```
212.            listob= importExcelUtilService.getBankListByExcel(in,file.get-
                   OriginalFilename());
213.            //该处可调用service相应方法将数据保存到数据库中,现只对数据输出
214.            Clazz clazz=new Clazz();
215.            clazz.setId(clazz_id);
216.            for (int i=0; i<listob.size(); i++) {
217.                List<Object>lo =listob.get(i);
218.                if(lo.get(0)==null||lo.get(0).equals("")||lo.get(1)==null
                      ||lo.get(1).equals(""))break;
219.                Student student=new Student();
220.                student.setLoginname((String)lo.get(0));
221.                student.setPassword((String)lo.get(0));
222.                student.setUsername((String)lo.get(1));
223.                student.setClazz(clazz);
224.                studentService.addStudent(student);
225.            }
226.
227.        in.close();
228.        mv.setViewName("student/success");
229.    }
230.    return mv;
231. }
232.
233. }
```

第 188 行 leadStudentExcel() 方法的作用是：首先应用 Spring MVC 中的 MultipartFile 类接收 Excel 文件，然后调用 importExcelUtilService 类中的 getBankListByExcel()方法将 Excel 文件中的学生信息转换为一个 List，最后将 List 中的学生信息存入 tb_student 学生表中。在页面中传递 Excel 文件的同时要将学生所在班级 id 一起发送过来。这样，在添加学生信息的同时，将学生所在班级的 id 同时加入数据库中。

（2）上传 Excel 文件页面。

上传 Excel 文档的 JSP 页面是 leadStudentExcel.jsp，页面显示如图 10-16 所示。

图 10-16　上传学生名单 Excel 文件页面

leadStudentExcel.jsp 代码如下。

```
1.  <form method="POST" enctype="multipart/form-data" id="form1" action=
    "${ctx}/student/leadStudentExcel">
2.
3.      <label>上传文件:</label>
4.      <input id="upfile" type="file" name="upfile"><br><br>
5.      <input type="hidden" name="flag" value="2"/>
6.      班级:
7.      <select name="clazz_id" style ="width:143px;">
8.                          <option value="0">--请选择班级
    --</option>
9.      <c:forEach items="${requestScope.clazzs}" var="clazz">
10.         <option value="${clazz.id }">${clazz.cname}</option>
11.     </c:forEach>
12.     </select>
13.     <input type="submit" value="提交" onclick="return checkData()">
14.  </form>
```

第7~12行是一个下拉列表框，显示所有班级信息，上传文件前要先选择班级，上传Excel文件的同时上传班级的id，处理请求映射是student/leadStudentExcel，对应的是StudentController类中的leadStudentExcel()方法，这个方法在前面已经讲解过，这里不再赘述。

学生管理模块的其他JSP页面部分与教师管理模块相似，读者可参照教师管理模块的JSP页面分析，此处不再列出。

10.4.4 课程管理模块

课程管理模块主要涉及tb_course表的增、删、改、查功能，本模块涉及的类除了实体类Course外，主要包括以下的接口和类。

- CourseDao接口：定义课程表的增、删、改、查方法。
- CourseService接口：定义业务逻辑的服务接口。
- CourseServiceImpl类：是CourseService的实现类。
- CourseController类：是课程管理模块的核心控制类，所有操作都由这个类分配完成。

这几个类之间的关系可通过下面的类图加以描述。课程管理模块的UML类图如图10-17所示。

从图10-17可以看出CourseController控制器与CourseService类之间的关系，CourseService类的对象courseService是CourseController的一部分，CourseService类实例由IoC容器自动注入控制器中。

同样，CourseDao的实例对象courseDao也是CourseServiceImpl类的成员变量，是其组成的一部分，CourseDao接口的实例化也是由IoC容器自动注入控制器中。

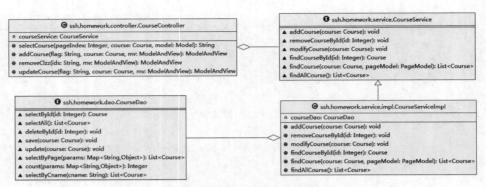

图 10-17　课程管理模块的 UML 类图

1. 数据访问层

数据访问层主要的组件是 CourseDao 接口，其中应用 MyBatis 的注解完成一些简单的数据库操作，涉及复杂的数据库操作，应用动态 SQL 语句完成，这部分代码写在相应的 provider 类中。

CourseDao 接口代码如下。

```
1.  package ssh.homework.dao;
2.
3.  import ssh.homework.dao.provider.CourseDynaSqlProvider;
4.  import ssh.homework.domain.Course;
5.  import static ssh.homework.common.HomeworkConstants.COURSETABLE;
6.
7.  public interface CourseDao {
8.      //根据 id 查询课程
9.      @Select("select * from " +COURSETABLE +" where id =#{id}")
10.     Course selectById(Integer id);
11.     //查询所有课程
12.     @Select("select * from " +COURSETABLE +" ")
13.     List<Course>selectAll();
14.
15.     //根据 id 删除课程
16.     @Delete(" delete from " +COURSETABLE +" where id =#{id} ")
17.     void deleteById(Integer id);
18.
19.     //插入课程
20.     @Insert("insert into " + COURSETABLE + " (cname) values (" + "#{cname})")
21.     @Options(useGeneratedKeys =true, keyProperty ="id")
22.     void save(Course course);
23.
```

```
24.        //更新课程
25.        @Update("update " +COURSETABLE +" SET CNAME=#{cname} WHERE id=#{id}")
26.        void update(Course course);
27.
28.        //动态查询
29.        @SelectProvider(type=CourseDynaSqlProvider.class,method= "select-
           WhitParam")
30.        List<Course> selectByPage(Map<String, Object>params);
31.
32.        //根据参数查询课程总数
33.        @SelectProvider(type=CourseDynaSqlProvider.class,method="count")
34.        Integer count(Map<String, Object>params);
35.        //根据课程名称精确查询课程
36.          @Select("select * from " +COURSETABLE +" where cname =#{cname}")
37.          List<Course> selectByCname(@Param("cname") String cname);
38.    }
```

第 30 行的 selectByPage()方法是一个动态查询,其参数是一个 Map,查询的条件是以键值对的方式存放在 params 参数中,动态查询的语句写在与之相应的 provider 文件中,这个动态 SQL 文件代码分析可参照教师管理模块中相应部分的讲解,此处不再列出。

2. 业务逻辑层

课程管理模块的业务层只有一个服务接口 CourseService,其代码如下。

```
1.    package ssh.homework.service;
2.
3.    import java.util.List;
4.
5.    import ssh.homework.domain.Course;
6.    import ssh.homework.tag.PageModel;
7.
8.    public interface CourseService {
9.      //添加课程
10.     void addCourse(Course course);
11.     //根据 id 删除课程
12.     void removeCourseById(Integer id);
13.     //修改课程
14.     void modifyCourse(Course course);
15.     //根据 id 查询课程
16.     Course findCourseById(Integer id);
17.     //根据 course 对象中属性的值动态查询课程表
18.     List<Course>findCourse(Course course,PageModel pageModel);
```

```
19.        //查询课程表中的所有课程
20.        List<Course> findAllCourse();
21.    }
```

CourseService 接口的实现类 CourseServiceImpl 的代码如下。

```
1.   package ssh.homework.service.impl;
2.
3.   import ssh.homework.dao.CourseDao;
4.   import ssh.homework.domain.Course;
5.   import ssh.homework.service.CourseService;
6.   import ssh.homework.tag.PageModel;
7.
8.   @Service("courseService")
9.   public class CourseServiceImpl implements CourseService {
10.      //操作 tb_course 表的 Dao 类,其实例由 IoC 容器注入
11.      @Autowired
12.      CourseDao courseDao;
13.      //添加课程
14.      @Override
15.      public void addCourse(Course course) {
16.          if(courseDao.selectByCname(course.getCname()).size()<=0)
17.              courseDao.save(course);
18.      }
19.      //根据 id 删除课程
20.      @Override
21.      public void removeCourseById(Integer id) {
22.          courseDao.deleteById(id);
23.      }
24.      //修改课程
25.      @Override
26.      public void modifyCourse(Course course) {
27.          courseDao.update(course);
28.
29.      }
30.      //根据 id 查询课程
31.      @Override
32.      public Course findCourseById(Integer id) {
33.          return courseDao.selectById(id);
34.      }
35.      //根据 course 对象中属性的值动态查询课程表
36.      @Override
```

```
37.    public List<Course>findCourse(Course course, PageModel pageModel) {
38.        /** 当前需要分页的总数据条数 */
39.        Map<String,Object>params=new HashMap<>();
40.        params.put("course", course);
41.        int recordCount=courseDao.count(params);
42.        pageModel.setRecordCount(recordCount);
43.        if(recordCount >0){
44.            /** 开始分页查询数据：查询第几页的数据 */
45.            params.put("pageModel", pageModel);
46.        }
47.        List<Course>clazzs=courseDao.selectByPage(params);
48.        return clazzs;
49.    }
50.    //查询课程表中的所有课程
51.    @Override
52.    public List<Course>findAllCourse() {
53.        return courseDao.selectAll();
54.    }
55. }
```

在 CourseServiceImpl 类中，第 11 行 @Autowired 注解通过 IoC 自动注入了 CourseDao 接口的对象。

3. Web 表示层

CourseController 类是课程管理模块的控制器，所有操作都需经过它来指挥完成，它是这个模块的核心类。这个类中主要有以下 4 个方法。

- addCourse()：添加课程方法，返回类型为 ModelAndView。
- removeCourse()：删除课程方法，返回类型为 ModelAndView。
- updateCourse()：更新课程方法，返回类型为 ModelAndView。
- selectCourse()：查询课程方法，返回类型为 String。

CourseController 控制器代码如下。

```
1.  package ssh.homework.controller;
2.
3.  import ssh.homework.domain.Course;
4.  import ssh.homework.service.CourseService;
5.  import ssh.homework.tag.PageModel;
6.  //有关课程请求的处理器类
7.  @Controller
8.  public class CourseController {
9.
10.     @Autowired
```

```java
11.     @Qualifier("courseService")
12.     //课程业务逻辑服务类,由IoC容器实例化并注入其对象
13.     private CourseService courseService;
14.     //处理查询课程的请求,查询条件由参数course中的属性决定
15.     @RequestMapping(value="/course/selectCourse")
16.     public String selectCourse(Integer pageIndex,
17.         @ModelAttribute Course course,
18.         Model model){
19.       PageModel pageModel=new PageModel();
20.       if(pageIndex!=null){
21.           pageModel.setPageIndex(pageIndex);
22.       }
23.       /** 查询用户信息   */
24.       List<Course> courses=courseService.findCourse(course, pageModel);
25.       model.addAttribute("courses", courses);
26.       model.addAttribute("pageModel", pageModel);
27.       return "course/course";
28.
29.     }
30.     //添加课程的请求处理
31.     @RequestMapping(value="/course/addCourse")
32.      public ModelAndView addCourse(
33.          String flag,
34.          @ModelAttribute Course course,
35.          ModelAndView mv){
36.        if(flag.equals("1")){
37.            //设置跳转到添加页面
38.            mv.setViewName("course/showAddCourse");
39.        }else{
40.            //执行添加操作
41.            courseService.addCourse(course);
42.            //设置客户端跳转到查询请求
43.            mv.setViewName("redirect:/course/selectCourse");
44.        }
45.        //返回
46.        return mv;
47.     }
48.     //删除课程的请求处理
49.     @RequestMapping(value="/course/removeCourse")
50.      public ModelAndView removeClzz(String ids,ModelAndView mv){
51.          // 分解id字符串
52.          String[] idArray=ids.split(",");
53.          try{
```

```
54.          for(String id : idArray){
55.              //根据id删除员工
56.              courseService.removeCourseById(Integer.parseInt(id));
57.          }
58.      }catch(Exception ex) {
59.          mv.addObject("error","课程不能删除!");
60.          mv.setViewName("error/error");
61.          return mv;
62.      }
63.      //设置客户端跳转到查询请求
64.      mv.setViewName("redirect:/course/selectCourse");
65.      //返回ModelAndView
66.      return mv;
67.  }
68.  //更新课程的请求处理
69.  @RequestMapping(value="/course/updateCourse")
70.   public ModelAndView updateCourse(
71.          String flag,
72.          @ModelAttribute Course course,
73.          ModelAndView mv){
74.      if(flag.equals("1")){
75.          //根据id查询用户
76.          Course target=courseService.findCourseById(course.getId());
77.          //设置Model数据
78.          mv.addObject("course", target);
79.          //返回修改员工页面
80.          mv.setViewName("course/showUpdateCourse");
81.      }else{
82.          //执行修改操作
83.          courseService.modifyCourse(course);
84.          //设置客户端跳转到查询请求
85.          mv.setViewName("redirect:/course/selectCourse");
86.      }
87.      //返回
88.      return mv;
89.  }
90. }
```

CourseController 类只有一个成员变量 courseService，这是课程的业务服务接口，由 IoC 负责创建实例，并将实例注入 CourseController 中。

课程管理模块的 JSP 页面部分与教师管理模块相似，读者可参照教师管理模块的 JSP 页面分析，此处不再列出。

10.4.5 习题管理模块

习题管理模块主要涉及 tb_exercise 表的增、删、改、查功能。由于 tb_exercise 表与 tb_teacher 表、tb_course 表都是多对一的关联，因此，对习题表的操作会涉及另外两个表，具体操作时会用到另外两个表的业务逻辑服务类，主要包括以下的类和接口。

- CourseDao 接口、CourseService 接口和实现类 CourseServiceImpl。
  由于 tb_exercise 表与 tb_course 表是多对一的关联，这组类的作用是实现对课程表的关联操作。
- ExerciseDao 接口、ExerciseService 接口和实现类 ExerciseServiceImpl。
  这组类实现与 tb_exercise 相关的所有操作。
- TeacherDao 接口、TeacherService 接口和实现类 TeacherServiceImpl。
  由于课程表 tb_exercise 与教师表 tb_teacher 是多对一的关联，因此这组类的作用是实现对教师表的关联操作。
- ExerciseController 类，这是习题管理模块的核心控制类，所有操作都经由这个类分配完成。

这几个类之间的关系可通过类图加以描述。习题管理模块的 UML 类图如图 10-18 所示。

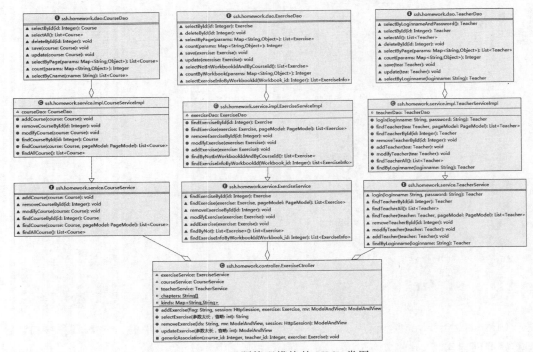

图 10-18　习题管理模块的 UML 类图

从图 10-18 可以看出 ExerciseController 控制器与 CourseService、TeacherService 和 ExerciseService 3 个组件相关联，这 3 个组件是 ExerciseController 的一部分，它们的实

例都是由 IoC 容器自动注入控制器中。

1. 数据访问层

数据访问层主要的组件是 ExerciseDao 接口,其中应用 MyBatis 的注解完成一些简单的数据库操作,涉及复杂的数据库操作,应用动态 SQL 语句完成,这部分代码写在相应的 provider 类中。

ExerciseDao 接口代码如下所示。

```
1.    package ssh.homework.dao;
2.
3.    import static ssh.homework.common.HomeworkConstants.EXERCISETABLE;
4.    import ssh.homework.dao.provider.ExerciseDynaSqlProvider;
5.    import ssh.homework.domain.Exercise;
6.    import ssh.homework.domain.ExerciseInfo;
7.    //tb_exercise 表的操作类
8.    public interface ExerciseDao {
9.        //根据 id 查询习题
10.       @Select("select * from "+EXERCISETABLE+" where id=#{id} ")
11.       @Results({
12.           @Result(id=true,column="id",property="id"),
13.           @Result(column="answer",property="answer"),
14.           @Result(column="chapter",property="chapter"),
15.           @Result(column="content",property="content"),
16.           @Result(column="tea_id", property="teacher",
17.              one=@One(select="ssh.homework.dao.TeacherDao.selectById",
18.               fetchType=FetchType.EAGER)),
19.           @Result(column="course_id", property="course",
20.              one=@One(select="ssh.homework.dao.CourseDao.selectById",
21.               fetchType=FetchType.EAGER))
22.       })
23.       Exercise selectById(Integer id);
24.
25.       //根据 id 删除习题
26.       @Delete(" delete from "+EXERCISETABLE+" where id=#{id} ")
27.       void deleteById(@Param("id") Integer id);
28.
29.       //动态查询
30.       @SelectProvider(type = ExerciseDynaSqlProvider.class, method =
              "selectWhitParam")
31.       @Results({
32.           @Result(id=true,column="id",property="id"),
33.           @Result(column="answer",property="answer"),
34.           @Result(column="chapter",property="chapter"),
35.           @Result(column="content",property="content"),
```

```
36.        @Result(column="course_id", property="course",
37.            one=@One(select="ssh.homework.dao.CourseDao.selectById",
38.             fetchType=FetchType.EAGER)),
39.        @Result(column="tea_id", property="teacher",
40.            one=@One(select="ssh.homework.dao.TeacherDao.selectById",
41.             fetchType=FetchType.EAGER))
42.        })
43.     List<Exercise>selectByPage(Map<String,Object>params);
44.
45.     //根据参数查询习题总数
46.     @SelectProvider(type=ExerciseDynaSqlProvider.class,method="count")
47.     Integer count(Map<String,Object>params);
48.
49.     //动态插入习题
50.     @SelectProvider( type = ExerciseDynaSqlProvider. class, method = "insert")
51.     void save(Exercise exercise);
52.     //动态更新习题
53.        @SelectProvider( type = ExerciseDynaSqlProvider. class, method = "update")
54.     void update(Exercise exercise);
55.
56.     //将已经布置到tb_assignment表中的习题从tb_exercise表中除去,这样已经
        //布置过的习题就不会出现在习题列表中
57.     @SelectProvider( type = ExerciseDynaSqlProvider. class, method = "selectByWorkbook")
58.     @Results({
59.        @Result(id=true,column="id",property="id"),
60.        @Result(column="answer",property="answer"),
61.        @Result(column="chapter",property="chapter"),
62.        @Result(column="content",property="content"),
63.        @Result(column="course_id", property="course",
64.            one=@One(select="ssh.homework.dao.CourseDao.selectById",
65.             fetchType=FetchType.EAGER)),
66.        @Result(column="tea_id", property="teacher",
67.            one=@One(select="ssh.homework.dao.TeacherDao.selectById",
68.             fetchType=FetchType.EAGER))
69.        })
```

```
70.        List< Exercise > selectNotInWorkbookIdAndByCourseId (Map < String,
           Object>params);
71.
72.        @SelectProvider( type = ExerciseDynaSqlProvider. class, method =
           "countByWorkbook")
73.        Integer countByWorkbook(Map<String, Object>params);
74.        //统计同一个作业中每个习题的错误次数
75.        @Select("SELECT exercise_id,count(*) as count "+
76.           "FROM student_workbook where workbook_id=#{workbook_id} and
           grade>score "
77.           +"group by exercise_id ")
78.        @Results({
79.           @Result(column="count",property="count"),
80.           @Result(column="exercise_id", property="exercise",
81.             one=@One(select="ssh.homework.dao.ExerciseDao.selectById",
82.               fetchType=FetchType.EAGER))
83.           })
84.        List<ExerciseInfo>selectExerciseInfoByWorkbookId(Integer Workbook_id);
85.     }
```

第 70 行的 selectNotInWorkbookIdAndByCourseId()方法是动态查询,其参数是一个 Map,查询的条件是以键值对的方式存放在 params 参数中,动态查询的语句写在与之相应的 provider 文件中。这个方法的功能是查询还没有在作业中出现的习题,查询这个动态 SQL 文件代码分析可参照教师管理模块中相应部分的讲解,此处不再列出。

2. 业务逻辑层

习题管理模块的业务逻辑层主要涉及 ExerciseService、TeacherService 和 CourseService 3 个接口。TeacherService 和 CourseService 接口前面已经介绍过,在此不再讲解。下面重点讲解 ExerciseService 接口及实现类。

(1) ExerciseService 接口。

ExerciseService 接口代码如下。

```
1.     package ssh.homework.service;
2.
3.     import ssh.homework.domain.Exercise;
4.     import ssh.homework.domain.ExerciseInfo;
5.     import ssh.homework.domain.Workbook;
6.     import ssh.homework.tag.PageModel;
7.
8.     public interface ExerciseService {
```

```
9.   /**
10.    * 根据id查询习题
11.    * @param id
12.    * @return 习题对象
13.    */
14.   Exercise findExerciseById(Integer id);
15.   /**
16.    * 根据exercise对象中属性的值动态查询习题
17.    * @return Exercise对象的List集合
18.    */
19.   List<Exercise> findExercise(Exercise exercise,PageModel pageModel);
20.   /**
21.    * 根据id删除习题
22.    * @param id
23.    */
24.   void removeExerciseById(Integer id);
25.
26.   /**
27.    * 修改习题
28.    * @param Exercise 习题对象
29.    */
30.   void modifyExercise(Exercise exercise);
31.
32.   /**
33.    * 添加习题
34.    * @param Exercise 习题对象
35.    */
36.   void addExercise(Exercise exercise);
37.   /**
38.    * 根据workbook查询不在当前作业中的习题
39.    * @param Workbook 作业对象
40.    * @return 习题对象的集合
41.    */
42.   List<Exercise> findByNotInWorkbookIdAndByCourseId(Workbook workbook,PageModel pageModel,String isNotIn);
43.   //统计同一个作业中每个习题的错误次数
44.   List<ExerciseInfo> findExerciseInfoByWorkbookId(Integer Workbook_id);
45. }
```

（2）ExerciseServiceImpl类。

ExerciseServiceImpl是ExerciseService的实现类，代码如下。

```java
1.  package ssh.homework.service.impl;
2.
3.  import ssh.homework.dao.ExerciseDao;
4.  import ssh.homework.domain.Exercise;
5.  import ssh.homework.domain.ExerciseInfo;
6.  import ssh.homework.domain.Workbook;
7.  import ssh.homework.service.ExerciseService;
8.  import ssh.homework.tag.PageModel;
9.  //ExerciseService 服务接口的实现类
10. @Service("exerciseService")
11. public class ExerciseServiceImpl implements ExerciseService {
12.     //tb_exercise 表的操作接口实例,由 IoC 注入
13.     @Autowired
14.     ExerciseDao exerciseDao;
15.     /**
16.      * 根据id查询习题
17.      * @param id
18.      * @return 习题对象
19.      **/
20.     @Override
21.     public Exercise findExerciseById(Integer id) {
22.
23.         return exerciseDao.selectById(id);
24.     }
25.     /**
26.      * 获得所有习题
27.      * @return Exercise 对象的 List 集合
28.      **/
29.     @Override
30.     public List<Exercise> findExercise(Exercise exercise, PageModel pageModel) {
31.         /** 当前需要分页的总数据条数 */
32.         Map<String,Object>params=new HashMap<>();
33.         params.put("exercise", exercise);
34.         int recordCount=exerciseDao.count(params);
35.         pageModel.setRecordCount(recordCount);
36.         if(recordCount >0){
37.             /** 开始分页查询数据:查询第几页的数据 */
38.             params.put("pageModel", pageModel);
39.         }
40.         List<Exercise>exercises=exerciseDao.selectByPage(params);
```

```
41.            return exercises;
42.        }
43.    //根据id删除习题表tb_exercise中的数据
44.        @Override
45.        public void removeExerciseById(Integer id) {
46.            exerciseDao.deleteById(id);
47.        }
48.    //根据exercise对象中的属性值更新习题表
49.        @Override
50.        public void modifyExercise(Exercise exercise) {
51.            exerciseDao.update(exercise);
52.
53.        }
54.    //添加习题
55.        @Override
56.        public void addExercise(Exercise exercise) {
57.            exerciseDao.save(exercise);
58.
59.        }
60.    //查询习题表中的习题,查询条件为只列出不在当前作业中的所有习题
61.        @Override
62.        public List<Exercise> findByNotInWorkbookIdAndByCourseId(Workbook
        workbook, PageModel pageModel, String isNotIn) {
63.            Map<String,Object>map=new HashMap<String,Object>();
64.            map.put("workbook",workbook);
65.            map.put("isNotIn", isNotIn);
        //此参数表示是根据exercise的id查询,还是根据不在exercise的id范围内查询
66.            List<Exercise>exercises=new ArrayList<Exercise>();
67.            Integer recordCount=exerciseDao.countByWorkbook(map);
68.            if(recordCount>0) {
69.                pageModel.setRecordCount(recordCount);
70.                map.put("pageModel", pageModel);
71.            exercises=exerciseDao.selectNotInWorkbookIdAndByCourseId(map);
72.            }
73.
74.            return exercises;
75.        }
76.    //根据某个作业统计每道题出错的次数,用于分析学生做作业的情况
77.        @Override
78.        public List<ExerciseInfo> findExerciseInfoByWorkbookId(Integer
        Workbook_id) {
```

```
79.
80.            return exerciseDao.selectExerciseInfoByWorkbookId(Workbook_
               id);
81.        }
82. }
```

第 78 行的 findExerciseInfoByWorkbookId() 方法查询的是 student_workbook 表，作用是统计同一个作业中每道题学生完成作业时出错的次数。通过此方法的统计数据可分析学生作业中出现的问题。这个方法的返回值是一个 List 集合，其元素是 ExerciseInfo 对象。

ExerciseInfo 类代码如下。

```
1. package ssh.homework.domain;
2.
3. public class ExerciseInfo {
4.     private Exercise exercise;
5.     private int count;            //某一习题出错误的次数
6. }
```

上面类中有两个成员变量：一个是实体类 Exercise 的对象；另一个是整型变量。两个成员变量的作用是：根据 Exercise 的对象统计学生做题出错的次数，并将数值存放在 count 变量中。

3. Web 表示层

ExerciseController 类是习题管理模块的核心，所有请求都需经过这个控制器，由控制器再调用相应的接口服务，完成相应的业务逻辑。ExerciseControlle 类主要有以下 5 个接口。

- addExercise()：添加习题方法，返回值是 ModelAndView。
- removeExercise()：删除习题方法，返回值是 ModelAndView。
- updateExercise()：更新习题方法，返回值是 ModelAndView。
- selectExercise()：查询习题方法，返回值是 String。
- genericAssociation()：将 Exercise 与 Course 进行关联，Exercise 与 Teacher 进行关联。

ExerciseController 类的代码如下。

```
1. package ssh.homework.controller;
2.
3. import ssh.homework.common.HomeworkConstants;
4. import ssh.homework.domain.Course;
5. import ssh.homework.domain.Exercise;
6. import ssh.homework.domain.Teacher;
```

```
7.    import ssh.homework.service.CourseService;
8.    import ssh.homework.service.ExerciseService;
9.    import ssh.homework.service.TeacherService;
10.   import ssh.homework.tag.PageModel;
11.
12.   @Controller
13.   public class ExerciseCtroller {
14.       @Autowired
15.       @Qualifier("exerciseService")
16.       ExerciseService exerciseService;
17.       @Autowired
18.       @Qualifier("courseService")
19.       CourseService courseService;
20.       @Autowired
21.       @Qualifier("teacherService")
22.       private TeacherService teacherService;
23.       //存放章的数组
24.       private static String[] chapters;
25.       //存放习题类型的数组
26.       private static Map<String,String>kinds;
27.       static{
28.           //初始化 chapters
29.           chapters=new String[20];
30.           for(int i=0;i<20;i++)chapters[i]=String.valueOf(i+1);
31.           //初始化 kinds
32.           kinds=new HashMap<String,String>();
33.           kinds.put("1","选择题");
34.           kinds.put("2","填空题");
35.           kinds.put("3","简答题");
36.       }
37.
38.       /**
39.        * 处理添加学生请求
40.        * @param String flag 标记,1表示跳转到添加页面,2表示执行添加操作
41.        * @param Exercise exercise 接收添加参数
42.        * @param ModelAndView mv
43.        **/
44.       @RequestMapping(value="/exercise/addExercise")
45.        public ModelAndView addExercise(
46.            String flag,
47.            HttpSession session,
```

```
48.              @ModelAttribute Exercise exercise,
49.              ModelAndView mv){
50.          if(flag.equals("1")){
51.              //查询教师和课程信息
52.              List<Course>courses=courseService.findAllCourse();
53.              List<Teacher>teachers=teacherService.findTeacherAll();
54.
55.              //设置Model数据
56.              mv.addObject("courses", courses);
57.              mv.addObject("teachers",teachers);
58.              mv.addObject("chapters",chapters);
59.              mv.addObject("kinds",kinds);
60.
61.              //返回添加习题页面
62.              mv.setViewName("exercise/showAddExercise");
63.          }else{
64.              //判断是否有关联对象传递,如果有,则创建关联对象
65.              //this.genericAssociation(course_id,teacher_id,exercise);
66.              //添加操作
67.              Teacher teacher=(Teacher)session.getAttribute
                        (HomeworkConstants.TEACHER_SESSION);
68.              exercise.setTeacher(teacher);
69.              exerciseService.addExercise(exercise);
70.
71.              //设置客户端跳转到查询请求
72.              mv.setViewName("redirect:/exercise/selectExercise");
73.          }
74.          //返回
75.          return mv;
76.
77.      }
78.
79.      /**
80.       * 处理查询请求
81.       * @param pageIndex 请求的是第几页
82.       * @param Integer course_id 课程id
83.       * @param Integer teacher_id 教师id*
84.       * @param exercise 模糊查询参数
```

```
85.        * @param Model model
86.        **/
87.       //处理习题查询,如果查询条件为空,则显示所有习题,否则根据所选条件进行查询
88.       @RequestMapping(value="/exercise/selectExercise")
89.         public String selectExercise(Integer pageIndex,
90.             Integer course_id,
91.             Integer teacher_id,
92.             String kind,
93.             String chapter,
94.             @ModelAttribute Exercise exercise,
95.             Model model){
96.
97.            //创建分页对象
98.            PageModel pageModel=new PageModel();
99.            pageModel.setPageSize(20);
100.            //如果参数 pageIndex 不为 null,则设置 pageIndex,即显示第几页
101.            if(pageIndex!=null){
102.                pageModel.setPageIndex(pageIndex);
103.            }
104.
105.            // 查询教师和课程信息
106.            List<Course>courses=courseService.findAllCourse();
107.            List<Teacher>teachers=teacherService.findTeacherAll();
108.            Course course=null;
109.            Teacher teacher=null;
110.            //下面的代码是为了保存每次查询的条件,分别是课程、教师姓名、题型、章
111.            if(exercise.getCourse()!=null&&exercise.getCourse().getId()!=null)
112.            course_id=exercise.getCourse().getId();
113.
114.            if(exercise.getTeacher()!=null&&exercise.getTeacher().getId()!=null)
115.                teacher_id=exercise.getTeacher().getId();
116.            if(course_id!=null) {
117.                course=courseService.findCourseById(course_id);
118.                exercise.setCourse(course);
119.            }
120.            if(teacher_id!=null) {
121.                teacher=teacherService.findTeacherById(teacher_id);
```

```
122.            exercise.setTeacher(teacher);
123.        }
124.        if(kind!=null)exercise.setKind(kind);
125.        if(chapter!=null)exercise.setChapter(chapter);
126.        //根据exercise查询习题信息
127.        List<Exercise>exercises=exerciseService.findExercise(exercise,
            pageModel);
128.        //设置Model数据
129.
130.        model.addAttribute("exercises", exercises);
131.
132.        model.addAttribute("courses", courses);
133.        model.addAttribute("teachers", teachers);
134.        model.addAttribute("chapters",chapters);
135.        model.addAttribute("pageModel", pageModel);
136.        model.addAttribute("course_id", course_id);
137.        model.addAttribute("teacher_id", teacher_id);
138.        model.addAttribute("course", course);
139.        model.addAttribute("teacher", teacher);
140.        model.addAttribute("kind",kind);
141.        model.addAttribute("chapter",chapter);
142.
143.        //返回习题页面
144.        return "exercise/exercise";
145.
146.    }
147.
148.    /**
149.     * 处理删除学生请求
150.     * @param String ids 需要删除的id字符串
151.     * @param ModelAndView mv
152.     **/
153.    @RequestMapping(value="/exercise/removeExercise")
154.    public ModelAndView removeExercise(String ids, ModelAndView mv,
        HttpSession session) {
155.        String[] idArray=ids.split(",");
156.        Teacher teacher = (Teacher) session.getAttribute(HomeworkConstants
            .TEACHER_SESSION);
157.        for(String id : idArray){
158.            try {
159.                //根据id删除习题
```

```java
160.                Exercise exercise=exerciseService
                        .findExerciseById(Integer.parseInt(id));
161.            //判断当前要删除的试题是不是当前教师所出
162.            if(!teacher.getId().equals(exercise
                        .getTeacher().getId()))
163.            {
164.                mv.addObject("error","不是自己出的习题不能删除!");
165.                mv.setViewName("error/error");
166.                return mv;
167.            }
168.            exerciseService.removeExerciseById(Integer.parseInt(id));
169.        }catch(Exception ep) {
170.            mv.addObject("error","习题不能删除!");
171.            mv.setViewName("error/error");
172.            return mv;
173.        }
174.    }
175.    //设置客户端跳转到查询请求
176.
177.    mv.setViewName("redirect:/exercise/selectExercise");
178.
179.    return mv;
180.
181. }
182. /**
183.  * 处理修改学生请求
184.  * @param String flag 标记,1表示跳转到修改页面,2表示执行修改操作
185.  * @param Integer course_id 课程 id
186.  * @Param Integer teacher_id 教师 id
187.  * @param Exercise exercise 要修改学生的对象
188.  * @param ModelAndView mv
189.  **/
190. @RequestMapping(value="/exercise/updateExercise")
191. public ModelAndView updateExercise(
192.        String flag,
193.        Integer course_id,
194.        Integer teacher_id,
195.        HttpSession session,
196.        @ModelAttribute Exercise exercise,
197.        ModelAndView mv) {
198.    if(flag.equals("1")) {
```

```
199.            //根据 id 查找要修改的学生
200.            Exercise target = exerciseService.findExerciseById(exercise
                .getId());
201.            // 查询教师和课程信息
202.            List<Course>courses=courseService.findAllCourse();
203.            List<Teacher>teachers=teacherService.findTeacherAll();
204.
205.            mv.addObject("exercise",target);
206.            mv.addObject("courses",courses);
207.            mv.addObject("teachers",teachers);
208.            mv.addObject("chapters",chapters);
209.            mv.addObject("kinds",kinds);
210.            mv.setViewName("exercise/showUpdateExercise");
211.
212.        }
213.        else
214.        {
215.            Teacher teacher=(Teacher)session
                .getAttribute(HomeworkConstants.TEACHER_SESSION);
216.            if(!teacher.getId().equals(teacher_id))
217.            {
218.                mv.addObject("error","不是自己出的习题不能修改!");
219.                mv.setViewName("error/error");
220.                return mv;
221.            }
222.            //封装关联对象
223.
224.            this.genericAssociation(course_id,teacher_id, exercise);
225.            //执行更新
226.            exerciseService.modifyExercise(exercise);
227.
228.            mv.setViewName("redirect:/exercise/selectExercise");
229.        }
230.
231.        return mv;
232.    }
233.    /**
234.     *将 exercise 对象通过 course_id、teacher_id 与 Course、Teacher 对象建立关联
```

```
235.    **/
236.    private void genericAssociation(Integer course_id,Integer teacher_
        id,Exercise exercise){
237.        if(course_id !=null){
238.            Course course=new Course();
239.            course.setId(course_id);
240.            exercise.setCourse(course);
241.        }
242.        if(teacher_id!=null)
243.        {
244.            Teacher teacher=new Teacher();
245.            teacher.setId(teacher_id);
246.
247.            exercise.setTeacher(teacher);
248.        }
249.
250.    }
251.
252. }
```

这个类是习题管理模块的核心类,有关习题操作的所有请求都由这个类处理。由于 tb_exercise 表与 tb_teacher 表、tb_course 表之间存在多对一的关联,而类中的很多操作涉及 tb_teacher 表和 tb_course 表,所以类中定义了 TeacherService 类的引用变量 teacherService 和 CourseService 类的引用变量 courseService。这两个变量的实例化是由 Spring 的 IoC 完成的。

习题管理模块的 JSP 页面部分与教师管理模块相似,读者可参照教师管理模块的 JSP 页面分析,此处不再列出。

10.4.6 作业管理模块

作业管理模块是整个系统的核心之一,几乎涉及所有表的查询,但只对 tb_workbook 表和 tb_assignment 表需要增、删、改、查功能,主要包括以下的类和接口。

- WorkbookService 接口:作业表 tb_workbook 的相关操作。
- ClazzService:班级表 tb_clazz 的相关操作。
- TeacherService:教师表 tb_teacher 的相关操作。
- CourseService:课程表 tb_course 的相关操作。
- AssignmentService:作业任务表 tb_assignment 的相关操作。
- ExcelService:Excel 文件的相关操作。
- ExerciseService:习题表 tb_exercise 的相关操作。
- StudentService:学生表 tb_student 的相关操作。
- StudentWorkbookService:学生作业表的相关操作。

- WorkbookController：这是作业管理模块的核心控制类，所有操作都由这个类分配完成。

这几个类的关系可通过类图加以描述。作业管理模块的 UML 类图如图 10-19 所示。

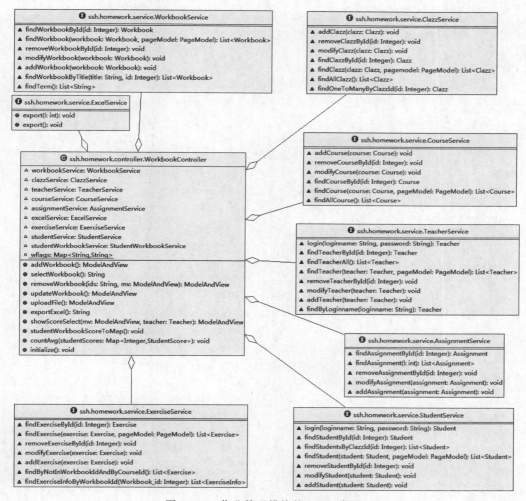

图 10-19　作业管理模块的 UML 类图

从图 10-19 可以看出，WorkbookController 控制器与 CourseService 等 9 个组件相关联，这 9 个组件是 WorkbookController 的一部分，它们的实例都是由 IoC 容器自动注入控制器中。

1. 数据访问层

数据访问层主要的组件是 WorkbookDao 接口，其中应用 MyBatis 的注解完成一些简单的数据库操作，涉及复杂的数据库操作，应用动态 SQL 语句完成，这部分代码写在相应的 provider 类中。

WorkbookDao 接口代码如下。

```java
1.  package ssh.homework.dao;
2.  
3.  import static ssh.homework.common.HomeworkConstants.WORKBOOKTABLE;
4.  import ssh.homework.dao.provider.WorkbookDynaSqlProvider;
5.  import ssh.homework.domain.Workbook;
6.  
7.  public interface WorkbookDao {
8.      //根据id查询作业
9.      @Select("select * from "+WORKBOOKTABLE+" where id=#{id}")
10.     @Results({
11.         @Result(id=true,column="id",property="id"),
12.         @Result(column="wflag",property="wflag"),
13.         @Result(column="term",property="term"),
14.         @Result(column="title",property="title"),
15.         @Result(column="fileName",property="fileName"),
16.         @Result(column="createDate",property="createDate",
                javaType=java.util.Date.class),
17.         @Result(column="teacher_id", property="teacher",
18.             one=@One(select="ssh.homework.dao.TeacherDao.selectById",
19.             fetchType=FetchType.EAGER)),
20.         @Result(column="clazz_id", property="clazz",
21.             one=@One(select="ssh.homework.dao.ClazzDao.selectById",
22.             fetchType=FetchType.EAGER)),
23.         @Result(column="course_id", property="course",
24.             one=@One(select="ssh.homework.dao.CourseDao.selectById",
25.             fetchType=FetchType.EAGER)
26.     })
27.     Workbook selectById(Integer id);
28.     
29.     //根据id删除作业
30.     @Delete(" delete from "+WORKBOOKTABLE+" where id =#{id} ")
31.     void deleteById(@Param("id") Integer id);
32.     
33.     
34.     
35.     //动态查询
36.     @SelectProvider(type = WorkbookDynaSqlProvider.class, method =
            "selectWhitParam")
37.     @Results({
38.         @Result(id=true,column="id",property="id"),
39.         @Result(column="wflag",property="wflag"),
40.         @Result(column="term",property="term"),
```

```java
41.        @Result(column="title",property="title"),
42.        @Result(column="fileName",property="fileName"),
43.        @Result(column="createDate",property="createDate",javaType=
           java.util.Date.class),
44.        @Result(column="teacher_id", property="teacher",
45.        one=@One(select="ssh.homework.dao.TeacherDao.selectById",
46.           fetchType=FetchType.EAGER)),
47.        @Result(column="clazz_id", property="clazz",
48.        one=@One(select="ssh.homework.dao.ClazzDao.selectById",
49.           fetchType=FetchType.EAGER)),
50.        @Result(column="course_id", property="course",
51.        one=@One(select="ssh.homework.dao.CourseDao.selectById",
52.           fetchType=FetchType.EAGER))
53.        })
54.    List<Workbook>selectByPage(Map<String,Object>params);
55.
56.    //根据参数查询作业总数
57.    @SelectProvider( type = WorkbookDynaSqlProvider. class, method = "count")
58.    Integer count(Map<String,Object>params);
59.
60.    //动态插入作业
61.    @SelectProvider( type = WorkbookDynaSqlProvider. class, method = "insert")
62.    void save(Workbook workbook);
63.    //动态更新作业
64.        @SelectProvider(type=WorkbookDynaSqlProvider.class,method=
           "update")
65.    void update(Workbook workbook);
66.
67.    //根据班级id查询关联的workbook
68.    @Select("select * from "+WORKBOOKTABLE+" where clazz_id=#{id}")
69.    @Results({
70.        @Result(id=true,column="id",property="id"),
71.        @Result(column="wflag",property="wflag"),
72.        @Result(column="term",property="term"),
73.        @Result(column="title",property="title"),
74.        @Result(column="fileName",property="fileName"),
75.        @Result(column="createDate",property="createDate",javaType=
           java.util.Date.class),
76.        @Result(column="teacher_id", property="teacher",
77.        one=@One(select="ssh.homework.dao.TeacherDao.selectById",
78.           fetchType=FetchType.EAGER)),
```

```
79.         @Result(column="clazz_id", property="clazz",
80.         one=@One(select="ssh.homework.dao.ClazzDao.selectById",
81.           fetchType=FetchType.EAGER)),
82.         @Result(column="course_id", property="course",
83.         one=@One(select="ssh.homework.dao.CourseDao.selectById",
84.           fetchType=FetchType.EAGER))
85.         })
86.      List<Workbook>selectWorkbooksByClazzId(Integer id);
87.      //根据班级id和作业名称查询
88.      @Select("select * from "+WORKBOOKTABLE+" where title=#{title}
         and clazz_id =#{id}")
89.      List<Workbook>selectWorkbookByTitle(@Param("title")String
         title,@Param("id") Integer id);
90.      //根据教师,课程id,班级id,已批改作业和学期查询作业
91.      @Select("select * from "+WORKBOOKTABLE+" where"
92.         +" teacher_id=#{teacher_id} and "
93.         +" course_id=#{course_id} and "
94.         +" clazz_id =#{clazz_id} and "
95.         +"wflag='2' and term=#{term}")
96.      List<Workbook>selectWorkbookByCourseIdAndClazzIdAndWflagAndTerm(
97.         @Param("teacher_id")Integer teacher_id,
98.         @Param("course_id")Integer course_id,
99.         @Param("clazz_id")Integer clazz_id,
100.        @Param("term")String term);
101.     //查询workbook表中的所有学期(去掉重复)
102.     @Select("select distinct term FROM "+WORKBOOKTABLE)
103.     List<String>selectTerm();
104.  }
```

第 27 行的 selectById()方法是根据 id 查询作业信息,由于 tb_workbook 表与 tb_teacher、tb_course、tb_clazz 3 个表相关联,因此,在注解中进行属性映射时应用 3 个 @One 注解实现了 tb_workbook 表与其他 3 个表的多对一的关联。

同样,第 54 行和第 86 行的两个方法的结果映射与 selectById()方法中的结果映射一样,都用到@One 注解与 3 个表进行了关联映射。WorkbookDao 对应的 provider 类是 WorkbookDynaSqlProvider,这里存放的是动态查询的 SQL 语句,读者可参照教师管理模块中这部分内容的分析,此处不再讲解。

2. 业务逻辑层

作业管理模块的业务逻辑层涉及的服务接口较多,这在本小节开头已经介绍了,这里主要介绍 WorkbookService 服务接口。

(1) WorkbookService 接口的代码。

```
1.    package ssh.homework.service;
2.
3.    import ssh.homework.domain.Workbook;
4.    import ssh.homework.tag.PageModel;
5.    //tb_workbook表的业务逻辑操作类
6.    public interface WorkbookService {
7.        /**
8.         * 根据id查询作业
9.         * @param id
10.        * @return 作业对象
11.        * */
12.       Workbook findWorkbookById(Integer id);
13.
14.       /**
15.        * 根据条件查询作业
16.        * @return Workbook对象的List集合
17.        * */
18.       List<Workbook>findWorkbook(Workbook workbook,PageModel pageModel);
19.
20.       /**
21.        * 根据id删除作业
22.        * @param id
23.        * */
24.       void removeWorkbookById(Integer id);
25.
26.       /**
27.        * 修改作业
28.        * @param Workbook 作业对象
29.        * */
30.       void modifyWorkbook(Workbook workbook);
31.
32.       /**
33.        * 添加作业
34.        * @param Workbook 作业对象
35.        * */
36.       void addWorkbook(Workbook workbook);
37.       //根据作业名称进行查找
38.       List<Workbook>findWorkbookByTitle(String title,Integer id);
39.       //根据教师id,课程id,班级id,已批改作业和学期查询作业
40.       List<Workbook>findWorkbookByCourseIdAndClazzIdAndWflagAndTerm(
41.           Integer teacher_id,Integer course_id,Integer clazz_id,String term);
42.       //查询workbook表中的所有学期(去掉重复)
43.       List<String>findTerm();
44.   }
```

（2）WorkbookServiceImpl 类的代码。

```java
1.   package ssh.homework.service.impl;
2.
3.   import ssh.homework.dao.WorkbookDao;
4.   import ssh.homework.domain.Workbook;
5.   import ssh.homework.service.WorkbookService;
6.   import ssh.homework.tag.PageModel;
7.
8.   //tb_workbook 表相关的业务逻辑服务类
9.   @Service("workbookService")
10.  public class WorkbookServiceImpl implements WorkbookService {
11.
12.      @Autowired
13.      //tb_workbook 表的操作类对象
14.      WorkbookDao workbookDao;
15.      @Override
16.      //根据 id 查询作业信息
17.      public Workbook findWorkbookById(Integer id) {
18.
19.          return workbookDao.selectById(id);
20.      }
21.
22.      @Override
23.      //根据 workbook 对象属性的值动态查询作业信息
24.      public List<Workbook> findWorkbook(Workbook workbook, PageModel pageModel) {
25.          /** 当前需要分页的总数据条数 */
26.          Map<String,Object> params=new HashMap<>();
27.          params.put("workbook", workbook);
28.          int recordCount=workbookDao.count(params);
29.          pageModel.setRecordCount(recordCount);
30.          if(recordCount>0){
31.              /** 开始分页查询数据：查询第几页的数据 */
32.              params.put("pageModel", pageModel);
33.          }
34.          List<Workbook> workbooks=workbookDao.selectByPage(params);
35.
36.          return workbooks;
37.      }
38.
39.
40.      @Override
41.      //根据 id 删除作业信息
42.      public void removeWorkbookById(Integer id) {
```

```java
43.            workbookDao.deleteById(id);
44.        }
45.
46.        @Override
47.        //修改作业信息
48.        public void modifyWorkbook(Workbook workbook) {
49.            workbookDao.update(workbook);
50.
51.        }
52.
53.        @Override
54.        //添加作业信息
55.        public void addWorkbook(Workbook workbook) {
56.            List< Workbook > workbooks = findWorkbookByTitle (workbook
                .getTitle(),workbook.getClazz().getId());
57.            if(workbooks.size()<=0)
58.            workbookDao.save(workbook);
59.
60.        }
61.
62.        @Override
63.        //根据作业名称和班级id查询作业信息
64.         public List< Workbook > findWorkbookByTitle (String title, Integer
                clazz_id) {
65.
66.            return workbookDao.selectWorkbookByTitle(title,clazz_id);
67.        }
68.        //根据教师,课程id,班级id,已批改作业和学期查询作业
69.        @Override
70.        public List<Workbook>findWorkbookByCourseIdAndClazzIdAndWflagAndTerm
           (Integer teacher_id,
71.            Integer course_id,
72.            Integer clazz_id, String term) {
73.        return workbookDao.selectWorkbookByCourseIdAndClazzIdAndWflagAndTerm(
74.                teacher_id, course_id, clazz_id, term);
75.        }
76.        //查询workbook表中的所有学期(去掉重复)
77.        @Override
78.        public List<String>findTerm() {
79.            return workbookDao.selectTerm();
80.        }
81.    }
```

3. Web 表示层

WorkbookController 类是作业管理模块的控制器,所有操作都需经过它来指挥完成,它是这个模块的核心类。这个类中主要有以下 12 个方法。

- addWorkbook():处理添加作业请求。
- addWorkbookSearch():将满足查询条件的学生成绩添加到 MAP 中。
- countAvg():处理计算学生作业平均成绩的请求。
- exportExcel():将作业输出到 Excel 文件。
- initialize():初始化 clazzs、courses、terms。
- removeWorkbook():处理删除作业的请求。
- searchStudentScore():处理生成查询作业成绩请求,并将结果呈现在浏览器。
- selectWorkbook():处理查询作业的请求。
- showScoreSelect():处理生成作业成绩页面请求。
- studentWorkbookScoreToMap():将 workbook_id 对应的作业成绩写到 studentScores 中。
- updateWorkbook():处理修改作业的请求。
- uploadFile():教师上传作业文件。

WorkbookContorller 控制器代码。

```
1.   package ssh.homework.controller;
2.
3.   import ssh.homework.common.HomeworkConstants;
4.   import ssh.homework.domain.*;
5.   import ssh.homework.service.*;
6.   import ssh.homework.tag.PageModel;
7.
8.   @Controller
9.   public class WorkbookController {
10.      @Autowired
11.      @Qualifier("workbookService")
12.      WorkbookService workbookService;
13.      @Autowired
14.      @Qualifier("clazzService")
15.      ClazzService clazzService;
16.      @Autowired
17.      @Qualifier("teacherService")
18.      TeacherService teacherService;
19.      @Autowired
20.      @Qualifier("courseService")
21.      CourseService courseService;
22.      @Autowired
```

```java
23.     @Qualifier("assignmentService")
24.     AssignmentService assignmentService;
25.     @Autowired
26.     @Qualifier("excelService")
27.     ExcelService excelService;
28.     @Autowired
29.     @Qualifier("exerciseService")
30.     ExerciseService exerciseService;
31.     @Autowired
32.     @Qualifier("studentService")
33.     StudentService studentService;
34.     @Autowired
35.     @Qualifier(value="studentWorkbookService")
36.     StudentWorkbookService studentWorkbookService;
37.     //存放作业状态的 Map
38.         private static Map<String,String>wflags;
39.         static {
40.             wflags=new HashMap<String,String>();
41.             wflags.put("0","未发布");
42.             wflags.put("1","发布");
43.             wflags.put("2","批改");
44.
45.         }
46.
47.     /**
48.      * 处理添加作业请求
49.      * @param String flag 标记,1表示跳转到添加页面,2表示执行添加操作
50.      *
51.      * @param Workbook workbook 接收添加参数
52.      * @param ModelAndView mv
53.      * */
54.     //添加作业
55.     @RequestMapping(value="/workbook/addWorkbook")
56.      public ModelAndView addWorkbook(
57.          String flag,
58.          @ModelAttribute Workbook workbook,
59.          HttpSession session,
60.          ModelAndView mv){
61.          if(flag.equals("1")){
62.             //查询班级信息
63.             List<Clazz>clazzs=clazzService.findAllClazz();
64.             List<Teacher>teachers=teacherService.findTeacherAll();
```

```
65.            List<Course>courses=courseService.findAllCourse();
66.
67.            //设置Model数据
68.            mv.addObject("clazzs", clazzs);
69.            mv.addObject("teachers",teachers);
70.            mv.addObject("courses", courses);
71.
72.            //返回添加作业页面
73.            mv.setViewName("workbook/showAddWorkbook");
74.        }else{
75.            //添加操作
76.            Teacher teacher = (Teacher) session.getAttribute(HomeworkConstants.TEACHER_SESSION);
77.            workbook.setTeacher(teacher);
78.            workbook.setCreateDate(new Date());
79.            workbookService.addWorkbook(workbook);
80.
81.            //设置客户端跳转到查询请求
82.            mv.setViewName("redirect:/workbook/selectWorkbook");
83.        }
84.     //返回
85.     return mv;
86.
87.    }
88.
89.    /**
90.     * 处理查询请求
91.     * @param pageIndex 请求的是第几页
92.     * @param workbook 模糊查询参数
93.     * @param Model model
94.     * @return String
95.     **/
96.    @RequestMapping(value="/workbook/selectWorkbook")
97.    public String selectWorkbook(Integer pageIndex,
98.         HttpSession session,
99.         @ModelAttribute Workbook workbook,
100.         Model model){
101.
102.        //创建分页对象
103.        Teacher teacher = (Teacher) session.getAttribute(HomeworkConstants.TEACHER_SESSION);
104.        PageModel pageModel =new PageModel();
105.        pageModel.setPageSize(40);
```

```java
106.        //如果参数pageIndex不为null,则设置pageIndex,即显示第几页
107.        if(pageIndex !=null){
108.            pageModel.setPageIndex(pageIndex);
109.        }
110.        //查询班级信息,用于模糊查询
111.        List<Clazz>clazzs=clazzService.findAllClazz();
112.        List<Teacher>teachers=new ArrayList<Teacher>();
113.        teachers.add(teacher);
114.        List<Course>courses=courseService.findAllCourse();
115.        //查询作业信息
116.        workbook.setTeacher(teacher);
117.        List<Workbook>workbooks =workbookService
                .findWorkbook(workbook,pageModel);
118.        //设置Model数据
119.        model.addAttribute("workbooks", workbooks);
120.        model.addAttribute("clazzs", clazzs);
121.        model.addAttribute("teachers", teachers);
122.        model.addAttribute("courses", courses);
123.        model.addAttribute("pageModel", pageModel);
124.        //返回作业页面
125.        return "workbook/workbook";
126.
127.    }
128.    /**
129.     * 处理删除作业请求
130.     * @param String ids 需要删除的id字符串
131.     * @param ModelAndView mv
132.     **/
133.    @RequestMapping(value="/workbook/removeWorkbook")
134.    public ModelAndView removeWorkbook(String ids,ModelAndView mv) {
135.        String[] idArray=ids.split(",");
136.        for(String id : idArray){
137.            try {
138.                //根据id删除作业
139.                workbookService.removeWorkbookById(Integer.parseInt(id));
140.            }catch(Exception ep) {
141.                mv.addObject("error","作业不能删除!");
142.                mv.setViewName("error/error");
143.                return mv;
144.            }
145.        }
146.        //设置客户端跳转到查询请求
147.        mv.setViewName("redirect:/workbook/selectWorkbook");
148.
```

```
149.            return mv;
150.
151.        }
152.        /**
153.         * 处理修改作业请求
154.         * @param String flag 标记,1 表示跳转到修改页面,2 表示执行修改操作
155.         * */
156.
157.        @RequestMapping(value="/workbook/updateWorkbook")
158.        public ModelAndView updateWorkbook(
159.            String flag,
160.            HttpSession session,
161.            @ModelAttribute Workbook workbook,
162.            ModelAndView mv) {
163.            if(flag.equals("1")) {
164.                //根据 id 查找要修改的学生
165.                Workbook target=workbookService.findWorkbookById(workbook.getId());
166.                //查询所有班级
167.                List<Clazz>clazzs=clazzService.findAllClazz();
168.                //List<Teacher>teachers=teacherService.findTeacherAll();
169.                List<Course>courses=courseService.findAllCourse();
170.
171.                mv.addObject("workbook",target);
172.                mv.addObject("clazzs", clazzs);
173.                //mv.addObject("teachers", teachers);
174.                mv.addObject("courses", courses);
175.                mv.addObject("wflags",wflags);
176.                mv.setViewName("workbook/showUpdateWorkbook");
177.
178.            }
179.            else
180.            {
181.                Teacher teacher=(Teacher)session.getAttribute(HomeworkConstants.TEACHER_SESSION);
182.                workbook.setTeacher(teacher);
183.                Assignment assignment=new Assignment();
184.                assignment.setWorkbook(workbook);
185.                List<Assignment>assignments=assignmentService
                        .findAssignment(assignment,new PageModel());
186.                Iterator<Assignment>it=assignments.iterator();
187.                //判断已存在的作业中的习题是否已经有了分值
```

```java
188.            if(assignments.size()<=0) {
189.                mv.addObject("error","作业题没有分值,请给每个题提供分值");
190.                mv.setViewName("/error/error");
191.                return mv;
192.            }
193.            while(it.hasNext()) {          //判断作业题是否有分值
194.                if(it.next().getGrade()==0) {
195.                    mv.addObject("error","作业题没有分值,请给每个题提供分值");
196.                    mv.setViewName("/error/error");
197.                    return mv;
198.                }
199.
200.            }
201.        workbookService.modifyWorkbook(workbook);
202.
203.        mv.setViewName("redirect:/workbook/selectWorkbook");
204.
205.        }
206.
207.        return mv;
208.    }
209.    //教师在作业管理时上传文件
210.    @RequestMapping(value="/workbook/uploadFile")
211.    public ModelAndView uploadFile(
212.        String flag,
213.        @ModelAttribute MultipartFile file,
214.        Integer id,                   //workbook的id
215.        ModelAndView mv,
216.        HttpSession session)throws Exception{
217.        if(flag.equals("1")){
218.            mv.addObject("id",id);          //将workbook的id保存到request中
219.            mv.setViewName("workbook/showUploadFile");
220.        }else{
221.            //上传文件路径
222.            String path=session.getServletContext().getRealPath(
223.                "/upload/");
224.            System.out.println(path);
225.            //上传文件名
226.            String fileName=file.getOriginalFilename();
227.                //将上传文件保存到一个目标文件中
228.            file.transferTo(new File(path+File.separator+fileName));
229.            //插入数据库
230.            Workbook workbook=workbookService.findWorkbookById(id);
```

```
231.            workbook.setFileName(fileName);
232.            //插入数据库
233.            workbookService.modifyWorkbook(workbook);
234.            //返回
235.            mv.setViewName("redirect:/workbook/selectWorkbook");
236.        }
237.        //返回
238.        return mv;
239.    }
240.    //将作业输出到Excel文件
241.    @RequestMapping(value="/workbook/exportExcel")
242.    public @ResponseBody String exportExcel(HttpServletResponse response,
243.            Integer id,
244.        ModelAndView mv)
245.    {
246.        response.setContentType("application/binary;charset=UTF-8");
247.        Workbook workbook=workbookService.findWorkbookById(id);
248.
249.        try{
250.            ServletOutputStream out=response.getOutputStream();
251.            try {
252.                //设置文件头：最后一个参数是设置下载的文件名
253.                response.setHeader("Content-Disposition",
                    "attachment;fileName=" +URLEncoder.encode(workbook
                    .getTitle()+".xls", "UTF-8"));
254.            } catch (UnsupportedEncodingException e1) {
255.                e1.printStackTrace();
256.                //mv.setViewName("redirect:/error/error");
257.                return "error";
258.            }
259.            String isNotIn="in";
260.            //从tb_exercise表中查找当前作业的习题
261.            PageModel pagemodel=new PageModel();
262.            pagemodel.setPageSize(50);
263.            List<Exercise>exercises=exerciseService.findByNot-
                InWorkbookIdAndByCourseId(workbook, pagemodel,
                isNotIn);
264.            //将exercises输出到Excel表中
265.            excelService.export(exercises, out);
266.            //mv.setViewName("redirect:/workbook/selectWorkbook");
267.            return "success";
268.        } catch(Exception e){
269.            e.printStackTrace();
```

```java
                    //mv.setViewName("redirect:/error/error");
                    return "error";
                }
            }

        //显示生成作业成绩的页面
        @RequestMapping(value="/workbook/showScoreSelect")
        public ModelAndView showScoreSelect(ModelAndView mv,
                @SessionAttribute( HomeworkConstants. TEACHER _ SESSION )
                Teacher teacher) {
            Set<Clazz>clazzs=new HashSet<Clazz>();       //当前教师所教的班级
            Set<Course>courses=new HashSet<Course>();//当前教师所教的课程
            Set<String>terms=new HashSet<String>();    //当前教师涉及的学期
            initialize(clazzs, courses, terms, teacher);
            mv.addObject("clazzs",clazzs);
            mv.addObject("courses",courses);
            mv.addObject("terms",terms);
            mv.setViewName("workbook/searchStudentWorkbookScore");
            return mv;
        }
        //将某班当前学期作业成绩输出到浏览器
        @RequestMapping(value ="/workbook/searchStudentScore")
        public  ModelAndView  searchStudentScore  ( HttpServletResponse response,
            @RequestParam Integer clazz_id, @RequestParam Integer course_id, @RequestParam String term,
            @SessionAttribute(HomeworkConstants.TEACHER_SESSION) Teacher teacher, ModelAndView mv) {
            response.setContentType("application/binary;charset=UTF-8");
            Course course=courseService.findCourseById(course_id);
            Clazz clazz=clazzService.findClazzById(clazz_id);
            String workbookInfo=clazz.getCname()+course.getCname()+"作业成绩";
            //从student_workbook表中查找当前班级所有学生的当前课程作业成绩并添
            //入map中
            Map<Integer, StudentScore>map=addWorkbookSearch(teacher, clazz_id, course_id, term);
            Collection<StudentScore>studentScores=map.values();
            Iterator<StudentScore>it=studentScores.iterator();
            int scoreCount=0;
            if(it.hasNext()) {
                scoreCount=it.next().getScores().size();    //作业数
```

```java
306.        }
307.        String scoreTitles[]=new String[scoreCount];     //列表表头
308.        int i;
309.        for(i=0;i<scoreCount-1;i++) scoreTitles[i]="作业"+(i+1);
310.        scoreTitles[i]="平均成绩";
311.        Set<Clazz>clazzs=new HashSet<Clazz>();           //当前教师所教的班级
312.        Set<Course>courses=new HashSet<Course>();        //当前教师所教的课程
313.        Set<String>terms=new HashSet<String>();          //当前教师涉及的学期
314.        initialize(clazzs, courses, terms, teacher);
315.        mv.addObject("clazzs",clazzs);
316.        mv.addObject("courses",courses);
317.        mv.addObject("terms",terms);
318.        mv.addObject("clazz",clazz);
319.        mv.addObject("course",course);
320.        mv.addObject("term",term);
321.        mv.addObject("scoreTitles",scoreTitles);
322.        mv.addObject("workbookInfo",workbookInfo);
323.        mv.addObject("studentScores",studentScores);
324.        mv.setViewName("workbook/searchStudentWorkbookScore");
325.        return mv;
326.
327.    }
328.
329.    //将某班的当前学期作业成绩输出到Excel文件
330.    @RequestMapping(value="/workbook/exportToScoreExcel")
331.    public @ResponseBody String exportScoreToExcel (HttpServletResponse response,
332.            Integer clazz_id,Integer course_id,String term,
333.            @SessionAttribute(HomeworkConstants.TEACHER_SESSION)
                    Teacher teacher,
334.            ModelAndView mv)
335.    {
336.        response.setContentType("application/binary;charset=UTF-8");
337.        Course course=courseService.findCourseById(course_id);
338.        Clazz clazz=clazzService.findClazzById(clazz_id);
339.        String workbookInfo=clazz.getCname()+course.getCname()+"作业
                成绩";
340.
341.        try{
342.            ServletOutputStream out=response.getOutputStream();
343.            try{
344.                //设置文件头：最后一个参数是设置下载的文件名
```

```
345.                    response.setHeader("Content-Disposition",
                           "attachment;fileName=" +URLEncoder.encode
                           (workbookInfo+".xls", "UTF-8"));
346.                } catch (UnsupportedEncodingException e1) {
347.                    e1.printStackTrace();
348.                    //mv.setViewName("redirect:/error/error");
349.                    return "error";
350.                }
351.                //从 student_workbook 表中查找当前班级所有学生的当前
                    //课程作业成绩并添入 map 中
352.
353.                Map<Integer,StudentScore>map=addWorkbookSearch
                        (teacher,clazz_id,course_id,term);
354.                //将 map 输出到 Excel 表中
355.                excelService.export(map,out,workbookInfo);
356.                //mv.setViewName("redirect:/workbook/selectWorkbook");
357.                return "success";
358.            } catch(Exception e) {
359.                e.printStackTrace();
360.                //mv.setViewName("redirect:/error/error");
361.                return "error";
362.            }
363.
364.        }
365.    //根据查询条件得到学生成绩的 Map
366.    public Map<Integer,StudentScore>addWorkbookSearch(Teacher
            teacher,Integer clazz_id,Integer course_id,String term) {
367.        //查询当前教师、所选班级、所选课程和所选学期对应的 workbook
368.        List<Workbook>workbooks=workbookService.findWorkbook-
            ByCourseIdAndClazzIdAndWflagAndTerm(
369.            teacher.getId(), course_id, clazz_id, term);
370.        Iterator<Workbook>it=workbooks.iterator();
371.        //Map 的键是学生的 id,值是学生每次作业的成绩
372.        Map<Integer,StudentScore>studentScores=new
            HashMap<Integer,StudentScore>();
373.        //所选班级的所有学生
374.        List<Student> students = studentService.findStudents-
            ByClazzId(clazz_id);
375.        Iterator<Student>stu_it=students.iterator();
376.        //利用 students 对 studentScores 进行初始化
377.        while(stu_it.hasNext()) {
```

```
378.                    Student student=stu_it.next();
379.                    List<Float>scores=new ArrayList<Float>();
                                                    //某个学生所有作业的成绩
380.                    StudentScore studentScore=new StudentScore();
381.                    studentScore.setStudent(student);
382.                    studentScore.setScores(scores);
383.                    studentScores.put(student.getId(), studentScore);
384.                }
385.                //遍历 workbooks
386.                while(it.hasNext()) {
387.                    Integer id=it.next().getId(); //作业 id
388.                    studentWorkbookScoreToMap(id,studentScores);
                                                    // 将每个作业
                                                    //成绩添加到 map 中
389.
390.                }countAvg(studentScores);       //计算每个学生的平均成绩
391.                return studentScores;
392.
393.            }
394.        //将 workbook_id 对应的作业成绩写到 studentScores 中
395.        public void studentWorkbookScoreToMap(Integer workbook_id,
            Map<Integer,StudentScore>studentScores) {
396.            //根据 workbook_id 从 studentWorkbook 表中查找所有做此作业
                //的学生,并将查询结果封装到 StudentInfo 对象中
397.            List<StudentInfo>studentInfos=studentWorkbookService
                .findStudentInfoGroupByStudentIdByWorkbookId(workbook_
                id);
398.            Set<Integer>set=studentScores.keySet();
399.            Iterator<Integer>it=set.iterator();
400.            //对 studentScores 进行遍历
401.            while(it.hasNext()) {
402.                Integer studentId=it.next();
403.                List<Float>list=studentScores.get(studentId)
                    .getScores();              //得到当前学生作业成绩列表
404.                Iterator<StudentInfo>sit=studentInfos.iterator();
405.                StudentScore studentScore=new StudentScore();
406.                int flag=0;                //标识学生是否有作业成绩
407.                //对 studentInfos 进行遍历,查找当前 student 对应的成绩,
                    //如果找到,则将当前作业成绩添加到 studentScores 中
408.                while(sit.hasNext()) {
409.                    StudentInfo studentInfo=sit.next();
```

```
410.
411.                    if(studentInfo.getStudent().getId()
                        .equals(studentId)) {//找到当前学生对应的成绩添加
                                              //到当前学生的成绩列表中
412.                        list.add(studentInfo.getScore());
413.                        studentScore.setStudent(studentInfo
                            .getStudent());
414.                        studentScore.setScores(list);
415.                        flag=1;
416.                        break;
417.                    }
418.
419.                }
420.                if(flag==0) {
                                //若该学生当前作业没有成绩,则在成绩列表中添入
421.                    list.add(0.0f);
422.                    Student student=studentService
                        .findStudentById(studentId);
                                    //此学生没有完成这次作业,重新查询学生信息
423.                    studentScore.setStudent(student);
424.                    studentScore.setScores(list);
425.                }
426.
427.                studentScores.put(studentId, studentScore);
                                              //将成绩写入 map
428.            }
429.        }
430.        //计算 studentScores 中每个学生的平均成绩
431.        public void countAvg (Map < Integer, StudentScore >
            studentScores) {
432.            Set<Integer>set=studentScores.keySet();
433.            Iterator<Integer>it=set.iterator();
434.            while(it.hasNext()) {
435.                Integer studentId=it.next();
436.                List<Float>list=studentScores.get(studentId)
                    .getScores();           //得到当前学生作业的成绩列表
437.                float sum=0;
438.                int count=list.size();
439.                for (int n =0; n <count; n++) {
440.                    sum +=list.get(n);    //每个学生的所有作业成绩之和
441.                }
442.                float x=(int)Math.ceil(sum / count);    //四舍五入取整
443.                list.add(x);             //将学生作业平均成绩添加到 list 中
```

```
444.                Student student=studentService.findStudentById
                   (studentId);
445.                StudentScore studentScore=new StudentScore();
                                                //构造 StudentScore
446.                studentScore.setStudent(student);
447.                studentScore.setScores(list);
448.                //将更新后的 studentScore 重新写入 studentScores 中
449.                studentScores.put(studentId, studentScore);
450.            }
451.        }
452.        //初始化 clazzs、courses、terms
453.         public void initialize(Set<Clazz> clazzs, Set<Course>
           courses, Set<String> terms, Teacher teacher) {
454.            Workbook workbook=new Workbook();
455.            workbook.setTeacher(teacher);
456.            PageModel pageModel=new PageModel();
457.            pageModel.setPageSize(100);
458.            List<Workbook>workbooks=workbookService
                   .findWorkbook(workbook, pageModel);
459.            Iterator<Workbook>it=workbooks.iterator();
460.            while(it.hasNext()) {
461.                Workbook wb=it.next();
462.                clazzs.add(wb.getClazz());
463.                courses.add(wb.getCourse());
464.                terms.add(wb.getTerm());
465.            }
466.        }
467.    }
```

WorkbookController 类中的方法比较多，实现的功能也较多。其实，这个类拆分为两个或 3 个类更容易管理。这个模块的 JSP 页面由于篇幅的关系不再列出，读者可以阅读源码自行分析。

10.4.7 批改作业模块

批改作业模块也是整个系统的核心之一，几乎涉及所有表的查询，但重点操作的是 student_workbook 表，需要增、删、改、查功能，主要包括的类和接口与作业管理模块相同，在此不再列出。

这几个类之间的关系可通过类图进行描述。批改作业模块的 UML 类图如图 10-20 所示。类图中为了简化，没有列出 CorrectWorkbookController 类的方法，只列出它的成员变量。

从图 10-20 可以看出 CorrectWorkbookController 控制器与 CourseService 等 9 个组件相关联，这 9 个组件是 CorrectWorkbookController 的一部分，它们的实例都是由 IoC

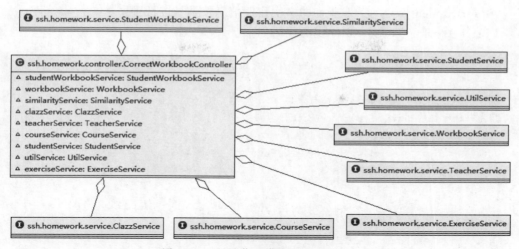

图 10-20　批改作业模块的 UML 类图

容器自动注到控制器中。

1．数据访问层

数据访问层主要的组件与作业管理模块基本相同，此处不再单独列出，可参照作业管理部分的内容。

2．业务逻辑层

批改作业模块的业务逻辑层涉及的服务接口较多，但基本与作业管理涉及的服务接口相同，此处不再列出。

3．Web 表示层

作业管理模块的业务逻辑层涉及的服务接口较多，这在本小节开头已经介绍了，这里主要介绍 WorkbookController 控制器类。

CorrectWorkbookController 类是批改作业模块的控制器，所有操作都需经过它来指挥完成，它是这个模块的核心类。这个类中主要有以下 9 个方法。

- correctWorkbook()：处理批改作业请求。
- downLoad()：处理下载学生作业文件请求。
- findExerciseInfoByWorkbookId()：处理查询学生作业习题出错的情况。
- findStudentWorkbookByWorkbookIdAndStudentId()：根据 workbookId 和 studentId 查询学生作业的情况。
- queryRep()：处理查询学生作业代码是否有抄袭。
- removeStudentWorkbook()：删除学生作业。
- runSimilarity()：根据选中的班级对学生作业进行查重。
- selectWrokbook()：处理请求查询当前教师作业列表。

- sortStudentInfos()：对查询的学生作业信息按学号进行排序。

CorrectWorkbookController 类的代码如下。

```
1.   package ssh.homework.controller;
2.
3.   import ssh.homework.common.*;
4.   import ssh.homework.domain.*;
5.   import ssh.homework.service.*;
6.   import ssh.homework.tag.PageModel;
7.   //处理批改作业请求的控制器
8.   @Controller
9.   public class CorrectWorkbookController {
10.      @Autowired
11.      @Qualifier(value="studentWorkbookService")
12.      StudentWorkbookService studentWorkbookService;
13.
14.      @Autowired
15.      @Qualifier("workbookService")
16.      WorkbookService workbookService;
17.
18.      @Autowired
19.      @Qualifier("similarityService")
20.      SimilarityService similarityService;
21.      @Autowired
22.      @Qualifier("clazzService")
23.      ClazzService clazzService;
24.
25.      @Autowired
26.      @Qualifier("teacherService")
27.      TeacherService teacherService;
28.
29.      @Autowired
30.      @Qualifier("courseService")
31.      CourseService courseService;
32.
33.      @Autowired
34.      @Qualifier("studentService")
35.      StudentService studentService;
36.
37.      @Autowired
38.      @Qualifier(value="utilService")
39.      UtilService utilService;
40.
```

```java
41.    @Autowired
42.    @Qualifier(value="exerciseService")
43.    ExerciseService exerciseService;
44.
45.    //Teacher teacher;
46.    //显示当前教师作业列表
47.    @RequestMapping(value="/correctWorkbook/selectCorrectWorkbook")
48.     public String selectWorkbook(Integer pageIndex,
49.         String isSimilarity,      //如果其值为1,则进入要批改的作业显示列表;
                                      //如果其值为2,则进入作业查重页面
50.         HttpSession session,
51.         @ModelAttribute Workbook workbook,
52.         Model model){
53.
54.        //创建分页对象
55.        Teacher teacher=(Teacher)session.getAttribute(HomeworkConstants
           .TEACHER_SESSION);
56.        PageModel pageModel=new PageModel();
57.        pageModel.setPageSize(40);
58.        //如果参数pageIndex不为null,则设置pageIndex,即显示第几页
59.        if(pageIndex!=null){
60.          pageModel.setPageIndex(pageIndex);
61.        }
62.        // 查询作业信息
63.        workbook.setTeacher(teacher);
64.        //workbook.setWflag("2");   //只查询处于批改状态的作业
65.        List<Workbook> workbooks=workbookService.findWorkbook(workbook,
           pageModel);
66.        //设置Model数据
67.        model.addAttribute("workbooks", workbooks);
68.        model.addAttribute("pageModel", pageModel);
69.        if(isSimilarity!=null&&isSimilarity.equals("1"))
70.        //返回查看作业页面
71.            return "correctWorkbook/correctWorkbook";
72.        else
73.        //返回查重页面
74.            return "correctWorkbook/showToSimilarityWorkbook";
75.
76.
77.    }
78.    //根据workbook_id和clazz_id显示当前作业所有学生的作业信息
79.
80.    @RequestMapping("/correctWorkbook/findStudentByWorkbookId")
```

```
81.    public ModelAndView findStudentByWorkbookId(
82.        Integer clazz_id,
83.        Integer workbook_id,
84.        Integer pageIndex,
85.        ModelAndView mv) {
86.        //根据workbook_id从studentWorkbook表中查找所有做此作业的学生,
           //并将查询结果封装到StudentInfo对象中
87.        List<StudentInfo>studentInfos1=studentWorkbookService
           .findStudentInfoGroupByStudentIdByWorkbookId(workbook_id);
88.        //得到当前班的所有学生
89.        List<Student>students=studentService
           .findStudentsByClazzId(clazz_id);
90.        //为了能显示当前作业班级的所有学生(不管是否已提交作业),构造对应的
           //StudentInfo数组
91.        List<StudentInfo>studentInfos=new ArrayList<StudentInfo>();
92.        for(int i=0;i<students.size();i++) {
93.            int j;
94.            for(j=0;j<studentInfos1.size();j++) {
95.                StudentInfo stInfo=studentInfos1.get(j);
96.                //如果是已做作业的学生,就添加到当前列表中
97.                if(stInfo.getStudent().getId().equals(students
                   .get(i).getId())) {
98.                    studentInfos.add(stInfo);
99.                    j=-1;
100.                   break;
101.               }
102.           }
103.           if(j!=-1) {
104.           StudentInfo studentInfo=new StudentInfo();
105.           studentInfo.setStudent(students.get(i));
106.           studentInfos.add(studentInfo);
107.           }
108.       }
109.
110.       //对studentInfos按学号进行排序
111.       sortStudentInfos(studentInfos);
112.       PageModel pageModel=new PageModel();
113.       pageModel.setPageSize(studentInfos.size());
114.
115.       pageModel.setRecordCount(studentInfos.size());
116.       if(pageIndex !=null){
117.           pageModel.setPageIndex(pageIndex);
```

```java
118.    }
119.    Workbook workbook=workbookService.findWorkbookById(workbook_id);
120.    mv.addObject("pageModel",pageModel);
121.    mv.addObject("workbook",workbook);
122.    mv.addObject("workbook_id",workbook_id);
123.    mv.addObject("studentInfos",studentInfos);
124.    mv.setViewName("correctWorkbook/showStudent");
125.    return mv;
126. }
127.
128. //根据 workbook_id 和 student_id 从 studentworkbook 表中删除某一个学生的作业
129. @RequestMapping("/correctWorkbook/removeStudentWorkbook")
130. public ModelAndView removeStudentWorkbook(
131.    Integer workbook_id,
132.    String ids,
133.    ModelAndView mv
134.    ) {
135.    String student_ids[]=ids.split(",");
                                                //得到要删除的学生作业的所有学生的 id
136.    for(String student_id:student_ids) {
137.    studentWorkbookService.removeStudentWorkbookByWorkbookIdAnd-
            StudentId(workbook_id, Integer.parseInt(student_id));
138.    }
139.    mv.addObject("workbook_id",workbook_id);
140.    mv.setViewName("redirect:/correctWorkbook/findStudentByWorkbookId");
141.    return mv;
142.
143.    }
144. //批改作业,并将结果保存到 student_workbook 表中
145. @RequestMapping("/correctWorkbook/correctWorkbook")
146. public ModelAndView correctWorkbook(String flag, //标识 flag 为 1,表
     //示显示批改作业页面;标识 flag 为 2,表示保存数据到 student_workbook 表中
147.    Integer workbook_id, Integer student_id,
148.    HttpSession session,
149.    @ModelAttribute("saveStudentWork") SaveStudentWorkbook
        saveStudentWorkbook, ModelAndView mv) {
150. Teacher teacher=(Teacher)session
        .getAttribute(HomeworkConstants.TEACHER_SESSION);
151.    String teacherName=teacher.getLoginname();
152.    //根据 workbook_id 和 student_id 从 studentWorkbook 表中查找所有做此作
        //业的学生,并将查询结果封装到 StudentInfo 对象中
153.    Student student=studentService.findStudentById(student_id);
154.    if (flag.equals("1")) {
```

```java
155.            Workbook workbook=workbookService.findWorkbookById(workbook_
                    id);
156.
157.            //根据workbook_id和student_id查询StudentWorkbook
158.            List<StudentWorkbook>studentWorkbooks =findStudent-
                WorkbookByWorkbookIdAndStudentId(workbook_id,
159.                student_id);
160.            //对学生作业按习题类别排序
161.            utilService.sortStudentWorkbook(studentWorkbooks);
162.            saveStudentWorkbook.setStudentWorkbooks(studentWorkbooks);
163.            mv.addObject("workbook", workbook);
164.            mv.addObject("student", student);
165.            mv.addObject("workbook_id", workbook_id);
166.            mv.addObject("student_id", student_id);
167.            mv.addObject("saveStudentWorkbook", saveStudentWorkbook);
168.            mv.setViewName("correctWorkbook/correctStudentWorkbook");
169.        } else {
170.            //得到批改作业前的studentWorkbook列表
171.            List < StudentWorkbook > studentWorkbooks = findStudentWorkbookBy-
                WorkbookIdAndStudentId(workbook_id,
172.                student_id);
173.            Iterator<StudentWorkbook>it=studentWorkbooks.iterator();
174.            //得到批改作业后的studentWorkbook列表
175.            Iterator<StudentWorkbook>it1 =saveStudentWorkbook.getStudent-
                Workbooks().iterator();
176.            //只对批改作业前的studentWorkbook列表中的grade和notes属性
                //进行更新,其余属性不变
177.
178.            while (it.hasNext()) {
179.                StudentWorkbook sw1=it.next();
180.                StudentWorkbook sw2=it1.next();
181.                sw1.setScore(sw2.getScore());
182.                sw1.setNotes(sw2.getNotes());
183.            }
184.            //对studentWorkbook表进行更新
185.            it=studentWorkbooks.iterator();
186.            StudentWorkbook studentWorkbook;
187.            while (it.hasNext()) {
188.                studentWorkbook=it.next();
189.                studentWorkbookService.addStudentWorkbook(studentWorkbook);
190.                workbook_id=studentWorkbook.getWorkbook().getId();
191.            }
```

```java
192.            mv.addObject("clazz_id",student.getClazz().getId());
193.            mv.addObject("workbook_id", workbook_id);
194.            mv.setViewName("redirect:/correctWorkbook/findStudentBy-
                    WorkbookId");
195.        }
196.        return mv;
197.    }
198.
199.    //下载学生作业文件
200.    @RequestMapping(value ="/correctWorkbook/downloadFile")
201.    public ResponseEntity<byte[]>downLoad(Integer workbook_id,
                                    //作业 workbook 的 id
202.             Integer student_id,        //学生 student 的 id
203.             HttpSession session) throws Exception {
204.        //得到 Workbook 对象
205.        Workbook workbook=workbookService.findWorkbookById(workbook_id);
                                    //得到当前 id 对应 workbook 对象
206.        //得到 Student 对象
207.        Student student=studentService.findStudentById(student_id);
208.        //创建 StudentWorkbook 对象,将 Workbook 和 Student 对象赋值给
            //StudentWorkbook 对象的相关属性
209.        StudentWorkbook studentWorkbook=new StudentWorkbook();
210.        studentWorkbook.setStudent(student);
211.        studentWorkbook.setWorkbook(workbook);
212.        //根据 StudentWorkbook 对象进行查询
213.        List<StudentWorkbook>studentWorkbooks=studentWorkbookService
            .findStudentWorkbook(studentWorkbook, new PageModel());
214.        Iterator<StudentWorkbook>it=studentWorkbooks.iterator();
215.        String fileName=new String();
216.        if(it.hasNext())
217.            fileName=it.next().getFileName();
218.        //下载文件路径
219.        String path=session.getServletContext().getRealPath(
220.                "/upload/");
221.
222.        File file=new File(path +File.separator +fileName);
223.        //创建 springframework 的 HttpHeaders 对象
224.        HttpHeaders headers=new HttpHeaders();
225.        //下载显示的文件名,解决中文名称乱码问题
226.        String downloadFielName=new String(fileName.getBytes("UTF-8"),
            "iso-8859-1");
227.        //通知浏览器以 attachment(下载方式)打开图片
228.        headers.setContentDispositionFormData ( " attachment ", download-
            FielName);
```

```
229.        //application/octet-stream: 二进制流数据(最常见的文件下载)
230.        headers.setContentType(MediaType.APPLICATION_OCTET_STREAM);
231.        //201 HttpStatus.CREATED
232.        return new ResponseEntity<byte[]>(FileUtils
            .readFileToByteArray(file), headers, HttpStatus.CREATED);
233.    }
234.    //根据选中的班级对学生作业进行查重
235.    @RequestMapping(value="/correctWorkbook/runSimilarity")
236.    public ModelAndView runSimilarity(String ids,
                                    //存放要查重的班级id
237.            float checkRate,
238.        ModelAndView mv) {
239.        String[] idArray=ids.split(",");
                                    //得到所要查重作业的workbook的id
240.        //将所选班级所有学生的作业信息放到下面的数组中
241.        List<StudentWorkbook> studentWorkbooks=new
            ArrayList<StudentWorkbook>();
242.        for (int i=0; i<idArray.length; i++) {
243.            Workbook workbook=new Workbook();
244.            workbook.setId(Integer.parseInt(idArray[i]));
245.
246.            StudentWorkbook studentWorkbook=new StudentWorkbook();
247.            studentWorkbook.setWorkbook(workbook);
248.            PageModel pageModel=new PageModel();
249.            pageModel.setPageSize(1000);
250.            List<StudentWorkbook> sw=studentWorkbookService.findStudent-
                Workbook(studentWorkbook,pageModel);
251.            studentWorkbooks.addAll(sw);
252.        }
253.        //对所选学生进行查重,并更新student_workbook
254.        similarityService.studentWorkbookSimilarity(studentWorkbooks,
            checkRate);
255.
256.        mv.addObject("isSimilarity", "2");
257.        mv.setViewName("redirect:/correctWorkbook/selectCorrectWorkbook");
258.
259.        return mv;
260.    }
261.    //查询学生作业习题错误的情况
262.    @RequestMapping(value="/correctWorkbook/findExerciseInfo-
        ByWorkbookId")
263.    public ModelAndView findExerciseInfoByWorkbookId(Integer workbook_id,
264.            ModelAndView mv) {
```

```
265.            Workbook workbook=workbookService.findWorkbookById(workbook_id);
266.            List<ExerciseInfo>exerciseInfos=exerciseService
                    .findExerciseInfoByWorkbookId(workbook_id);
267.            mv.addObject("exerciseInfos",exerciseInfos);
268.            mv.addObject("workbook",workbook);
269.            mv.setViewName("/correctWorkbook/showWorkbookErrorCount");
270.            return mv;
271.        }
272.
273.
274.    //根据workbook_id和student_id查找studentWorkbook
275.    public List<StudentWorkbook>findStudentWorkbookByWorkbookIdAnd-
        StudentId(Integer workbook_id,Integer student_id){
276.
277.        StudentWorkbook studentWorkbook=new StudentWorkbook();
278.
279.        Workbook workbook=workbookService.findWorkbookById(workbook_id);
280.        Student student=studentService.findStudentById(student_id);
281.        studentWorkbook.setStudent(student);
282.        studentWorkbook.setWorkbook(workbook);
283.        if(workbook==null||student==null)return null;
284.        PageModel pageModel=new PageModel();
285.        pageModel.setPageSize(50);
286.        List<StudentWorkbook>studentWorkbooks=student-
            WorkbookService.findStudentWorkbook(studentWorkbook,
            pageModel);
287.
288.        return studentWorkbooks;
289.    }
290.    //对查询的student_workbook根据student_loginname(学号)进行排序
291.    public void sortStudentInfos(List<StudentInfo>studentInfos) {
292.        Collections.sort(studentInfos, new Comparator<StudentInfo>(){
293.            /*
294.             * int compare(StudentInfo p1, StudentInfo p2) 返回一个基本
                   类型的整型,
295.             * 返回负数表示:p1 小于p2,
296.             * 返回0 表示:p1 和p2 相等,
297.             * 返回正数表示:p1 大于p2
298.             */
299.            public int compare(StudentInfo p1, StudentInfo p2) {
300.                //按照年龄升序排列
301.                if(p1.getStudent().getLoginname().compareTo( p2
                        .getStudent().getLoginname())>0){
302.                    return 1;
```

```
303.                    }
304.                    if(p1.getStudent().getLoginname().compareTo( p2.getStudent()
                        .getLoginname())==0){
305.                        return 0;
306.                    }
307.                    return -1;
308.                }
309.            });
310.        }
311.    }
312. }
```

批改作业模块的主要功能有两个：一个是批改学生作业；另一个是对学生作业进行查重，并将查重的结果保存在 student_workbook 表中。JSP 页面可参考源代码，这里不再列出。

10.4.8 学生端作业管理模块

学生端作业管理模块主要包括学生查看作业并完成作业、查看已经完成的作业和修改个人信息 3 部分。操作的核心表为 student_workbook，涉及的接口和类有以下 7 个。

- StudentWorkbookService：学生作业表 student_workbook 的服务接口。
- StudentService：学生表 tb_student 的服务接口。
- WorkbookService：作业表 tb_workbook 的服务接口。
- ClazzService：班级表 tb_clazz 的服务接口。
- UtilService：这是一个工具类接口。
- AssignmentService：发布作业表 tb_assignment 的服务接口。
- StudentWorkbookController：处理学生作业的所有请求的控制器，它也是该模块的核心控制类，所有的请求都首先通过控制器分配后，才能到达映射的方法进行处理。

以上几个接口和类的关系可以通过以下的 UML 类图展示，如图 10-21 所示。

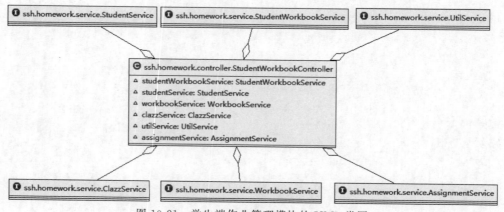

图 10-21 学生端作业管理模块的 UML 类图

1. 数据访问层

```
1.    package ssh.homework.dao;
2.
3.    import static ssh.homework.common.HomeworkConstants.STUDENTWORKBOOKTABLE;
4.
5.    import ssh.homework.dao.provider.StudentWorkbookDynaSqlProvider;
6.    import ssh.homework.domain.ExerciseInfo;
7.    import ssh.homework.domain.StudentInfo;
8.    import ssh.homework.domain.StudentWorkbook;
9.
10.   public interface StudentWorkbookDao {
11.       //根据id查询学生作业内容
12.       @Select("select * from "+STUDENTWORKBOOKTABLE+" where id =#{id}")
13.       @Results({
14.           @Result(id=true,column="id",property="id"),
15.           @Result(column="studentAnswer",property="student-
                  Answer"),
16.           @Result(column="studentAnswer",property="student-
                  Answer"),
17.           @Result(column="grade",property="grade"),
18.           @Result(column="score", property="score"),
19.           @Result(column="notes", property="notes"),
20.           @Result(column="rate",property="rate"),
21.           @Result(column="studentRate",property="studentRate"),
22.           @Result(column="instructions",property="instructions"),
23.           @Result(column="fileName",property="fileName"),
24.           @Result(column="workbook_id", property="workbook",
25.             one=@One(select="ssh.homework.dao.WorkbookDao
                  .selectById",
26.              fetchType=FetchType.EAGER)),
27.           @Result(column="exercise_id", property="exercise",
28.             one=@One(select="ssh.homework.dao.ExerciseDao
                  .selectById",
29.              fetchType=FetchType.EAGER)),
30.           @Result(column="student_id", property="student",
31.             one=@One(select="ssh.homework.dao.StudentDao.selectById",
32.              fetchType=FetchType.EAGER))
33.       })
34.       StudentWorkbook selectById(Integer id);
35.
36.       //根据id删除学生作业内容
37.       @Delete(" delete from "+STUDENTWORKBOOKTABLE+" where id =#{id} ")
```

```
38.        void deleteById(@Param("id") Integer id);
39.
40.        //根据 workbook_id 和 student_id 删除学生作业内容
41.        @Delete(" delete from "+STUDENTWORKBOOKTABLE+" where workbook_
           id=#{workbook_id} "
42.            +"and student_id=#{student_id} ")
43.        void deleteByWorkbookIdAndStudentId(@Param("workbook_id")
           Integer workbook_id,
44.            @Param("student_id") Integer student_id);
45.
46.        //动态查询
47.        @SelectProvider(type=StudentWorkbookDynaSqlProvider.class,
           method="selectWhitParam")
48.        @Results({
49.            @Result(id=true,column="id",property="id"),
50.            @Result(column="studentAnswer",property=
               "studentAnswer"),
51.            @Result(column="grade",property="grade"),
52.            @Result(column="score", property="score"),
53.            @Result(column="notes",property="notes"),
54.            @Result(column="rate",property="rate"),
55.            @Result(column="studentRate",property="studentRate"),
56.            @Result(column="instructions",property="instructions"),
57.            @Result(column="fileName",property="fileName"),
58.            @Result(column="workbook_id", property="workbook",
59.            one=@One(select="ssh.homework.dao.WorkbookDao.selectById",
60.              fetchType=FetchType.EAGER)),
61.            @Result(column="exercise_id", property="exercise",
62.            one=@One(select="ssh.homework.dao.ExerciseDao
               .selectById",
63.              fetchType=FetchType.EAGER)),
64.            @Result(column="student_id", property="student",
65.            one=@One(select="ssh.homework.dao.StudentDao
               .selectById",
66.              fetchType=FetchType.EAGER))
67.        })
68.        List<StudentWorkbook> selectByPage(Map<String, Object>
           params);
69.
70.        //根据参数查询学生作业内容总数
71.        @SelectProvider(type=StudentWorkbookDynaSqlProvider.class,
           method="count")
```

```
72.            Integer count(Map<String, Object>params);
73.
74.            //动态插入学生作业内容
75.            @SelectProvider(type=StudentWorkbookDynaSqlProvider.class,
               method="insert")
76.            void save(StudentWorkbook studentWorkbook);
77.            //动态更新学生作业内容
78.            @SelectProvider(type=StudentWorkbookDynaSqlProvider.class,
               method="update")
79.            void update(StudentWorkbook studentWorkbook);
80.
81.
82.            //根据 workbook_id 按 student_id 统计每个学生的成绩、最大查重率和
               //习题数
83.            @Select("SELECT student_id,fileName,sum(score) as score,
               max(rate) as rate,count(*) as count "+
84.            "FROM student_workbook where workbook_id=#{workbook_id} group
               by student_id ")
85.            @Results({
86.                @Result(column="fileName",property="fileName"),
87.                @Result(column="score",property="score"),
88.                @Result(column="rate",property="rate"),
89.                @Result(column="count", property="count"),
90.                @Result(column="student_id", property="student",
91.                    one=@One(select="ssh.homework.dao.StudentDao.selectById",
92.                      fetchType=FetchType.EAGER))
93.            })
94.            List<StudentInfo>selectStudentInfoGroupByStudentId(Integer
               workbook_id);
95.      }
```

StudentWorkbookDao 类主要完成对 student_workbook 表的操作，这个表中存放的是与学生作业相关的详细信息，包括作业习题内容、学生作业答案、学生作业每道题的得分、学生作业查重情况等。

第 94 行是类中最后一个方法，其作用是根据作业 id 查询每个学生在作业题中的查重情况，查询的结果保存在一个对象中，这个对象是 StudentInfo 类的实例，其代码如下：

```
1.    package ssh.homework.domain;
2.
3.    public class StudentInfo {
4.
5.        private Student student;              //学生信息
6.        private float score;                  //作业题得分
```

```
7.     private float rate;            //设置查重率
8.     private int count;             //满足查重率的学生人数
9.     private String fileName;       //学生上传作业的文件路径
10.    //setter and getter 方法
11. }
```

2. 业务逻辑层

批改作业模块的业务逻辑层涉及的服务接口较多,这在本小节开头已经介绍了,这里主要介绍 StudentWorkbookService 服务接口。

(1) StudentWorkbookService 接口的代码如下。

```
1.  package ssh.homework.service;
2.
3.  import ssh.homework.domain.StudentInfo;
4.  import ssh.homework.domain.StudentWorkbook;
5.  import ssh.homework.tag.PageModel;
6.  //处理学生作业相关业务逻辑服务接口
7.  public interface StudentWorkbookService {
8.      /**
9.       * 根据 id 查询学生作业
10.      * @param id
11.      * @return 学生作业对象
12.      */
13.     StudentWorkbook findStudentWorkbookById(Integer id);
14.
15.     /**
16.      * 获得所有学生作业
17.      * @return User 对象的 List 集合
18.      */
19.     List<StudentWorkbook> findStudentWorkbook(StudentWorkbook student-
            Workbook,PageModel pageModel);
20.
21.     /**
22.      * 根据 id 删除学生作业
23.      * @param id
24.      */
25.     void removeStudentWorkbookById(Integer id);
26.
27.     /**
28.      * 根据 workbook_id 和 student_id 删除学生作业
29.      * @param workbook_id
30.      */
31.     void removeStudentWorkbookByWorkbookIdAndStudentId(Integer workbook_
            id,Integer student_id);
```

```
32.    /**
33.     * 修改学生作业
34.     * @param StudentWorkbook 学生作业对象
35.     **/
36.    void modifyStudentWorkbook(StudentWorkbook studentWorkbook);
37.
38.    /**
39.     * 添加学生作业
40.     * @param StudentWorkbook 学生作业对象
41.     **/
42.    void addStudentWorkbook(StudentWorkbook studentWorkbook);
43.    //
44.    /**
45.     * 根据 workbook_id 按 student_id 统计每个学生的成绩、最大查重率和习题数
46.     * @param workbook_id
47.     * @return StudentInfo 对象的集合 List
48.     **/
49.    List<StudentInfo> findStudentInfoGroupByStudentIdByWorkbookId(Integer workbook_id);
50.
51. }
```

（2）StudentWorkbookServiceImpl 类的代码如下。

```
1.  package ssh.homework.service.impl;
2.
3.  import ssh.homework.dao.StudentWorkbookDao;
4.  import ssh.homework.domain.StudentInfo;
5.  import ssh.homework.domain.StudentWorkbook;
6.  import ssh.homework.service.StudentWorkbookService;
7.  import ssh.homework.tag.PageModel;
8.
9.  //StudentWorkbookService 的实现类
10. @Service("studentWorkbookService")
11. public class StudentWorkbookServiceImpl implements StudentWorkbookService {
12.     @Autowired
13.     private StudentWorkbookDao studentWorkbookDao;
14.     @Override
15.     //根据学生作业 id 查询学生作业信息
16.     public StudentWorkbook findStudentWorkbookById(Integer id) {
17.
```

```
18.        return studentWorkbookDao.selectById(id);
19.    }
20.    //根据学生 id 和作业 id 查询某个学生的所有作业内容
21.    @Override
22.    public List<StudentWorkbook> findStudentWorkbook(StudentWorkbook studentWorkbook, PageModel pageModel) {
23.        Map<String,Object>params=new HashMap<String,Object>();
24.        params.put("studentWorkbook",studentWorkbook);
25.
26.        int recordCount=studentWorkbookDao.count(params);
27.        pageModel.setRecordCount(recordCount);
28.        if(recordCount>0)
29.            params.put("pageModel",pageModel);
30.        List<StudentWorkbook>stus=studentWorkbookDao
            .selectByPage(params);
31.
32.        return stus;
33.    }
34.
35.    @Override
36.    //根据 id 删除学生作业
37.    public void removeStudentWorkbookById(Integer id) {
38.        studentWorkbookDao.deleteById(id);
39.
40.    }
41.    //学生重做或修改作业
42.    @Override
43.    public void modifyStudentWorkbook(StudentWorkbook studentlWorkbook) {
44.        studentWorkbookDao.update(studentWorkbook);
45.
46.    }
47.
48.    @Override
49.    //第一次做作业时,执行添加作业
50.    public void addStudentWorkbook(StudentWorkbook studentWorkbook) {
51.        //查找是否有重复提交的作业,查找重复提交的条件是
            //workbook_id,student_id,assignment_id 不能同时相同
52.        List<StudentWorkbook>list=findStudentWorkbook(studentWorkbook, new PageModel());
53.        if(list.size()<=0)
54.        studentWorkbookDao.save(studentWorkbook);
55.        else
```

```
56.            studentWorkbookDao.update(studentWorkbook);
                                       //如果存在这条记录,就执行更新操作
57.
58.        }
59.        @Override
60.        //根据 workbook_id 按 student_id 统计每个学生的成绩、最大查重率和习题数
61.        public List<StudentInfo>findStudentInfoGroupByStudentIdByWorkbook-
           Id(Integer workbook_id) {
62.
63.            return studentWorkbookDao.selectStudentInfoGroupByStudentId
                   (workbook_id);
64.        }
65.
66.        @Override
67.        //根据 workbook_id 和 student_id 删除学生作业
68.        public void removeStudentWorkbookByWorkbookIdAndStudentId(Integer
           workbook_id, Integer student_id) {
69.
70.            studentWorkbookDao.deleteByWorkbookIdAndStudentId(workbook_
               id, student_id);
71.        }
72.    }
```

3. Web 表示层

学生端作业管理模块的业务逻辑层涉及的服务接口较多,这在本小节开始已经介绍了,这里主要介绍 StudentWorkbookController 类。

StudentWorkbookController 类是学生端作业管理模块的控制器,所有操作都需经过它来指挥完成,它是这个模块的核心类。这个类中主要有以下 6 个方法。

- doStudentWorkbook():处理学生提交作业请求。
- downLoad():处理学生下载文件请求。
- findStudentWorkbook:查询某一个学生的所有作业。
- modifyPassword():处理修改学生的登录密码请求。
- selectStudentWorkbookByPage():处理学生查询作业请求,这里要调用 findStudentWorkbook()方法。
- uploadFile():处理学生上传作业文件请求。

(1) StudentWorkbookController 控制器类代码如下:

```
1. package ssh.homework.controller;
2.
3. import static ssh.homework.common.HomeworkConstants.STUDENT_SESSION;
4.
```

```java
5.    import ssh.homework.domain.*;
6.    import ssh.homework.service.*;
7.    import ssh.homework.tag.PageModel;
8.    //处理学生端作业请求的控制器类
9.    @Controller
10.   public class StudentWorkbookController {
11.       //相关的service属性定义省略
12.       //显示当前学生的所有作业
13.       @RequestMapping(value="/studentWorkbook/selectStudentWorkbook")
14.       public String selectStudentWorkbookByPage(
15.           Integer pageIndex,
16.           String isFinish,    //1表示显示已发布但未批改的作业,2表示已批改的作业
17.           HttpSession session,
18.           Model model) {
19.       //从session中得到登录学生的信息
20.       Student student=(Student)session.getAttribute(STUDENT_SESSION);
21.       //重新查询Student,得到Student关联对象的信息
22.       student=studentService.findStudentById(student.getId());
23.       //得到学生所在班级
24.       Clazz clazz=student.getClazz();
25.       //查询关联的workbook
26.       clazz=clazzService.findOneToManyByClazzId(clazz.getId());
27.
28.       Workbook workbook=new Workbook();
29.       workbook.setClazz(clazz);
30.       if(isFinish.equals("1"))         //发布的作业
31.           workbook.setWflag("1");
32.       else
33.           workbook.setWflag("2");      //在批改的作业
34.
35.       PageModel pageModel=new PageModel();
36.       if(pageIndex!=null)pageModel.setPageIndex(pageIndex);
37.
38.       //得到当前学生的所有作业列表
39.
40.       List<Workbook>workbooks=workbookService.findWorkbook(workbook,
            pageModel);
41.       pageModel.setPageSize(workbooks.size());
42.
43.       model.addAttribute("workbooks",workbooks);
44.       model.addAttribute("pageModel",pageModel);
```

```java
45.          model.addAttribute("isFinish",isFinish);
                            //1表示显示已发布但未批改的作业,2表示已批改的作业
46.
47.          return "studentWorkbook/studentWorkbook";
48.
49.     }
50.     //学生完成作业
51.     @RequestMapping(value="/studentWorkbook/doStudentWorkbook")
52.     public ModelAndView doStudentWorkbook(
53.          String flag,
54.          Integer workbook_id,
55.          @ModelAttribute("isFinish")String isFinish,    //1表示显示已发布但
          //未批改的作业,2表示已批改的作业
56.          @ModelAttribute("saveStudentWork") SaveStudentWorkbook saveStudent-
          Work,
57.          HttpSession session,
58.          ModelAndView mv) {
59.     //提示信息
60.     Student student=(Student)session.getAttribute(STUDENT_SESSION);
61.     if(flag.equals("1")) {                                    //进入完成作业的页面
62.          PageModel pageModel=new PageModel();
63.          Workbook workbook=workbookService.findWorkbookById(workbook_id);
64.          pageModel.setPageSize(100);
65.          Assignment assignment=new Assignment();
66.          assignment.setWorkbook(workbook);
67.          //根据workbook,student在student_workbook表中查找学生作业
68.          List<StudentWorkbook> sw=findStudentWorkbook(workbook,student,
          pageModel);
69.          if(sw.size()>0) {    //说明student_workbook表中已经有记录,学生已
                    //经交过这个作业了,将查询得到的作业内容保存到saveStudentWork中
70.
71.              saveStudentWork.setStudentWorkbooks(sw);
72.          }
73.      else {
74.     //这个学生是第一次提交作业,根据当前作业查找相应的习题添加到
        //saveStudentWork中
75.          List<Assignment>assignments=assignmentService
                .findAssignment(assignment, pageModel);
76.          for(int i=0;i<assignments.size();i++) {
77.              StudentWorkbook studentWorkbook=new StudentWorkbook();
78.              studentWorkbook.setExercise(assignments.get(i).getExercise());
79.              studentWorkbook.setGrade(assignments.get(i).getGrade());
80.              studentWorkbook.setWorkbook(workbook);
```

```
81.             studentWorkbook.setStudent(student);
82.             saveStudentWork.getStudentWorkbooks().add(studentWorkbook);
83.
84.         }
85.     }
86.     //对学生作业按习题类别排序
87.     utilService.sortStudentWorkbook(saveStudentWork
        .getStudentWorkbooks());
88.
89.     mv.addObject("saveStudentWork",saveStudentWork);
90.
91.     mv.addObject("workbook",workbook);
92.     if(isFinish.equals("1"))
93.     mv.setViewName("studentWorkbook/doStudentWorkbook");
                                            //isFinish 为 1 表示完成作业
94.     else
95.         mv.setViewName("studentWorkbook/viewStudentWorkbook");
                                            //isFinish 为 2 表示查看作业
96.
97.     }else
        //如果 flag 不等于 1,则执行保存操作,将作业内容提交给 student_workbook 表
98.     {
99.     List<StudentWorkbook>studentWorkbooks=saveStudentWork
        .getStudentWorkbooks();
100.      Iterator<StudentWorkbook>it=studentWorkbooks.iterator();
101.      while(it.hasNext()) {
102.         StudentWorkbook st=it.next();
103.         //只要不是简答题,就判断习题答案与学生的答案是否一样
104.
105.         if(st.getExercise().getAnswer().trim().toLowerCase()
106.             .equals(st.getStudentAnswer().trim().toLowerCase())&&!st
                .getExercise().getKind().toLowerCase().equals("3"))
107.         st.setScore(st.getGrade());
                //如果学生回答正确,则将成绩添加到 studentWorkbook 的 score 字段中
108.         studentWorkbookService.addStudentWorkbook(st);
                                            //对 student_workbook 表执行添加操作
109.
110.      }
111.      mv.setViewName("redirect:/studentWorkbook/selectStudentWorkbook");
112.      }
113.      mv.addObject("isFinish",isFinish);
114.
```

```
115.        return mv;
116.    }
117.    /**
118.     * 处理修改学生密码的请求
119.     * @param String flag 标记,1表示跳转到修改页面,2表示执行修改操作
120.     * @param String clazz_id 班级id号
121.     * @param Student student 要修改学生的对象
122.     * @param ModelAndView mv
123.     **/
124.    @RequestMapping(value="/studentWorkbook/modifyPassword")
125.    public ModelAndView modifyPassword(
126.            String flag,
127.            @ModelAttribute Student student,
128.            HttpSession session,
129.            ModelAndView mv) {
130.        if(flag.equals("1")) {
131.            //从session中得到登录学生的信息
132.            student=(Student)session.getAttribute(STUDENT_SESSION);
133.            //重新查询Student,得到Student关联对象的信息
134.            student=studentService.findStudentById(student.getId());
135.
136.
137.            mv.addObject("student",student);
138.            mv.setViewName("/studentWorkbook/modifyPassword");
139.
140.        }
141.        else
142.        {
143.            //执行更新
144.            studentService.modifyStudent(student);
145.            mv.addObject("isFinish","1");
146.
147.            mv.setViewName("redirect:/studentWorkbook/selectStudentWorkbook");
148.        }
149.
150.        return mv;
151.    }
152.
153.    //学生在作业管理时上传文件
154.    @RequestMapping(value="/studentWorkbook/uploadFile")
155.    public ModelAndView uploadFile (String flag, @ModelAttribute
        MultipartFile file, Integer id,                    // Workbook的id
```

```java
156.        ModelAndView mv, HttpSession session) throws Exception {
157.            if (flag.equals("1")) {
158.                mv.addObject("id", id);        //将Workbook的id保存到request中
159.                mv.setViewName("studentWorkbook/showStudentUploadFile");
160.            } else {
161.                String fileName1=file.getOriginalFilename();
162.
163.                //上传文件路径
164.                String path = session.getServletContext().getRealPath("/upload/");
165.                //从session中得到登录学生的信息
166.                Student student = (Student) session.getAttribute(STUDENT_SESSION);
167.                //重新查询Student,得到Student关联对象的信息
168.                student=studentService.findStudentById(student.getId());
169.                Workbook workbook=workbookService.findWorkbookById(id);
                                        //得到当前id对应的workbook对象
170.                StudentWorkbook studentWorkbook=new StudentWorkbook();
171.                //根据作业和学生得到所有的作业中的习题
172.                List<StudentWorkbook>studentWorkbooks=
                    findStudentWorkbook(workbook, student, new PageModel());
173.    //如果studentWorkbooks.size()==0,则说明此学生还没有提交当前作业,不能
       //上传文件
174.                if (studentWorkbooks.size()==0) {
175.                    mv.addObject("error","没有提交作业前不能上传文件");
176.                    mv.setViewName("error/error");
177.                    return mv;
178.                }
179.                //如果文件有后缀,则查找后缀,并在文件名后加上后缀,否则不加
180.                String fileName;
181.                if (fileName1.indexOf(".") >0) {
182.                    String suffix=fileName1.substring(fileName1.lastIndexOf("."));
183.                    //上传文件名由作业名称+学生学号+学生姓名组成
184.                    fileName=workbook.getTitle().trim() +student
                        .getLoginname().trim() +student.getUsername().trim() +
                        suffix;
185.
186.                } else
187.                    fileName = workbook.getTitle().trim() +student
                        .getLoginname().trim() +student.getUsername().trim();
188.
189.                //将上传文件保存到一个目标文件中
190.                if (file.getSize()>1048576)
                        //如果上传文件超出文件大小限制(这里是1MB),则返回error.jsp
```

```
191.            {
192.                mv.addObject("error", "文件上传失败,可能是文件太大(文件大小
                        不超过 1MB),请将 lib 中的 JAR 包删除!");
193.                mv.setViewName("/error/error");
194.                return mv;
195.            }
196.            file.transferTo(new File(path +File.separator +fileName));
197.            //根据作业和学生查询更新数据库
198.            Iterator<StudentWorkbook> it =studentWorkbooks.iterator();
199.            while (it.hasNext()) {
200.
201.                studentWorkbook =it.next();
202.                studentWorkbook.setFileName(fileName);
203.                //更新数据库
204.                studentWorkbookService.modifyStudentWorkbook(studentWorkbook);
205.            }
206.            //返回
207.            mv.setViewName("redirect:/studentWorkbook/selectStudentWorkbook");
208.            mv.addObject("isFinish", "1");
209.        }
210.        //返回
211.        return mv;
212.    }
213.
214.    //学生下载作业文件
215.
216.    @RequestMapping(value="/studentWorkbook/downloadFile")
217.    public ResponseEntity<byte[]>downLoad(Integer id,
                                    //作业 workbook 的 id
218.        HttpSession session) throws Exception {
219.        Workbook workbook=workbookService.findWorkbookById(id);
                                    //得到当前 id 对应的 workbook 对象
220.        String fileName=workbook.getFileName();
221.        //下载文件路径
222.        String path=session.getServletContext().getRealPath(
223.                "/upload/");
224.
225.        File file=new File(path +File.separator +fileName);
226.        //创建 springframework 的 HttpHeaders 对象
227.        HttpHeaders headers=new HttpHeaders();
228.        //下载显示的文件名,解决中文名称乱码问题
229.        String downloadFielName=new String(fileName.getBytes("UTF-8"),
        "iso-8859-1");
```

```
230.      //通知浏览器以attachment(下载方式)打开图片
231.      headers.setContentDispositionFormData("attachment",
          downloadFielName);
232.      //application/octet-stream：二进制流数据(最常见的文件下载)
233.      headers.setContentType(MediaType.APPLICATION_OCTET_STREAM);
234.      //201 HttpStatus.CREATED
235.      return new ResponseEntity<byte[]>(FileUtils
          .readFileToByteArray(file), headers, HttpStatus.CREATED);
236.  }
237.  //根据workbook和student在student_workbook中查找此学生是否有当前的作业
238.  public List<StudentWorkbook> findStudentWorkbook(Workbook workbook,
      Student student,PageModel pageModel){
239.      StudentWorkbook studentWorkbook=new StudentWorkbook();
240.      studentWorkbook.setStudent(student);
241.      studentWorkbook.setWorkbook(workbook);
242.      List<StudentWorkbook> sw=studentWorkbookService.find-
          StudentWorkbook(studentWorkbook, pageModel);
243.      return sw;
244.  }
245.
246.  }
```

（2）学生完成作业的JSP页面doStudentWorkbook.jsp。

```
1.   <body>
2.     <table>
3.       <!--当前作业内容数据展示区-->
4.   <tr><td><table width="100% " border="1" cellpadding="5" cellspacing
     ="0" style="border:#c2c6cc 1px solid; border-collapse:collapse;">
5.       <tr class="main_trbg_tit" align="center">
6.       <td width="100% ">作业标题：${workbook.title} 班级:${workbook
         .clazz.cname}
7.       课程：${workbook.course.cname}<p></td>
8.       </table></td></tr>
9.     <tr valign="top">
10.    <td height=" 20 " width =" 100% " > < form action =" ${ctx}/
       studentWorkbook/doStudentWorkbook" method="post">
11.    <input type="hidden" name="flag" value="2">
12.       <input type="hidden" name="isFinish" value="${isFinish}">
13.    < table class =" main _ locbg font2" width =" 100% " border =" 1"
       cellpadding="5" cellspacing="0" style="border:#c2c6cc 1px solid;
       border-collapse:collapse;">
14.    <tr><td >序号</td><td >题型 </td><td align="center">题干</td></tr>
```

```jsp
15.          <c:forEach items="${requestScope.saveStudentWork
         .studentWorkbooks}" var="studentWorkbook" varStatus="stat">
16.           <tr id="data_${stat.index}" class="main_trbg" align="center" >
17.             <td width="2% " >
18.             <input type="hidden"
19.             name="studentWorkbooks[${stat.index}].id" value=
             "${studentWorkbook.id}" >
20.             <input type="hidden"
21.             name="studentWorkbooks[${stat.index}].exercise.id" value
             ="${studentWorkbook.exercise.id}" >
22.             <input type="hidden"
23.             name="studentWorkbooks[${stat.index}].workbook.id" value
             ="${studentWorkbook.workbook.id}" >
24.             <input type="hidden"
25.             name="studentWorkbooks[${stat.index}].student.id" value="
             ${studentWorkbook.student.id}" >
26.               <input type="hidden"
27.             name="studentWorkbooks[${stat.index}].exercise.answer"
             value="${studentWorkbook.exercise.answer}" >
28.               <input type="hidden"
29.             name="studentWorkbooks[${stat.index}].exercise.kind" value=
             "${studentWorkbook.exercise.kind}" >
30.               <input type="hidden"
31.             name="studentWorkbooks[${stat.index}].grade" value=
             "${studentWorkbook.grade}" >
32.          ${stat.index+1} </td>
33.             <td ><c:if test="${studentWorkbook.exercise.kind==1}">
34.                 选择题
35.               <c:set var="rows1" value="10"></c:set>
36.                 <c:set var="cols1" value="180"></c:set>
37.                   <c:set var="rows2" value="1"></c:set>
38.                   <c:set var="cols2" value="180"></c:set>
39.                   <c:set var="maxlength" value="200"></c:set>
40.               </c:if>
41.              <c:if test="${studentWorkbook.exercise.kind==2}">
42.                 填空题
43.                 <c:set var="rows1" value="10"></c:set>
44.                 <c:set var="cols1" value="180"></c:set>
45.                 <c:set var="rows2" value="1"></c:set>
46.                 <c:set var="cols2" value="180"></c:set>
47.                 <c:set var="maxlength" value="200"></c:set>
48.               </c:if>
```

```
49.                    <c:if test="${studentWorkbook.exercise.kind==3}">
50.                    简答题
51.                    <c:set var="rows1" value="10"></c:set>
52.                    <c:set var="cols1" value="180"></c:set>
53.                    <c:set var="rows2" value="40"></c:set>
54.                    <c:set var="cols2" value="180"></c:set>
55.                     <c:set var="maxlength" value="5800"></c:set>
56.                    </c:if>
57.                    (${studentWorkbook.grade})
58.                </td>
59.                <td width="80% ">
60. < textarea rows ="${rows1}" cols ="${cols1}" readonly =" readonly ">
    ${studentWorkbook.exercise.content}</textarea></td></tr><tr>
61.                <td align="left" ></td>
62.            </tr>
63.            <tr><td colspan="2">     答案:</td><td width="80% ">
64.                <textarea name =" studentWorkbooks [ ${ stat. index }].
    studentAnswer"
65.                rows="${ rows2 }" cols ="${ cols1 }" maxlength ="
    ${maxlength}">${studentWorkbook.studentAnswer}
66.                </textarea>
67.            </td>
68.        </tr>
69.    </c:forEach>
70.    <tr><td align="right"><input type="submit" value="提交">
    </td></tr>
71.        </table></form>
72.          </td>
73.      </tr>
74.    </table>
75.    <div style="height:10px;"></div>
76.  </body>
```

在 Spring MVC 中，无法直接从 Controller 传递集合类型的数据，如 List 类型的数据无法通过 Controller 直接进行传递。但可以采用间接的方法，可以将一个 List 作为一个对象的属性，然后将对象保存在 Model 中，这样可以在 JSP 页面中通过 EL 表达式得到这个集合。第 15 行 <c:forEach> 标签中的 items 属性的值是一个 List，这个 List 是 saveStudentWorkbook 对象的属性 studentWorkbooks。这个集合的内容是当前学生作业的内容，也就是学生作业中的习题。学生完成作业时提交作业，同时会将作业答案也一同提交到 student_workbook 表中，提交作业时会对客观题进行评阅打分，而简答题则要由教师评阅。

10.5 单元测试

在 Spring 环境中可以很方便地对系统做单元测试，只要在创建测试中使用以下两个注解即可。
- @RunWith(SpringJUnit4ClassRunner.class)：让测试运行于 Spring 测试环境。
- @ContextConfiguration Spring 整合 JUnit 4.0 测试时，使用注解引入多个配置文件。

在测试类中只要有以上两个注解，就可为每个类创建测试类。在测试类中采用依赖注入方式实例化类的属性。下面以创建课程表 CourseDao 类的测试类为例，演示如何在 Spring 环境下进行单元测试。

CourseDao 类的测试类 CourseDaoTest 代码如下。

```
1.   package ssh.homework.test;
2.
3.   import ssh.homework.dao.CourseDao;
4.   import ssh.homework.domain.Course;
5.   @RunWith(SpringJUnit4ClassRunner.class)
6.   @ContextConfiguration(locations ={ "classpath:applicationContext.xml" })
7.   //CourseDao 类测试类
8.   public class CourseDaoTest {
9.       @Autowired
10.      CourseDao courseDao;
11.  //测试根据 id 查询课程 selectById()方法
12.      @Test
13.      public void testSelectById() {
14.          Course course=courseDao.selectById(1);
15.          assertNotNull("查询失败",course);
16.          System.out.println(course.getCname());
17.      }
18.  //测试根据 id 删除 deleteById()方法
19.      //@Test
20.      public void testDeleteById() {
21.          courseDao.deleteById(5);
22.      }
23.  //测试 selectByPage()方法
24.      @Test
25.      public void testSelectByPage() {
26.          Map<String,Object>map=new HashMap<String,Object>();
27.          Course course=new Course();
```

```
28.         course.setCname("设计");
29.         map.put("course",course);
30.         List<Course>list=courseDao.selectByPage(map);
31.         System.out.println(list.size());
32.     }
33.     //测试count()方法
34.     //@Test
35.     public void testCount() {
36.         Map<String,Object>map=new HashMap<String,Object>();
37.         Course course=new Course();
38.         course.setCname("VC 程序设计");
39.         map.put("course",course);
40.         System.out.println(courseDao.count(map));
41.     }
42.     //测试save()方法
43.     //@Test
44.     public void testSave() {
45.         Course course=new Course();
46.         course.setCname("VB 程序设计");
47.         courseDao.save(course);
48.     }
49.     //测试update()方法
50.     //@Test
51.     public void testUpdate() {
52.         Course course=new Course();
53.         course.setCname("VC 程序设计");
54.         course.setId(7);
55.         courseDao.update(course);
56.     }
57. }
```

第 5 行的@RunWith 注解指定了运行测试的类,@ContextConfiguration 注解指定 Spring 配置文件的位置信息,并解析这个配置文件,这样可以通过@Autowires 注解将 clazzDao 实例注入测试类中。以上只给出一个课程的测试类,其他的测试类可参照此类进行设计,此处不再讲解。

10.6 发布运行系统

首先将项目发布到 Tomcat 8.0 服务器中,当然,前提是数据库已经建好。发布正常后可以启动 Tomcat 服务器。在如下地址栏中访问 Web 服务:

```
http://localhost/homework
```

首先登录页面，以系统管理员角色登录后出现系统管理员页面，如图 10-22 所示。

图 10-22　系统管理员页面

如果以教师角色登录，则出现如图 10-23 所示的教师管理作业页面。

图 10-23　教师管理作业页面

如果以学生角色登录，则出现如图 10-24 所示的学生管理作业页面。

图 10-24　学生管理作业页面

参 考 文 献

[1] 刘伟. Java 设计模式[M]. 北京：清华大学出版社，2018.
[2] 史胜辉，王春明，陆培军. JavaEE 基础教程[M]. 北京：清华大学出版社，2010.
[3] 史胜辉，王春明，沈学华. JavaEE 轻量级框架 Struts2＋Spring＋Hibernate 整合开发[M]. 北京：清华大学出版社，2014.
[4] 杨开振，周吉文，梁华辉，等. JavaEE 互联网轻量级框架整合开发[M]. 北京：电子工业出版社，2017.
[5] 黑马程序员. JavaEE 企业级应用开发教程[M]. 北京：人民邮电出版社，2017.
[6] 疯狂软件. Spring＋MyBatis 企业应用实战[M]. 北京：电子工业出版社，2019.